THE MEANINGS OF THE GENE

RHETORIC OF THE HUMAN SCIENCES

The Meanings of the Gene

PUBLIC DEBATES ABOUT HUMAN HEREDITY

CELESTE MICHELLE CONDIT

THE UNIVERSITY OF WISCONSIN PRESS

The University of Wisconsin Press
2537 Daniels Street
Madison, Wisconsin 53718

3 Henrietta Street
London WC2E 8LU, England

5 4 3 2 1

Printed in the United States of America

Library of Congress Cataloging-in-Publication Data
Condit, Celeste Michelle, 1956–
The meanings of the gene : public debates about human heredity / by Celeste Michelle Condit.
342 pp. cm. — (Rhetoric of the human sciences)
Includes bibliographical references and index.
ISBN 0-299-16360-1 (cloth: alk. paper)
ISBN 0-299-16364-4 (paper: alk. paper)
1. Genetics—Social aspects—United States—History—20th century. 2. Genetics—Public opinion—History—20th century. 3. Genetics in mass media—History—20th century. I. Title. II. Series.
QH438.7.C65 1999
304.5—dc21 99-6274

For
Bruce Railsback

small drops, well laden
sweet love
builds fantastic pillars

CONTENTS

Acknowledgments ix

Introduction 3

PART ONE — CLASSICAL EUGENICS: 1900–1935 25

Chapter 1 Breeding the Human Stock 27

Chapter 2 Challenges to Eugenics 46

PART TWO — FAMILY GENETICS: 1940–1954 63

Chapter 3 The Transformation to Genetics 65

Chapter 4 The Family Gene 82

PART THREE — GENETIC EXPERIMENTATION: 1956–1976 97

Chapter 5 Recoding Humanity 99

Chapter 6 Genetic Counseling 124

Chapter 7 Ethical Challenges and Biohazards 140

PART FOUR — GENETIC MEDICINE AND BEYOND: 1980–1995 157

Chapter 8 Blueprints for Genetic Commerce 159

Chapter 9 Toward a Personal Health Genetics 178

Chapter 10 Ethical Challenges for Genetic Health 188

Conclusions and Speculations 209

Appendix 1: Theory and Methodology 249
Appendix 2: Quantitative Data 269
Appendix 3: Coding Sheet and Instructions 274
Notes 280
References 296
Index 314

ACKNOWLEDGMENTS

This book is one product of a six-year-long journey that was made possible by dozens of individuals and institutions. Whatever merit this project holds would not have been possible without each of their contributions, and I am enormously grateful to them all. The journey began in the early nineties, when I began to notice that genetic selection was a growing issue within the reproductive issues I was studying. In order to learn some basic genetics, I applied to the University of Georgia's Study in a Second Discipline program, a marvelous fellowship developed by Vice President for Academic Affairs William Prokasy. I was fortunate to receive the support of this fellowship and the support of Sidney Kushner, who allowed me to work in his lab during 1993–1994 with helpful and supportive guidance from his postdocs, Eileen O'Hara and Caroline Ingle, along with many friendly and helpful graduate students. I thank you all for making my stay in the lab both enormously informative and enjoyable. I also took classes from the UGA genetic faculty, to whom I give my warmest thanks, especially to Jan Westpheling and Rich Meagher, who extended my genetic education by talking with me outside the classroom as well and who welcomed me warmly and continue to be supportive of my efforts. I also wish to thank Jonathon Arnold and the members of the Fungal Genome Research Group at UGA, who have been constantly supportive and helpful, allowing me to try out my ideas on them and to learn from them by observation.

As I began my fellowship, a colleague in my field, Barbara Biesecker, heard me talk about my interest in genetic discourse, and she graciously helped me to contact Barbara Bowles Biesecker at the National Institutes of Health. Bowles Biesecker helped me to apply for support as a visiting investigator at NIH and then served as my primary host for my research stay there. She not only arranged for me to observe genetic counseling sessions, to sit in on genetic training for medical students, and to observe numerous other meetings and clinical conferences, but she was also gracious, warm, and enthusiastic in the face of her own enormous workload. Don Hadley was similarly helpful and supportive, and to both of them I owe my enthusiasm for research in this area and the strength to forge ahead.

Somewhere in this process, I applied for a University of Georgia

Senior Faculty Research Grant to perform the pilot study that shaped the project of this book, and was again fortunate to receive that assistance. The work from the pilot study enabled me to apply for a grant from the Ethical, Legal, and Social Implications Branch of the Human Genome Project, and I received NCHGR, NIH Grant 5 R01 HG01362-02, which paid for the coders for the research and allowed me some time away from teaching to do the work. Sally Caudill and Kris Sheedy worked on the grant and did a fantastic job with much of what they call the grunt work and the coding. They have been a real delight in my life. Tahra Broderson also contributed to the coding, and Caitlin Wills assisted with final manuscript preparation. Nneka Ofulue worked with me on the issues of determinism and did her share of "grunt" work as well. Doyle Srader, John Hyatt, and Mary Jane Mahan are also thanked for their help with the early phases of development of the coding instrument.

As the project was nearing its completion, Roxanne Parrott linked me up with three marvelous undergraduate students who were also interested in genetics and communication. Holly Gooding did reference checking for the project and contributed the study of the visual images in the magazine articles to which I refer at several points. Stephanie Palmer did the detailed research on the transition from the term *germ-plasm* to *gene.* Beth O'Grady did the work on the term *mutation.*

Once a first draft of the manuscript was completed, I benefited from the support, critique, and suggestions of an army of thoughtful people: Betty Jean Craige and Peter Conrad graciously read and commented on the first draft of the manuscript. Edward Larson lent his expertise to late versions of the chapters on the classical eugenic era. James Crowe not only read the manuscript with enormous care; he also recalculated my chi-square tests using the Fisher exact test and reread portions of the manuscript after I had made initial revisions. David Depew likewise read not only the original manuscript but also subsequent revisions and has given me a great deal beyond the manuscript to think about. John Lyne, John Peters, John Campbell, and Ruth Schwartz Cowan all also thoughtfully read the manuscript and provided suggestions, advice, and support. They have made me wish I could start all over again and do this perfectly, but I doubt the NIH would fund a fresh start, so I have adopted as many of their suggestions as I could. In the last stage, Robin Whitaker provided careful, thoughtful assistance in producing a final draft through her expert copy editing, for which I am also thankful.

This journey also could not have been completed without the warm support of my department chair, Don Rubin, who did his best to protect me from my inability to say no to departmental and university service work and who manifested a warm faith in the importance of my research, a faith that has been of more significance to me than he knows. I am also continually indebted to Bruce Railsback, who makes life happy and comfortable for me, and to V'ger, who provided the impetus for many study breaks. I also have to admit that the project might not have been completed without Godiva dark chocolate and contemporary CD technology, both of which kept my energy levels up for this very long haul.

Finally, I would like to thank the people of the state of Georgia for their support of my position as a professor at a research institution and my fellow citizens of the nation for their grant support. I have done my best to fulfill their trust, and I hope that my contributions are worthy of our combined efforts.

THE MEANINGS OF THE GENE

INTRODUCTION

When Kenneth Burke (1968: 16) wrote that humanity is "rotten with perfection," he called to mind a characteristic that is simultaneously the glory of the species and its greatest peril. The linguistically embedded drive toward perfection, which manifests itself in all spheres of human endeavor, has created such marvels as lofty cathedrals, grand opera, subtle poetry, and the opportunity to view our planet from its moon. The obverse side of this same drive—to eradicate the imperfect—also has routinely encouraged anorexia, environmental devastation, war, and occasionally even genocide. The dualistic potential of the human drive for self-perfection—the potential it harbors for great good and great harm—has been most powerfully evident during this century in the development of genetic research and technology.

It would be no exaggeration to identify the twentieth century as the period in human social evolution when we began to exert conscious control over the hereditary characteristics of our species. The century opened with the rediscovery of Mendelian genetics and the accompanying rise of eugenics, a discourse which has been linked to the coercive sterilization of hundreds of thousands of persons, primarily at the hands of the Nazis but in smaller numbers around the globe. The century came to a close with genetics as the hottest scientific research arena, promising a hundred revolutionary miracles in agriculture, industry, and human health.

The revolutions promised for the next century of genetics are ever more far-reaching. A panel of genetic scientists and other experts met in 1997 and concluded that genetic enhancement technologies are inevitable and that physicians will be the arbiters of their use (Vogel, 1997). Using the devices of gene therapy, these experts concluded that in the not too distant future, well-to-do people might make themselves stronger, taller, bustier, or hairier, if not smarter, kinder, and more beautiful. They might even make themselves into something that is outside our traditional understandings of "human" as the implanting of nonhuman genes in humans becomes possible. At the least it seems increasingly likely that we will eventually clone someone.

While we debate the future contours of what it will mean, biologically, to be a human being, genetic screens to test for "defective" fetuses have already become the general rule for pregnant women.

Those who object to these tests are increasingly being treated as deviants. Sociologists argue that the entry of eugenics by the front door at the beginning of the twentieth century may have been repudiated, but it has now firmly entrenched itself in our national homes via the back door of parental choice and medical manipulation (Duster, 1990).

The Relationship of Eugenics and Genetics

Is eugenics, then, the inevitable consequence of our understanding of genes? Hundreds of books and articles have sought to shed light on that anxiety-laden question. Many of these works have raised particular policy issues or provided management strategies to ensure that genetic medicine avoids its negative potentials through pragmatic precautions. These works have addressed various issues: How do you keep DNA records private? and what if you can't? How do you get informed consent? How do you provide insurance for all in an era of genetic self-knowledge? How do you certify genetic counseling? (See, e.g., Andrews et al., 1994; Bartels, Leroy, and Caplan, 1992; Blank, 1990; Holtzman, 1989; Kevles and Hood, 1992; Kitcher, 1996; Marteau and Richards, 1996.) All of these are important questions, but they are in a sense pragmatic issues with a relatively limited focus.

Another set of writings has attempted to chronicle the historical aspects of genetics. Several studies of the scientific development of genetics are available (Carlson, 1991; Cravens, 1988; Gribbin, 1985; Judson, 1979; Olby, 1994; Portugal and Cohen, 1972; Sturtevant, 1965). And several histories of the eugenic movement have already clearly depicted the historical individuals, events, and ideas that made eugenics a social force of serious consequence in the early part of the twentieth century (Bannister, 1979; Kevles, 1995; Larson, 1995; Ludmerer, 1972; Pickens, 1968). Lily Kay has combined the social and scientific issues in offering an analysis of the social forces supporting the development of molecular biology (Kay, 1993). These wide-ranging and often thorough studies have outlined in some detail the individuals who supported eugenic discourses or genetic research and the socioeconomic and political conditions that made it possible for eugenic propaganda to have some prominence. They have also traced the way a core theory of eugenics—understood diversely as focused on sterilization (Larson, 1995; Reilly, 1991) or immigration control (Hasian, 1996)—manifested itself in a variety of venues, including the public realm and state legislatures, as well as private letters and elite books and grants. These researches have not, however, focused clearly on the specific fashion in which the public took up the issues of heredity and the gene beyond the context of the eugenic movement.

It is, of course, a crucial point of debate whether the term *genetics* can ever escape the social implications of *eugenics* (or whether it should do so). The resolution of this debate depends in large measure upon what we mean by *eugenics.* If we understand the term in its broadest sense—the desire to have children who are "well-born," that is, healthy, happy, and perhaps good moral participants in the life of the species—then most human societies are eugenic. Moreover, human beings in many societies have taken an enormous variety of steps to increase the likelihood of bringing children into the world who are well-born. They have married into the right families to attain or retain nobility, slept on the right side instead of the left to produce a boy child rather than a girl, eaten the right foods when they were pregnant, taken AZT, and exercised (or not exercised, depending on the conventional wisdom of the moment).

None of these activities, however, has historically been associated with the term *eugenics.* That term was coined by Francis Galton, who proclaimed, "Eugenics is the science which deals with all influences that improve the inborn qualities of a race; also with those that develop them to the utmost advantage" (Galton, [1909] 1985: 35). The combination of Galton's science and his public rhetoric launched a social movement that gave specific historical meanings to the term *eugenics.*

For many years, two features were identified as crucial to the distinctive and problematic character of eugenics. Eugenics was understood as coercive control of individual procreation and as reproductive decision making that was focused on the human gene pool as a whole, rather than on the good of individuals. As Diane Paul (1994: 145–48) has noted, however, neither of these definitional components carves out a precise territory. Many of the most active theorists and promoters of eugenics in the classical era of eugenics (1900–1930) did not advocate coercion but rather voluntary adherence to eugenic goals. Additionally, even when parents make reproductive choices on an individual basis for their own offspring, these individual decisions, taken in aggregate, will have a collective impact on the human gene pool. Given that individual decisions are always structured by social institutions and cultural norms, the distinction between individual outcomes and social outcomes may be little more than a mirage.

Underlying these contested components of the definition of *eugenics* is the largely unexplored assumption that eugenics is about the manipulation of genes. Scrutiny of that linkage creates some difficulties. The term *eugenics* was coined before the turn of the century, before the rediscovery of Mendel's work on the hereditary characteristics of peas. Galton and other eugenicists employed the term *germ-plasm* rather than *gene* even after 1905 when the term *genetics* was coined (the term *gene* was not coined until 1909; Portugal and Cohen, 1972: 118; see also Carlson, 1991). The term *gene* was not used in the public discussions of eugenics in magazines and newspapers until the midtwenties and was not dominant until the thirties and forties, when eugenics as a public movement had clearly waned. More crucially, the terms *gene* and *genetics* have outlived the term *eugenics* in practical applications by many decades. Consequently, for the next century, the question about what eugenics has meant will in some ways be surpassed in importance by the broader question of what our shifting understandings of the gene can mean.

Three Problems for Genetics

This shift in focus from eugenics to genetics does not erase serious attention to social problems. In addition to those works that address narrow pragmatic issues such as informed consent and laboratory quality, a vast and growing literature addresses the problems associated with the reconceptualization of ourselves and our societies that occurs when we begin to have access to genetic self-knowledge and to apply that knowledge and to talk about it collectively. Many such issues have been identified by critical scholars (Andrews et al., 1994; Cranor, 1994; Katz Rothman, 1998; Kaye, 1997; Murphy and Lappé, 1994; Pellegrino, Harvey, and Langan, 1990; Rothenberg and Thomson, 1994; K. Smith, 1992; Spallone, 1992; Weir, Lawrence, and Fales, 1997; etc.), but three of these issues have received the most substantial attention. The first is the potential that genetic knowledge has for making us all into biological determinists. Dorothy Nelkin and Susan Lindee (1995), Ruth Hubbard and Elijah Wald (1993), Ted Peters (1997), José van Dijck (1998), and Abby Lippman (1993) have all suggested that contemporary discourse about genetics encourages us to conceive of ourselves as biological automatons. As Nelkin and Lindee put it: "The images and narratives of the gene in popular culture reflect and convey a message we will call genetic essentialism. Genetic essentialism reduces the self to a molecular entity, equating human beings, in all their social, historical, and moral complexity, with their genes" (1995: 2). Abby Lippman identifies the same concern, though she labels it geneticization. Her fear is that popular discourse about genetics designates the gene as the determiner of all facets of human character, encouraging us to ignore the impact of social structures, especially access to nutrition and health care, factors she deems more important than genes in shaping what each person comes to be.

The concern that popular discourse about genetics is deterministic is motivated in substantial part by a second concern, that such genetically deterministic beliefs lead to discriminatory attitudes.[1] Critics of genetic research and technology argue that if we conceive of ourselves as primarily determined by our genes, and if we conceive of different sets of genes as superior and inferior, we will strengthen inappropriate and destructive prejudices against persons of different races, classes, genders, and abilities. Even the narrowest and most medically motivated genetic choice—to abort a fetus with Down syndrome[2] or to select for implantation an embryo that lacks the extensive CAG repeats of Huntington's chorea—entails a sense of what is a better genetic configuration and what is worse. It therefore can be taken to entail a judgment about the value of the living people who bear these genetic configurations.

Once such a hierarchy among persons is set up, it is difficult to draw hard and fast lines between which genetic configurations are worth selecting against and which are within the "normal" variation. Can we explain why we would select against Huntington's disease but not dwarfism? Can we explain why we would select against dwarfism but not growth hormone deficiency? If we select against growth hormone deficient children do we select against those who have short stature of a "normal" variety? If we select on the basis of height and gender, can we avoid selecting on the basis of skin color and perfect pitch? Moreover, isn't it likely that the rich will have access to more selections than the poor? Will that establish hereditary biological castes (Nelkin, 1992)? And what will it do to social support for disabilities? Will we blame those who have disabilities (or their parents) and therefore refuse to provide assistance to those with special needs?

The issue of discrimination leads readily to a third concern: What attitudes toward our children, collectively and individually, will the prospect and the practice of genetic choice encourage? Scholars and public advocates have worried that genetic choice leads almost inevitably to perfectionist attitudes toward our children. Dorothy Wertz and James Sorenson (1989) have made the most explicit argument about this concern. They have suggested that the evolution of public norms in regard to genetics is correlated with a rise in perfectionist expectations with regard to offspring. Although Wertz has sometimes qualified her use of the word *perfect* (Wertz and Fletcher, 1993: 174), the basic argument runs that in the closing decades of the century a variety of social forces have encouraged parents to pursue perfect offspring, whereas in earlier eras parents were less concerned about the quality of their offspring. Wertz and Sorenson (1989: 556) suggest that parents today are in search of the "perfect child" and are more eager to "spend large sums of money and time in enriching its life" than in previous eras. Genetic technologies provide one important avenue for such expenditures.

Wertz and Sorenson's claim is not uncontested. Nelkin and Lindee (1995: 139–43) advance an apparently opposite claim, that genetic discourse is exhibiting a tendency to release parents from responsibility for the ways in which they raise their children. They argue that parents can blame the problems of their children upon the children's genes, over which parents do not have direct control. Therefore, they suggest that parents may be predisposed by public talk that is genetically essentialist to spend less effort on the difficult task of raising their children. Given these conflicting claims of observers about trends in social norms, the questions remain open as to what degree perfectionist discourse is evident in talk about reproduction that is associated with genetic discourse and how such responsibility is apportioned among parents and other biological and social agencies.

Explorations of all three issues—determinism, discrimination, and perfectionism—have tended to be presentist and absolutist. The leading critics have assumed or implied (if only by neglect) that before this century, human beings did not collectively examine our heritage or factor heredity as a component of our world view. But this assumption is not true. People have long thought about what it has meant, what it can mean, to say that we are "related" to one another in a physical sense. Attributions of one's failings or strengths to family lineage or one's "blood" extend throughout many cultures and deep into history. Nor is it true, as Hubbard and Wald (1993: 9) imply, that contemporary discourse in the United States merely replays the tunes of the classical eugenic era in contrast with a golden era that focused exclusively on nonbiological, social factors in human life. Instead, the research grounding this book indicates that consistently across the century, human beings have grappled with their nature as biological creatures, physically linked to other human beings. Nor has this reflection varied dramatically with regard to degree or levels of genetic determinism across the century (see appendix 2, table 1). Even at the height of the eugenic movement, attention to genetics was not exclusive. Eugenics shared the scientific and public stage with behaviorism, just as today medical genetics shares the public stage with the environmental movement and cultural and social accounts of race, intelligence, and other human behaviors. The research presented in this book indicates that claims that genetic determinism is dominant in the culture and dramatically increasing are overstated. The absolutist portrait of contemporary treatments of genetics as uniquely determinist are readily denied by careful measures of degree. But ultimately, matters other than the question of degree turn out to be more interesting and consequential.

The assumption that molecular genetics brought with it a radically new degree of determinism, discrimination, and perfectionism arises in part from the belief that discourse about genetics is relatively uniform, that it generally has opposed sociocultural explanations of human action, and that it was therefore aligned with anti-egalitarian perspectives. More careful scrutiny indicates that the views expressed about heredity in U.S. public discourse during this century have been variable and multidimensional rather than simply "good" or "evil," "racist" or "egalitarian," "genetically deterministic" or "socially deterministic." Several scholars have already noted that eugenics was endorsed by political activists of both the Left and the Right (Paul, 1995; Pickens, 1968). More directly to the point, there has been an increasing, if still incomplete, success across the century in elaborating a view of genetics that is compatible with a humane and substantially egalitarian world view. Public discourse in the latter half of the century presented some models of gene-environment interaction that allowed more latitude for environmental influence than did those of the first half of the century. This latitude derived from more open metaphors about the fundamental nature of the gene. Such open, interactive perspectives have not become predominant—both critics of genetics and many geneticists have reified the narrower alternatives—but more diverse models are increasingly available and ever more thoroughly articulated. More recent discourse has also been more likely to promote egalitarian values and consideration of ethical issues (see appendix 2, tables 2, 3, 4, 5). Consequently, the crucial questions about the influence of genetics on social life go beyond the issue of how much influence has been attributed to genes; they also inquire about the range of models of the interactions between genetics and other variables that have been featured at different times in the public realm.

Working from a negative, critical approach, Diane Paul has defended the need for more multiperspectival approaches to genetics, arguing that "a history of eugenics that is sensitive to its complexities alerts us to the fact that genetic technologies present more than one kind of danger—and that if we are not very careful, we may avoid one only to court another" (Paul, 1995: 135). This statement might be supplemented by the more positive observation that the existence of genetic influences on human characteristics is a material fact, but one whose cultural interpretation is variable.[3] Consequently, it is only if those who are interested in promoting humane visions do the work necessary to explore those facets of heredity and genetics that support progressive agendas that we shall be able to articulate a compelling, substantively correct, and progressive view of what it means to be, in part, biological creatures linked to each other through genes.

Methodology

As a prelude to such efforts, we need a more broadly informed understanding of the range of public meanings about genetics than has so far been available. What I offer here is both a quantitative account and a critical interpretation of such public discourse. This should not be confused with a study of "popular" discourse. Studies of popular culture and everyday life have swept contemporary disciplines ranging from history to sociology to English to communication studies, and have received interdisciplinary attention through the vehicle of cultural studies. I do not, however, propose here to describe how discourse about genetics functioned in everyday life or to provide a sampling of the range of popular, individual opinions about genetics that were available at various moments in historical time (a kind of data that would be interesting but which is not available to us). Instead of the term *popular,* I follow a tradition established in the discipline of rhetorical studies and employ the term *public* to distinguish those discussions that were overtly intended to describe or influence collective behaviors and which appeared in widely accessible venues such as newspapers, magazines, and television.[4] In addition, public rhetoric includes those discourses which overtly defined the public motives behind legislation (evident in congressional debates and hearings).

This study thus presents a third alternative, something between the notions of the role of popular eugenics in everyday life and historical notions of eugenics as a practice guided by cohesive intellectual theories and by prominent individuals and instantiated in particular events. This rhetorical approach to the understandings of heredity produces a different account from earlier studies because public discourse is rarely orderly, consistent, and philosophically well developed. Whatever intellectuals like Galton or Weismann, Pearson or Davenport, might have meant by *eugenics,* whatever they might have thought about human heredity or genetics, they had only partial influence in public discourse about the subject; they did not have control over the public interpretations of their theories. By and large, few people read books by Galton or Popenoe. Highly educated professional-class persons probably formed most of the readership for books about eugenics and genetics, yet these figures and their texts have assumed a central role in previous explorations of what *eugenics* meant. In contrast, newspapers and, to a lesser extent, magazines were read both by the middle class and even by a fair proportion of the semieducated working and farming class (especially in the early part of the era). In the later eras, television obviously became the most widely attended public medium. The many and varied accounts in newspapers and magazines and on television did not faithfully and consistently adopt the rhetorics of the learned elites. Consequently, studies focusing on these different types of documents give us very different sets of meanings about genetics. Both types of studies are important for generating thorough understanding of the role of genetic science in public life. Until very recently, however, the scholarly emphasis has been on the leaders of the eugenic movement and their books and writings; even studies incorporating more popular media have been heavily influenced by the lens constituted by the elite-centered eugenic movement. What I recount here instead is the public usage of the concepts of genetics as manifested in the more widely accessible public media. Eugenics as understood by the eugenic theorists and activists was a part of this discourse and was quite influential, but these sources did not fully define the public discourse about genetics and heredity.

Most scholars have tended to focus on the discourse of intellectual leaders in part because it is more coherent and thus easier to write about and describe. In contrast, ideology as it appears in public media is necessarily truncated and incompletely elaborated. Moreover, popular media rarely attempt to present a singular, unified theory of the form that typically occurs in either professional journals or academic books. Instead, public media present discourses with a mix of challenges and responses. They present attractive fragments of theories and blend them with attractive features of other discourses. Thus, for example, eugenic discourse in books and scientific articles might have been defined by its focus on biological heredity, but in the public realm eugenics and euthenics were jumbled together as pieces of the same agenda.[5] Good heredity was viewed as a product of familial contributions to both the biology and the environment of children. In academic accounts of the era of eugenics, with their emphasis on the derivation of eugenics from genetic science, that linkage has been obfuscated. This study of public discourse, therefore, does not search for a single unifying theme that defines a principle such as eugenics. To explore public discourse requires, rather, that we read across the entire score of the public orchestra (or, quite often, cacophony) to identify the range, dispersal, and clustering of various ideas, many of which might not appear to belong together in a rigid, unified philosophical system, but which nonetheless tend to "go together" as a matter of empirical fact in the public discourse of an era.

Taking public discourse seriously also mandates that one avoid the tendency to view the inevitably messy, multifaceted, and truncated public appropriations of scientific theories as though they were deficient versions of the presentation of pure science that could be corrected simply by better scientific training. The common tendency of academics to prefer academic-style discourse tends to subordinate the function of public discourse to that of science and to hierarchize science as superior to public discourse. Such a hierarchy misleads us because public discourse guides and supports (or prevents) science every bit as much as the reverse (Kay, 1993; Latour and Woolgar, 1979). Anyone who would deny this tether between science and the public needs to consider carefully why it is that we have a Human Genome Project but no superconducting supercollider. A growing literature on the rhetoric of biology has begun to trace and elaborate the complex interactions between the biological sciences and the public sphere (Campbell, 1987, 1990; Lyne, 1990; Lyne and Howe, 1990, 1992; Turney, 1998).[6] As most of these works presume, the function of the public arena is not to produce or reproduce factual truths, but to come to shared determinations of how we should live together. Scientifically produced truths may play a role in that process, but the process of shared governance does not and cannot wait upon the production of scientific truths, nor are scientific truths sufficient for achieving public goals (Lyne, 1990). That being said, public discourse uses and affects science because the two are interrelated, though partly demarcated, spheres (Lyne, 1990; Taylor, 1996). We should not, therefore, expect public discourse merely to reproduce science in full and faithful ways. Instead, public discourse should be understood as appropriating science for its own ends, ends which are perfectly valid and necessary—ultimately more necessary than science itself. From this perspective, "public rhetoric" is not what is leftover after the science, logic, action, or other substance is left out, as is implied by the frequent misuse of the phrase *mere rhetoric.* Rather, public rhetoric consists of the set of communicative interactions through which members of a community share with each other their good reasons for choosing courses of action together, where these good reasons include evidence and logic, but also, necessarily, social values and affective relationships and identities (see Brock, Scott, and Chesebro, 1989: 14; Farrell, 1990: 83).

In this spirit, we can understand the arrival of eugenics in America not as the adoption of a genetic account of human characteristics, but as something more complicated than that. Instead, a shift, influenced by scientific conceptions of genetics, occurred in the understanding of what human heredity means, but these scientific conceptions were also influenced by the public discourse, and the public discourse about heredity was influenced by dozens of major social factors including world wars, the arrival of computers, changing population demographics, and so forth. Our vision of human heredity in this century is thus a rich smorgasbord of alternatives rather than a simple menu of bread and water.

Two recent books have contributed to the description of this smorgasbord. José van Dijck (1998) has offered a critical reading of popular and public discourse about genetics from the fifties to the present. Her postmodern approach mixes analyses of movies, mass magazines, and advertising circulars with academic criticism. Although she emphasizes popular rather than elite thought, the approach still features single significant texts, supplementing readings of Watson's *Double Helix* with analyses of particular science fiction films and novels. Differences in conclusions will be addressed at relevant points in this narrative.

Jon Turney (1998) has provided a study that extends in yet a different direction. He surveys both American and British discourse—scientific, literary, and public in both the nineteenth and twentieth centuries—on the broader topics of biological science and genetics. Turney's work is insightful, especially with regard to particular subtopics such as vivisection and in vitro fertilization, and the story told below will note areas where our interpretations buttress each other.

In contrast with these two broad-ranging accounts of popular genetics, this study offers a relatively systematic set of comparisons across the twentieth century of three particular facets of public genetic discourse—determinism, discrimination, and perfectionism. To achieve a systematic comparison without losing the unique meaningfulness that is a core attribute of discourse seemed to require a combination of critical and quantitative methods, applied to an appropriate set of public discourse. Although a wide range of primary and secondary sources have been consulted, the heart of the research rests in a systematically selected sample of American magazines, newspapers, congressional discourse, and television news coverage of eugenics and genetics from 1900 to 1995. This sample was constructed according to social scientific assumptions, but I then read the selected discourse employing the humanistic assumptions of rhetorical criticism.[7] (An extended discussion of the theoretical grounding and methodology is provided in appendix 1.) My critical reading of the more than a thousand documents in this sample produced a series of critical statements about the kinds of discourse present in various periods. These critical statements were then assessed as hypotheses using the quantitative methods of social science. A team of three coders recorded the appearance of each type of discourse across these time periods. The social scientific data thereby generated supported the critical reading in broad outlines, but they also provided rich additional details, some surprising modifications, and some intersubjective verification. With these data in hand, I returned as a critic to the discourse, rereading it all and factoring in the data from the quantitative study, trying to provide an account that captures the variability and richness of the public discourse, but which also seeks to explore some coherent threads in our understanding of heredity and genetics and to interpret, or at least explore, their implications.

I use the term *rhetorical formations* to designate the relatively co-occurrent sets of discourse—metaphors, narratives, values, and so on—that the data analysis and critical readings have located in each time period. In contrast with those who employ the term *discursive formation,* I do not presume that the discursive dynamics of an era are defined by an underlying episteme that uniformly permeates all talk and all practices. Instead, by using the term *rhetorical formation* I intend to indicate that most eras are dominated by change processes, that is, by the attempt to establish different ways of seeing the world, rather than by a static, monolithic perspective that blankets the public space. Because discourse is not a thing but a continual process of persuading people to see the world in particular ways and to employ and establish particular practices and institutions, rhetorics are always more or less contested, and at most times there is a range of discourses operating within the public sphere (see appendix 1). Finally, because a rhetorical formation is usually not dominated by a unified theory or principle, it is important to investigate the particularities of various parts of the discourse: the topics, metaphors, story lines, and images that frequently recur. From such a rhetorically oriented perspective, it matters whether one describes a gene as "beads on a string" or as "a coded message." For each time period, therefore, this study describes (1) the set of related themes, metaphors, stories, and values that appear more frequently than other themes, metaphors, stories, and values, (2) any contests among different sets of values, and (3) the transformation processes that reconstitute the clusters of rhetorical units into new formations in different time periods.

Such a means of defining public discourse assumes that the ways in which publics think and speak are not tightly organized around single principles such that, once one adopts a principle, all other ideas are generated from that principle. Instead, public advocates have multiple simultaneous agendas that are not tightly unified. Moreover, public advocates speak to attract the assent and identification of multiple audience members with multiple agendas of their own. Thus the rhetoric that gets created and repeated is multidimensional. It consists of a complex lacework of connected ideas, but it may also include internal contradictions as well as involve disputes among various advocate factions. In a substantial sense, therefore, the entire rhetorical complex of an era is unique. The exact patterns of discourse are unlikely to be reproduced elsewhere. Consequently, it is worthwhile to distinguish between the rhetorical formation of the classical eugenic era—which is the totality of discourse about heredity in circulation between 1900 and 1935—and discursive fragments that were used in that era, which thus might be understood as "eugenic" in origin but which are reknit into different patterns in later eras.

This account does not pretend to be comprehensive. I have not tried to provide a full narrative of the history of genetics in the American public arena; attention to prominent individuals, key historical events, literary treatments, and the development of the science of genetics are all virtually absent from these pages. Detailed description of the three major themes of determination, discrimination, and perfection in magazines, newspapers, and television news easily consumed all the space available to a book of reasonable size. Perhaps later efforts will be able to incorporate the results of this survey of public meanings into a more comprehensive account that integrates scientific history, the eugenic movement, and public consciousness.

Preview

THE CLASSICAL ERA OF EUGENICS

In the meantime, the following chapters will suggest that there have been four major rhetorical formations about genetics in this century. In the classical era of eugenics, which lasted from the opening of the century to approximately 1935, the dominant metaphor was that of stock breeding. In that era, well-educated members of the professional classes imagined a world where there was no poverty, no alcoholism, and far less crime. They hoped that they could empty the mental institutions and produce a world of intelligent, healthy, and happy people, much like themselves. They pursued this vision under the banner of the Progressive Era in many forms, but because of new research in biology, the eugenic movement constituted one of their central endeavors. The active and powerful minority that supported the major strands of the eugenic agenda imagined that, simply by controlling human reproduction, they could eliminate the problems of people who were "unfit" for their urban-industrial world and for the high culture that these elite groups wanted to promote.

As almost everyone now realizes, Adolf Hitler's implementation of this dream revealed it to be a nightmare of a proportion never previously realized on earth. Because it was highly classist, racist, and culturally exclusionist, this vision could not encompass humanity without doing violence to it. The stock-breeding metaphor spoke all too clearly of the perils of this imaginary world. The preferred children of the eugenic dream were "superior" ones, and children were directly compared to prize bulls at a fair. The metaphor also required a clear distinction between the superior breeders who would exercise control and a class of humans that were animallike, the to-be-bred. Huge groups of people, including the majority of racial and cultural groups, along with the lower classes of the dominant Northern European group, were deemed "unfit."

Rather than being genetically or even biologically deterministic, however, the discourse of public eugenics mixed home life and genes as causes of the superior and inferior characteristics of children. On this account, feeblemindedness was traceable to heredity, specifically to parents, but whether it was due to their inability to provide a nurturing home life or to their bad germ-plasm was not of first significance, at least not for public policy. In either case, it was deemed that the so-called unfit should be prevented from producing more unfit. In keeping with the merging of nature and nurture in one account of heredity, the key qualities of lack of fitness were not limited to the physical characteristics of animals, but rather they emphasized the moral and mental features of human individuals.

This rhetorical formation was widely visible but firmly contested, and the alternative views, based largely on individual reproductive rights, forestalled most eugenic policies in the United States. The challenging, anti-eugenic views gained dominance in the public discourse during the midthirties and early forties because of the combined influence of two factors. The rise and failings of the Third Reich provided the sociohistorical force for delegitimating eugenics. Simultaneously, a scientific delegitimation of eugenics occurred because increasing sophistication in genetic research shifted the fundamental scientific understanding of heredity from the eugenic concept of a unified, totalized "germ-plasm" to the more multiplicitous, variegated concept of genes.

FAMILY GENETICS

The rhetorical formation of classical eugenics was replaced by a vision of family-oriented genetics, which lasted from the midthirties to the midfifties. This era was characterized by strong reactions against key components of eugenic thought. Public writers sought to reformulate thinking about human heredity in ways that would not support the evils of eugenics. Chief among these reactions were overt attempts to sever the linkages between racism and hereditarian thought by positing an egalitarian account of human heredity and also a repudiation of government control in favor of control by individuals in collaboration with medical professionals.

Further embodying the reaction against eugenics, the era also featured as its dominant metaphor a model of the gene in control of human beings. The flamboyant dream of humans in control of their genes that had animated the eugenic era was reversed to a fatalist vision in which genes controlled human characteristics in a heavy-handed fashion. Drawing on the vocabulary of the world wars, genes were described as commanding, directing, or determining human life. Genes were modeled as powerful and unchanging particles—beads on a string—deposited in a being at conception, and the individual's personal genetic legacy could not thereafter be altered by any factor. The process of heredity in which the genes were dealt was imagined as a great lottery where the fates played tricks with the destinies of the lives of families.

Contemporary critics have assumed that high degrees of determinism correlate with anti-environmentalist initiatives, but this era demonstrates otherwise. Because genes were understood as impervious to human control, public advocates urged individuals to attend first to what they could control—family environment. Advocates proclaimed that genes set the boundaries of human abilities and that environment only determined where within those boundaries one might ultimately come to rest. Because environment was the most malleable component, however, caring parents would attend to providing good environments for their children.

There was one respect, however, in which families were told that they could address the biological component of their reproductive legacy. Because information about genetics indicated that the lottery was rigged in certain ways, wise persons were advised to play the odds when it came to major diseases. The game, however, was an individual rather than a collective one. The public vision no longer encompassed the assumption that the rigged odds could be subverted on a large scale for the race or nation as a whole. Individuals could choose not to reproduce if their personal odds were particularly bad, but the society as a whole could not make such a choice. The house of fate always won out against the species in the end.

Public rhetoric thus no longer focused thoughts about bettering heredity on the grand scale of races and nations, and no formal governmental controls were needed or wanted. Now it was the hopes and fears of individual families that mattered. Research in genetics and heredity offered to serve the smaller, but still poignant, dreams of families that they might increase the odds that they would have only children who were "normal" rather than defective or abnormal (which meant subnormal). The parameters of normality were now confined to medically defined health rather than to a broader range of abilities, and the medical profession was nominated as the rational agent who would assist the family in applying the complicated knowledge about genes and heredity to their own reproductive choices. The public discourse told families that to pursue the vision of familial hereditary normality they should know their family medical history, consult the new profession of genetic counseling, and consider making decisions that would positively influence their reproductive legacy by familial choice. Careful application of scientific knowledge and rational, strategic self-control promised a happier future for the family.

In this era the dream of producing superior children or people disappeared. In fact, the castigation of such aspirations proved to be a major mechanism for promoting the new vision of genetics as opposed to eugenics. The aspiration for superiority was defined as a central goal of Nazi eugenics. Genetics was separated from that goal and given a positive valence by describing it as a key point of contrast with the United States' new international enemy, the Soviet Union. The predominance of Lysenkoism among the communists was defined not only as opposition to classical genetics but also as opposition to free inquiry. Genetic research was thus transformed from an adjunct of eugenics to a central exemplar of freedom. This vision was virtually uncontested. Hence, the rhetorical formation regarding heredity in the era was surprisingly uniform, but it also lacked substantial social salience in terms of public policy.

EXPERIMENTAL GENETICS

No cataclysmic event such as a world war provided a neat demarcation for the end of the second era. Indeed, the rhetorical formation of familial genetics did not so much end as it became subsumed in fast-moving extensions of the vision. The change toward broader dreams and concerns in a more visible and colorful rhetorical formation can be conveniently demarcated by Watson and Crick's discovery of the structure of DNA in 1953. Though it took more than half a decade from that discovery to develop, the third rhetorical formation about genetics in the century might reasonably be labeled the era of genetic experimentation. Featuring a frontier spirit of wild exploration and eager enthusiasm in both science and public discourse, this era lasted from the last half of the fifties to the midseventies.

The most flamboyant discursive experimentation in this era consisted of the first broad and sustained speculations about what a genetically malleable future might bring to the human species. Uniquely in this era, the public imagination was loosed to contemplate what might be if the dreams offered by the genetic scientists—of total control over genes—were realized and applied in the realm of the everyday. Fantastic possibilities ranging from artificial reproduction and surrogacy to chimeras, cyborgs, and clones were considered in enthusiastic, even awestruck, discussions that both celebrated and feared the new possibilities, half of which would become reality before the close of the century.

The second prominent set of discursive experiments in the era was less flamboyant and more immediately pragmatic. This discourse responded to scientific experiments that directly applied molecular genetic technologies to human beings for the purposes of genetic selection. As a result of both technical developments and the legalization of abortion, the use of amniocentesis and maternal serum screening developed in this era. The public discourse rapidly normalized and legitimized these practices, placing them within the medical establishment. Discourse of this genre experimented with the vision of genetic counseling as a major social practice.

This fast-flowing scientific and discursive experimentation was explicated through the metaphor of the genetic code. Because of its enthusiasm about genetics, the era featured a higher level of attention to the gene itself than during any other time period. In that respect, the era was slightly more deterministic than any other era. Accounts of the role of environment were often truncated. However, because the metaphor of a code conjured up an image of an immaterial series of discursive fragments that could be easily "edited" or "rewritten," the outcomes of heredity—the products of genes—were imagined as more plastic than ever before. The dominant model of the interaction between environment and heredity still described heredity as setting the boundaries, but these boundaries were often now envisioned as ample and generous rather than tightly restrictive. Moreover, it was imagined that though heredity might deal humans a particular hand through their genes, as the symbol-wielding species, humans could readily rewrite this fundamental biological code to suit our own desires. Once we understood the code of life, it was presumed, it would be a relatively simple matter to recode all biological life to fit human needs and desires.

Because it was a relatively progressive era, these needs and desires included several egalitarian facets, including the shift from a concern with "normal" children to an interest in healthy ones, an increased attention to ethical issues, and a continued effort to establish more egalitarian visions of genetics, especially with regard to issues of race. The era ended because of the success of the scientific experimentation. A debate over the release of the organisms produced by the new recombinant DNA technologies displaced debate over the long-range future implications of the technologies, and the economic demands and values of commercialization displaced both scientists and genetic counseling.

MEDICAL GENETICS

The century closed with the public discourse focused on the nascent commercialization of genetics and individual health in medical genetics. In this era, the blueprint provided the dominant metaphor for genetics. The blueprint metaphor accompanied a refocusing of scientific vistas from the individual genes to the "genome," and this involved a more holistic look at the complicated role of DNA in creatures and cells. The blueprint metaphor was understood by many not only as more holistic than the fragments of bits of codes imagined in the previous era but also as more materially tied to the environment. Blueprints provided only starting points; real buildings were distinctive and individual, and this distinctiveness was a product of the specific materials employed in their initial construction as well as in their history of add-ons. The new metaphor was thus part of a discourse that was slightly less deterministic because it offered a new model for the relationship of genes and environment based on normative starting points that could be altered by both genes and environment, rather than on a starting point that was overdetermined by genes and subsequently affected by environment.

This more balanced, quasi-developmental account of the role of genes in human outcomes was accompanied by substantially expanded ethical debates. These debates included a focus on social structure that transcended individual and familial concerns. Simultaneously, however, other progressive components of the vision of heredity became more heavily contested. Sustained attention was given to a version of evolutionary psychology that offered retreaded eugenic images to account for putatively inborn deficiencies of both women and criminals. There was also a sustained but compartmentalized set of discourse that established the social conditions necessary for commercialization of genetic technologies, especially patenting rights and regulation through a process of "science-based" bureaucracy that excluded the public except in the role of consumers. The hidden costs of the vision of familial and personal genetics included the ceding of control to commercial enterprises.

One final, tentative change evident in the period may eventually prove to be the most momentous. In the previous eras, heredity had been conceived of as the circulation of genes, first within racial groups, later within families, and finally within the complicated developmental circuit of the biological individual. As the century closed, however, "genetics" and "heredity" were being disarticulated from each other. Not only was a personal health genetics replacing reproductive genetics as the preferred public focus (though it was not yet a widely practical option), but also the more that control could be exerted over the genes of every creature, the less was the form of any species or individual being fated by its heredity. To whatever extent genes or their products could be altered in the living individual, to that extent was a hereditary version of genetics less significant.

The ultimate trick of the genetic dream was thus to sever its connection to its origins. The dream of the power of genetics had arisen from a desire for and a vision of hereditary control. As the genetic vision became more powerful it also began to overrun the boundaries of its origin and to take on significance independent of its original hereditary formulation. The meaning of the gene in Western civilization had originated as a dependent fragment of a concern about heredity and evolution. After a century of turmoil, it promised to transcend the issue of heredity and become a mechanical part of the commercial process, subsumed in talk of genes as "factories" and "blueprints" and personal health options.

Dancing an Attitude toward Genetics

Across the century, therefore, major shifts have occurred in our public discussions of human heredity. The effort to describe the shifting versions of heredity in this century, to explore their possible implications, to encourage scrutiny, might be read by some as a condemnation of genetics. If all genetic discourse has potential negative consequences, then it might seem attractive simply to ban genetic research or public discussion of genetics. Such an urge to banish all discourse about heredity is fairly common in the critical literature about genetics (even when it is not overtly expressed). Such an urge is understandable, for indeed, discourse about heredity is "dangerous," in Foucault's sense of that term. But Foucault's lesson was that all powerful social discourse is dangerous. In *Politics, Philosophy, Culture* (1990: 169), Foucault made it clear that the dangerousness of any social discourse does not mean either that we can do without it or that some other discourse exists that is inherently not dangerous.

Earlier, nongenetic versions of heredity also had their negative potentials. Outside of Darwinism, popular Western discourse about heredity has often focused on God's will in ways that can be taken to be deterministic and discriminatory. The popular phrasing of this, borrowed from the Christian Bible, is that the sins of the father will be visited upon the sons for many generations. Even the concept of original sin can be interpreted in a highly deterministic fashion.[8] Pregenetic biological views also had this dangerous potential because they posited the homunculus, the little preformed man in the sperm who already had all his characteristics even before insemination (a dream which cloning unfortunately now revives and threatens to enact). Even environmentalist notions are not necessarily and inherently positive and guaranteed to generate positive outcomes: "spare the rod and spoil the child" comes to mind as a popular saying of an environmental bent that many would not endorse.

I find it impossible to specify an account of the causes of human characteristics that guarantees problem-free social applications in every possible situation.[9] Equally, however, I think that a social account which ignores the role of genes and heredity in the creation of human beings is prone to serious errors. Visions of untempered social environmentalism are amaterialistic in a fashion that ignores some real if inchoate needs and qualities of human beings. I have seen too many suffering and dying children in too many hospitals whose genetic configurations speak powerfully to the independent substantiality of human biology, of which genetics is one factor. Consequently, rather than being accusatory or celebratory, this account of the varying interpretations of genetics is offered in an effort to emulate what followers of Kenneth Burke have called the tragicomic spirit (Carlson, 1988). That is a world view which recognizes that human beings are powerful and flawed symbol-using animals. We invent symbolic and nonsymbolic tools that are always capable of carrying us over the edge of the abyss. If there is to be salvation from our own combined power and imperfection, it lies in our ability to recognize what havoc our symbol systems urge us to create, to understand that those potentials are always with us, and, through shared understanding of our comic position, to dampen our more virulent tendencies with the varied tools of laughter, awe, disgust, and insight.

Throughout the century, explorations of the implications of biological heredity and of our genes have raised fundamental issues of identity, social connectedness, and human potential. These issues are powerful and dangerous ones. That does not mean these are issues we can or should avoid. Perhaps if we understand the range of ways in which we have addressed these issues in the past we will be better equipped to craft better understandings for the future. Only perhaps.

PART ONE

CLASSICAL EUGENICS

1900–1935

As the nineteenth century gave way to the twentieth, a wave of eugenic sentiment swept across much of the Western world. In the United States, the rhetorical formation concerning the concept of heredity was focused on a debate over the beliefs, goals, and policies offered by eugenics. As chapter 1 indicates, pro-eugenics discourse was conveyed in a vivid and sustained fashion through a stock-breeding metaphor that had recently gained increased credibility through the appropriation of new statistical and cellular research into biological heredity. This new research was not yet called genetics, and eugenics did not neatly separate nature and nurture, but the statistical and biochemical research being conducted on the "germplasm" encouraged social reformers of many political hues to imagine that the regulation of human reproduction might provide an effective control lever for eliminating the social ills of the rapidly urbanizing nation. As chapter 2 will indicate, however, the pro-eugenics discourse was heavily contested by advocates who manifested a strong preference for individual choice in reproductive matters and who offered religious and scientific objections to eugenics. Consequently, eugenic discourse (at least narrowly defined) enjoyed limited policy success in the United States and was rapidly discredited when the Nazi uses of eugenics made the implications of such policies all too clear.

1 BREEDING THE HUMAN STOCK

In Nazi Germany, over 300,000 people were sterilized to improve the human breed (Kevles, 1995: 117; Paul, 1995: 89). When compulsory sterilization proved too slow and inefficient a method for selective breeding, around 6 million more people were killed outright to purify the race. The Aryan Nazis judged other persons unfit to contribute their genes to the common germ-plasm on grounds of their religion, their race, their sexual orientation, and their physical disabilities. The discourse of eugenics, operating in the particular sociopolitical environment of Nazi Germany, legitimated and encouraged this mass extermination and sterilization.

In the United States the offenses traceable to eugenics were far less extensive. Eugenics was bound to a different political system and modulated by other regnant discourses. Approximately 30,000 U.S. citizens were sterilized in the name of eugenics during the classical eugenic era (i.e., 1900–1935; Kevles, 1995: 117). Additionally, in 1924, partly because of eugenic arguments, ethnically biased immigration restrictions were imposed to control what kind of "stock" could achieve American citizenship. These concrete outcomes do not, however, fully summarize the social meaning of eugenics in the United States. Eugenics was a prominent and visible, if not widely popular, discourse during the first three decades of this century. It was significant not only for the actions taken in its name, but also because of its deep influence on the ways in which human beings thought about themselves, each other, and their common heredity. To understand this way of thinking about ourselves, this chapter introduces the manner in which eugenicists selectively appropriated the infant science of genetics to give enhanced plausibility to nascent eugenic discourses, and it explores the workings of the stock-breeding metaphor around which coalesced much of the public thought about eugenics.

The Formation of Public Eugenics

As Martin Pernick (1996: 14) has noted, "Eugenics had many different meanings," and today the boundaries of eugenic discourse remain highly disputed (Ludmerer 1972: 2; Paul, 1995: 3–4). Eugenic rhetoric came from many sources and featured many branches and variations, ranging from radical socialists to avid capitalists to Iowa farmwives. Some who considered themselves eugenicists were merely racists who saw the eugenic movement as a convenient ally for achieving racist social agendas. Others used the banner "eugenics" to argue for better regulation of rental housing (merging eugenics and euthenics as part of a single cause). Still others (like Margaret Sanger) used eugenics as an entering wedge to permit the discussion of free love or birth control. To the extent that there was a consensus definition, however, it was ably summarized by Robert De C. Ward in the *North American Review* in 1910. "To put it bluntly," he concluded, "eugenics has to do with breeding human beings for the betterment of the human race" (56).

Central to this notion of "betterment" was the idea that some people are superior and some are inferior. In that era and region of the world, the public discourse had long identified the "Anglo-Saxon" as the superior and everyone else, from the Irish to Italians, Africans, and Asians, as the inferior. Hence, the most fervent eugenicists (themselves self-identified with the Anglo-Saxon "race") claimed that saving Anglo-Saxon civilization required that they themselves breed more frequently and that the unfit be discouraged from increasing their proportion in the civilization. The unfit were those from other lineages, who also were primarily from lower classes, given the circumstances of the day. Prevention of their propagation was to be accomplished by government-controlled sterilization, institutional segregation, immigration controls, and/or education and voluntary initiative.

These tenets of eugenics are captured in an article in the *New York Times,* 22 March 1930, entitled "Fears Moron Types Will People Nation: Dr. Wiggam Says Democracy Will Not Outlast Century Unless the Intelligent Reproduce." The article reported a speech by Dr. Wiggam, a "scientist," at the New York Association of Biology Teachers. In his address to this gathering of the propagators of science, Wiggam reportedly warned: "Morons are multiplying much faster than college professors, business men or skilled workmen." Thus, he continued, "civilization is making the world safe for stupidity. Not only have the intelligent classes given up the family idea but they are going out of their way to help the physically and mentally unfit, through charitable institutions, prolong their lives and propagate others of their kind. It seems that we are doing everything we can think of to weaken the race." The concerned doctor further noted: "American intelligence is bound to decline and, when intelligence declines, moral character sinks with it. Society is dying at the top and democracy cannot continue, nor can civilization of any kind."[1]

Eugenic statements of this type were widely disseminated in U.S. public discourse in the first third of this century. Not only were they visible in newspapers and magazines, but also at state fairs, in high school textbooks, and at public museums (Hasian, 1996; Selden, 1989). The rise to prominence of this type of talk represented something new in the social matrix, even though most of the ideas central to eugenics had been available for several decades in isolated fragments. In the second half of the nineteenth century, at least since Darwin's writings, versions of eugenic discourse had been articulated and eugenic societies launched. Until the twentieth century, however—that is, until the formalization of scientific research on genetics—these were sporadic and largely unsuccessful efforts. Without genetic science, some crucial element was missing from the eugenic vision. What difference, then, did the science of genetics make?

First, it is useful to isolate what genetics did not contribute to the eugenic formation. Racism clearly played a central role in eugenics. But racism had been a staple of Anglo-American thought for centuries (Jordan, 1977). In Western thought, all species, including human beings, had been envisioned as part of a hierarchical chain of being that clearly demarcated inferior and superior. Anglo-Saxons had generally placed themselves at the top of this hierarchy, next to the white-feathered angels. This hierarchical vision of human beings was dependent on neither Darwinian nor Mendelian concepts of biological heredity. Before and after Darwin and Mendel, the diffuse concept of "blood" relationship and shared bloodlines was sufficient to justify racial hierarchies to many persons and groups.

The idea that modern civilization inhibited natural selection was also not a product of genetic science, if we understand the distinctive characteristics of the new science of genetics as the use of statistical, chemical, and eventually molecular methods to explore the heredity of biological organisms. The belief in the tension between civilization and natural selection had become widely discussed throughout the second half of the nineteenth century as a core element of social Darwinism. Social historians have observed that social applications of Darwinism were well established as a thread in popular American consciousness long before the term *genetics* was coined in 1905 and before Mendel's genetics was rediscovered in 1900. Kenneth Ludmerer (1972: 11) reports that almost 400,000 copies of Herbert Spencer's social Darwinist works had been sold by the turn of the century in America, and he notes the pervasiveness of the social Darwinist creed in the last half of the nineteenth century. Even if it was not enthusiastically endorsed by the populace at large, social Darwinism was broadly familiar and prominent in the public discourse, and it helped make familiar the concepts of evolution developed by Darwin.[2]

Though the mechanisms he offered for the transmission of hereditary traits were different from those of twentieth-century geneticists, Darwin's work had focused attention on heredity and had linked human beings more tightly to the animal world. This linkage had been implicit in Western thought, as far back as Aristotle's definition of human beings as animals who featured the specific traits of reason, language, and virtue (e.g., in *Politics* and *Ethics*). But Christianity had served to heighten the distance between the rational animal and its peers by demarcating the rational capacity as a "soul" with special status, linking it to God and the angels. Darwin shifted the balance back, moving human beings closer to the animal world, downplaying the differences between the not-so-rational animal and what naturalistic observation was demonstrating to be its ever-more-rational-appearing peers in the animal world.

While Darwin merely tightened the analogy between humans and other animals, social Darwinism expanded and modified that analogy to warn that civilization should not tamper with natural mechanisms of selection at penalty of self-destruction caused by the multiplication of the unfit. Social Darwinism thus created or supported a eugenic urge. Conditions at the turn of the century, such as heavy migration to cities and immigration of new ethnic groups, probably intensified that urge. However, based as it was in a model of the natural world, social Darwinism lacked an image of a control lever. Social Darwinists could warn against government activities that reversed the natural order, but few people wanted to return to life in the natural order. The escalating research on genetics at the turn of the century provided an alternative. Such research made it possible to envision mass control because it provided a concrete image of a control lever, accessible to human manipulation, within human civilization.

AN EFFECTIVE CONTROL LEVER

Genetic research involved the conscious and systematic manipulation of animals. Where Darwin had gone out into the natural world and observed nature's operations, geneticists actively experimented with rats and flies and salamanders within the confines of the laboratory, bringing nature within the bounds of civilization.

Genetic experiments were not, of course, the first instance of human manipulation of animals. Plant and animal breeding predated genetic experimentation as a model for the manipulation of animals. Moreover, in part because it was closer to the popular experience of the still heavily rural citizenry, stock breeding would form the basis of the popular imagery of eugenics (and many geneticists were active in stock-breeding activities). There were, however, insufficiencies in the common model of stock breeding that genetic research modified.

First, stock breeding made the application of stock-breeding images to human beings more tenable because it claimed to offer accurate prediction. One of the most distinctive features of twentieth-century genetics was its application of statistical techniques for prediction and explanation. Mendel's ([1865] 1959) genius lay in his ability to track, through rigorous numerical methods, particular features of his plants as they were manifested in progressive generations. Diagrams of genetic "crosses" that rested on these numerical analyses became the dominant visual icons of early genetics (in the mass media, in textbooks, and to some extent in genetic journals as well). The rediscovery and expansion of such statistical analyses at the beginning of the twentieth century were socially significant because these statistical methods purported to allow reasonably accurate (albeit probabilistic) prediction of the outcome of particular breeding plans.

Stock breeders had long known what particular characteristics they preferred, but they also experienced breeding largely as a matter of chance. One might cross a larger than ordinary bull with a larger than ordinary cow and still get only an ordinary calf. By quantifying those chances and explaining them, Mendelian genetics enhanced the image of control held by the breeder. Instead of taking unknown chances, breeders now could expect more systematic aggregate effects, explainable through the rational template provided by Mendelian genetics (or biometrics). The infant science of genetics thus suggested that one could breed animals in an orderly fashion with specific intended results, rather than in the haphazard, ad hoc fashion that had historically been typical of stock breeding. This vision of rational control heightened the appeal of eugenics for mass human applications in a crucial manner. As long as the outcomes of the control of human reproduction were merely the unpredictable product of chance, one could hardly justify such manipulation of human beings. But if one could claim to predict the outcomes in a rational and reliable manner, then the interests of the individual could begin to be weighed against a putatively specific and measurable benefit to future generations.

The new genetics also helped tip the balance away from the individual by directing attention to the chemical and subcellular realm. Classical stock breeding had been envisioned as the manipulation of particular individuals. From the relatively few animals one had at hand, one chose as breed stock those who were bigger or gave more milk. It was the biological individual that stock breeders manipulated, and it was difficult to imagine the direct manipulation of human individuals in the same fashion. The growing research in the new biochemically based genetics, however, subtly shifted the focus from the human individual to a microscopic and transindividual object, the "germ-plasm."

One of the distinctive facets of genetic research at the turn of the century was its intense focus on microscopic cellular subcomponents. By 1902, the chromosomes (originally understood simply as distinctively staining components within cells) had been linked by C. E. McClung to sex determination. Across the next decade a small group of scientists pursued the linkage of outward characteristics to particular chromosomes. As Portugal and Cohen (1972: 130) report, "By 1915, more than two hundred factors had been located on the X chromosome of *Drosophila* [fruit flies]." The increasing focus on the nucleus of the cell that characterized the new biology (Olby, 1994) and the rapid discoveries related to chromosomes meant that instead of imagining themselves as manipulating human individuals, eugenicists could focus attention on the manipulation of invisible "factors" embedded in a germ-plasm that was part of a common property belonging to all of a species. This vision arose in large part from August Weismann's pivotal theory of the continuity of the germ-plasm, which emphasized that the germ-plasm "is always passed on from the germ-cell in which an organism originates in direct *continuity* to the germ-cells of the succeeding generations" (1893: 9, emphasis in original). He noted further: "I have attempted to explain heredity by supposing that in each ontogeny, a part of the specific germ-plasm contained in the parent egg-cell is not used up in the construction of the offspring, but is reserved unchanged for the formation of the germ-cells of the following generation" (1891: 170). By identifying chromosomes as bearers of hereditary characteristics, the microscopic subcellular research gave the older Darwinian notions of heredity an apparently objective existence in a putatively real, material entity.[3] Thus the "germ-plasm" and the "unit factors" embedded in it were a central contribution of the new biology to eugenics; this combination of Mendel's and Weismann's ideas offered concrete and neutral objects for manipulation, which had been lacking in earlier, Darwinian accounts of heredity.

The move from the human individual to the transcendent and permanent mass of the familial or even racial germ-plasm made it easier to imagine the transfer of stock-breeding techniques to humans. Hence, when humans were mentioned as the objects of proposed manipulation in eugenic discourse, it was most frequently in the form of collectivities—extended families or large stigmatized groups.[4] To be sure, human individuals still appeared occasionally in public discourse about heredity, but generally they did so as the products of the germ-plasm rather than as the direct objects for manipulation. The way in which popular discourse used this focus on microscopic units to make discussion of the manipulation of human traits inoffensive can be seen in a description of biological heredity offered by French Strother to the audience of *World's Work* in 1924. Strother specifically cited Weismann's theory and then reported:

> Each of these chromosomes is, in effect, a string of beads; and each of these beads is the determiner of a specific characteristic of the adult human being who will grow from the cell in which it is housed. Thus, there is a bead that determines that this adult shall have red hair; a second, that he shall be color-blind; a third, that he shall be high tempered, and so on. (1924: 170)

With such an account, eugenicists were able to propose simple pairings and selections among chromosomes or hereditary factors that could sort out and eliminate the unwanted beads. Rather than discussing the manipulation of sacred human individuals with rights and dignity, such discourse was able to propose the manipulation of microscopic beads of collectively shared germ-plasm. Indeed, Strother's article then promoted the "building" of a "human race free from idiocy and imbecility and feeble-mindedness" as the long-term object of eugenics (174). His essay overtly and deliberately sidestepped the implications to individual human beings, declaring that those were "not any of his business" (170).

Early genetic research thus helped spur or consolidate eugenics in two important ways. First, the successful use of statistics in the growing scientific exploration and understanding of genetics seemed to offer the possibility of rational, planned breeding with predictable outcomes. Second, highlighting the germ-plasm allowed eugenicists to discuss controlled breeding without talking about controlling human individuals. It shifted the focus of discussion on the symbolic, if not practical, level from the manipulation of individual persons to the manipulation of an impersonal and invisible mass belonging to the race as a whole. This new symbolic focus depersonalized the process of controlled breeding and made its benefits appear permanent.

The findings of early genetic research were thus appropriated to provide what appeared to be a serviceable control lever for effecting the goals of earlier social Darwinist ideals. Early genetic research thereby contributed to the movement of Darwinian ideas out of the natural world into the realm of human civilization and control. This intellectual shift, occurring in a hospitable social environment, produced a relatively rapid escalation in the development and then popularization of eugenic ideas. The dominant metaphor for eugenics in the public discourse became this scientized vision of stock breeding.

The Stock-Breeding Metaphor

"Breeding Better Men," an article published in the magazine *World's Work* in 1908, illustrates the vision of human breeding that was thus created. The article opened with an extended comment on the accomplishments in plant and animal breeding that had taken place, noting how planned breeding could be added to environmental manipulations such as weeding and watering. It then argued that these same techniques should be equally applied to humans: "If physical and mental traits are inherited, it is of prime importance for the welfare of state or nation that those stocks which are on the whole endowed with the best traits should contribute more, many more, individuals to the next generation than should those stocks whose characteristics are on the whole bad" (Pearl, 1908: 9821).

The metaphor that human beings could be bred like plant and animal stock constituted the dominant rhetorical figure of the hereditarian discourse of this era. It appeared overtly in over a third of all articles written on genetics or eugenics in the era, regardless of their topic, focus, or slant (see appendix 2, table 2). This is an overwhelming presence for a single metaphor in any set of discourse.[5] By the end of the century, however, such talk about breeding human beings would come to carry a diffusely repugnant connotation. It requires some exploration, therefore, to understand the striking fit between the breeding metaphor and the rhetorical situation of its usage.

Those who used the metaphor most frequently and comfortably were members of the professional class.[6] Legislative successes and research activities in eugenics were undertaken and articulated by a relatively small group of well-educated Protestants of Northern European ancestry (Paul, 1995). This group can be reasonably identified in shorthand as the "professional" class of the time (though many professionals, of course, did not agree with eugenics). The readership of magazines, and especially of newspapers, was broader than this class, and eugenic discourse was heard widely outside this group. Moreover, different isolated facets of eugenics appealed to different groups, sometimes large sectors of the populace. However, the professional class constituted the major force promoting eugenics. Doctors, lawyers, scientists, and business persons generated the bulk of eugenic discourse, and it reflected their ideals and their concerns.

The turn of the century was an anxious era for the relatively well-to-do Americans whose ancestors had come from Europe and who had dominated the continent for 200 years. Massive migration from rural areas to cities was occurring. From 1880 to 1920 the population in the United States shifted from overwhelmingly rural (75 percent) to majority urban (51 percent) (*Historical Statistics,* 1970). This migration was creating and amplifying what we now know to be the standard problems of urban poverty in industrializing areas. However, the scope and scale of these problems were probably both new and shocking to the members of the society. While absolute numbers of those involved in crime and poverty might not have increased (Kevles 1995: 129), the proximity to crime, poverty, prostitution, and other social ills that accompanies life in a city environment was nonetheless capable of creating distress. The shock was exacerbated by the immigration of a variety of different ethnic groups with a vast array of languages, cultural values, and cultural practices different from those preferred by the people derived from Northern European ancestry and culture. As Diane Paul (1995: 103) has noted, a third of the population in 1920 consisted of immigrants and their children, and the proportions were higher in cities. The descendants of the Northern Europeans who had gained dominance by displacing the native population 200 years earlier were now faced with losing that dominance in turn.

A variety of initiatives were launched at the turn of the century to address the new social conditions.[7] These initiatives operated from the assumptions that elites could appropriately reform groups who were deficient (or who were experiencing socially enforced deficiencies, depending on one's political views) and that government should play a role in these efforts. Accordingly, institutions for the insane were built, reformatories for young people with problems were established, public educational initiatives were expanded, and labor protection laws such as child labor restrictions and the eight-hour work day were passed. Once social Darwinists appropriated key concepts from genetic research, a vision of eugenics that could fit into this model of top-down control developed. As Martin Pernick (1996: 44–46) has pointed out, this new biochemically based eugenic vision was particularly attractive because it claimed to displace many expensive social care-taking institutions by providing a solution that was permanent. Both left-wing and right-wing groups used the metaphor of stock breeding to envision a world in which poverty, mental illness, and moral depravity could be permanently bred out of the population.

The stock-breeding metaphor embodied the model of elite control of others and polarized the hereditary potential of the two extremes of the racial stock—the superior ("genius") and the unfit. In contrast with the discourse of genetics in later eras, pro-eugenic articles did not favor the average or normal person. Eugenics gave virtually no attention to what would later be understood as the numerically dominant middle sort of human being. Instead, the most common sets of terms used to designate the qualities that were sought in human offspring were those describing a small elite; the term *superior* itself and the term *genius* were most frequently employed, along with rare positive uses of the term *perfect.* In no other era were these the dominant terms used to describe the characteristics wanted in future generations of humans (these differences are statistically significant: see appendix 2, table 6).

The focus on superior humans often included detailed and extensively developed analogies between humans and animals. The genius was like the prize bull at the state fair. Indeed, state fairs featured "better baby" and "fitter family" contests with blue ribbons awarded for the best in show. But stock breeders also worked actively to cull the herd of the unfit. The terms used most frequently to designate a group of human beings who should not breed were *the unfit* and *the feebleminded.* These terms were identified variously with those who were poor, those who were associated with criminal activity, and/or those who engaged in behaviors disapproved of by the Northern European public culture, such as prostitution and high alcohol consumption. Because these terms were economically and culturally defined, they were further associated with those from other ethnic groups or races, especially the immigrants from Southern Europe, who were arriving in large numbers, bringing their varied cultures but scant economic resources. One of the major instabilities in the usage of the metaphor was the issue of just how many people and which groups constituted the unfit. The candidate groups ranged from the overwhelming majority of the population, to a commonly cited 10 percent, to the tiny sliver of the institutionalized who were actually sterilized. As chapter 2 will note, identifying who should count as unfit constituted one of the insurmountable difficulties for the pragmatic and rhetorical success of the metaphor.

Whatever the number of human beings consigned to each of its categories, the use of the stock metaphor shaped an evaluation of human heredity in a highly polarized and totalized form. One's heredity, in the form of one's germ-plasm, was treated as a singular entity, either wholly superior or wholly unfit. No gradations were considered, and no diversity was recognized within this framework. Human heredity was to be judged, and it was to be judged on absolute and dichotomous criteria. Within this vision, individual persons were not unique assemblages of multiple characteristics. They were simply excellent or unworthy. As such, they were imagined as members of biological castes.

This bifurcated vision was associated with the eugenicists' sustained preference for the term *germ-plasm* as opposed to the term *genes. Germ-plasm* was the older term, dating from a period when biological scientists knew that the "germ cells" were involved in the transmission of biological features, but when they were still uncertain about the components and mechanisms of physiological inheritance. As a singular and relatively abstract term, *germ-plasm* encouraged a vision of a unified whole. Weismann himself had begun his account of the germ-plasm by describing the regeneration of lost body parts. He took this ability as a starting point for understanding the germ-plasm, noting, "As a matter of fact, groups of units removed from an organism possess the power of constructing the whole anew; and we are thus obliged to admit that the tendency to take a specific form is present in all parts of the organism" (1893: 2). Each characteristic of an individual was thus substantively identified with the total character of that individual. The whole was implicit in each part. This holistic understanding of the germ-plasm lent itself to a totalistic, singular vision of heredity. The alternative was the vision of the "genes" as multiple and varied carriers of many complexly interacting bits of information. The foundation of this alternative vision was available in biological studies as early as 1910; the work of Nilsson-Ehnle and East clearly showed that multiple genes were involved in producing a single characteristic (Portugal and Cohen, 1977: 119).[8] Many studies were exploring the variability and combinations of genes during the time, and an account based in these biological differences would eventually encourage emphasis on the variable and multiple features of every individual rather than reducing individuals to singular qualities—superior or inferior.

As the new scientific data became available, however, public advocates for eugenics did not rapidly revise their word choices or their vision. The term *gene* did not appear in the sample set of public discourse used in this study until the midtwenties, and it did not replace *germ-plasm* as the dominant term until the midthirties. The transition from *germ-plasm* to *genes* in the public media was associated with an increase in anti-eugenic articles and with articles that were making the transition away from eugenics. Those who held the most enthusiastic views toward popular eugenics generally seemed also to have been most likely to persist in the use of the term *germ-plasm* rather than converting early to the concept of the genes.[9] This shift was also associated with a generational turnover in geneticists. As Cravens has noted (1988: 175), those who entered the profession before 1910 (when *germ-plasm* summed up the dominant understandings) were more likely to be active eugenicists, whereas those who entered after 1910 were more likely to reject eugenics or at least to stick closely to their workbench.

The notion of human heredity as a unified, coherent whole that was either consistently superior or uniformly unfit had a strong appeal that led to its endurance in public discourse long beyond its disappearance from the scientific cutting edge. One strong strand of this appeal was its ability to help the scientifically literate deal with what Diane Paul (1995: 22) has called "evolutionary anxieties." Thinking of oneself as just another animal (instead of, for example, the reflected image of God) might produce substantial existential angst. The stock-breeding metaphor allowed those who accepted the Darwinian account to distance themselves from the implications of that theory. Yes, human beings were animals, but it was others—those who were poor or lived according to different cultural rules—who most clearly manifested this animal nature. Promoters of eugenics could interpret themselves as superior to these others; they could envision themselves as the breeders, not the to-be-bred.[10]

The polarized character of the stock-breeding metaphor also functioned effectively to compress the burgeoning diversity in the nation into a simple, controllable hierarchy. Faced with an enormous variety of new cultures, values, and life-styles on the part of the new migrants, promoters of eugenics were able to use this metaphor to delegitimate the diversity and to avoid confronting and exploring it—to avoid even assessing it on any basis of particularity. Instead, the diversity was neatly compressed into a dualistic hierarchy of inferior and superior. The superior—the familiar ways, their ways, the homogeneous—was to be propagated. The inferior—all new ways—was to be eradicated under the banner of "lack of fit." It was, of course, no coincidence that the biological castes imagined with the stock-breeding metaphor precisely coincided with the cultural castes battled by those who employed the metaphor.

The stock-breeding metaphor also facilitated eugenic aims by deflecting attention from issues of individual reproductive rights. Animals obviously had no such rights. To the extent that unfit people were envisioned as mere animals, their individual hopes, dreams, and fears for the propagation of their selves and their ways of life could be discounted. The framework and potency of the metaphor thus helped to focus attention away from the issues of individual rights, which otherwise constituted a strong basis for rejecting eugenic visions (see chapter 2). Instead, the eugenicist discourse focused attention on the good of the herd, defined variously as nation, race, or society. This tendency was reflected in the percentage of articles whose key theme was controlling individual reproduction for the benefit of society. Forty-two percent of the articles in this period had the good of the race as a central theme, whereas only 16 percent focused on the reproductive interests of the individual, and no articles focused on the health of individuals (see appendix 2, table 3). Simeon Strunsky articulated these values in the *Atlantic Monthly* when he said that "society has the right to guard itself against the evil workings of the iron law of heredity" (1912: 850).

The society to be protected was uniformly envisioned along the lines desired by the Northern European–derived professional class. The *World's Work* ("Husbanding the Nation's Manhood," 1910: 13470) was starkly direct about this goal when it declared, "There is in the study of immigration and birth statistics a plain lesson for the old American stock: It is Extinction or Eugenics." What was good for society so defined included lower taxes (one needn't pay for educating and warehousing the unfit), but also less crime, alcoholism, and prostitution and more of the high culture favored by elitists such as Havelock Ellis. This focus on culture, however, clearly stretched the reach of the stock breeding metaphor, which called to mind animal characteristics rather than human morals and higher abilities.

Biology and Environment in Eugenic Discourse

Although eugenicists often took physical appearance as a sign of overall fitness (Pernick, 1996), the dominant traits eugenicists desired for humans were not the physical traits identified with most of the animal models used in the stock-breeding metaphor.[11] Instead, American eugenicists generally indicated that human beings were to be bred primarily for intellectual qualities, including the moral or behavioral qualities that they deemed to be linked to intellectual abilities. In all four eras in this century, physical characteristics were linked to heredity or genetics in approximately two-thirds of the popular press articles (see appendix 2, table 3). However, only in this early era did two-thirds of the articles link mental characteristics to heredity in this manner. By the end of the twentieth century, that percentage had dropped to one-third. Similarly, behavioral (or moral) characteristics were linked to genetics in close to half (43 percent) of the articles in this classical era of eugenics, whereas by 1995, the frequency had dropped to 14 percent. French Strother's list of characteristics linked to specific genes provides an example of the flavor of the early assertions about pervasive and simple genetic linkages to human outcomes:

> . . . some of those now known to be hereditary: absence of pigment from the eyes (albinism), six-fingered hands and six-toed feet, color of hair, color of eyes, shape of the facial features, double cow-lick, color blindness, inability to see at dusk, musical talent, mathematical talent, inventive genius, literary talent, high temper, cheerful disposition, feeble-mindedness, epilepsy, certain types of insanity, and certain types of criminality. (1924: 174)

In spite of the smooth transitions among different types of characteristics presented by this list, the tight linkage among intellect, moral behavior, and heredity that dominated the classical eugenic era actually belied the animal model. There were some efforts to demonstrate the inheritance of behavioral patterns in animals (especially by using models of dog behaviors such as herding, swimming, or hunting). However, the eugenic models clearly went far beyond the bounds of the animal-breeding metaphor in this regard. Yet this was necessary to the eugenic program. No one was particularly interested in breeding humans for appearance alone. Concerns over breeding for physical strength and stature received some minor attention only in articles relating to the eugenic effects of war. Physical health and rigor received passing and generalized mention as desirable characteristics in the racial stock and as indicators of mental qualities. However, there was no persistent concern for the physical health of the "less fit." The fact that the lives of the poor were physically short, nasty, and brutish was not perceived as a national crisis by the eugenicists. The crisis was located in the intellectual and associated moral qualities of the populace.

This lack of fitness was usually summarized in the term *feeble-minded.* Feeble-mindedness was not constituted solely by a low IQ (though it was often measured that way). It was a way of living, particularly the way of living common to the lower classes. Eugenic field-workers "diagnosed" individuals as feeble-minded merely by observing the homes in which they lived, the apparel they wore, and their manner of speech (J. D. Smith, 1985). Feeble-mindedness, although defined as a lack of intellectual development, was thus inseparable from moral and behavioral inadequacy. The feeble-minded, it was argued, did not have the intellectual ability to avoid poverty and moral delinquencies such as alcoholism, prostitution, and crime. The *Literary Digest* of 1921 noted, for example, that "TOO MANY ARE FEEBLE-MINDED. The mentally below par fill the segregated rooms in our schools and grow up to be vagrants and criminals. . . . these are societies [*sic*] future jailbirds and prostitutes" ("Our Drift toward Degeneracy," 1921).

The linkage between intellectual ability and behavior allowed the discrediting of immigrants and their cultures in a stunningly effective fashion. If the differences in behavior manifested by those from immigrant cultures were the product of inferior intellect, then the new immigrants were intellectually incapable of assimilating by learning the dominant culture. Simultaneously, because it was the product of inferior intellect, the immigrant's culture was not worth welcoming or even exploring. These sentiments were expressed by Representative Albert Johnson of Washington in the debate over the 1924 immigration act. He warned, "If we continue to admit great numbers of those who fail to assimilate, we will be in danger of ourselves being assimilated. Our customs will be changed for us" (*Congressional Record*, 1 March 1924: 3424).[12]

The linkage of intellect and moral culture to each other and to heredity was of central significance to the political usages to which eugenics was put. Without this linkage, the eugenicists could not have used arguments from heredity to justify the stringent immigration restrictions on non–Northern European groups, or even the segregation and sterilization of the poor. Thus, even though arguments about moral and intellectual character were not strictly drawn from the stock-breeding metaphor and actually ran counter to it by emphasizing nonanimalistic characteristics, these arguments were dominant in the eugenic discourse.

That genetic research lacked any evidence linking something like animal intellect or animal moral behavior to genes in this era is obvious. Moreover, statistical analyses of the intellectual and moral characteristics of human families were inherently incapable of separating nature from nurture. Cadres of researchers from sites such as the Eugenics Records Office, headed by the conservative biologist Charles Davenport and funded by the wealthy and politically liberal Mary Harriman (see Kevles, 1995: 44–56), gathered information on physical and social traits that ran in families. Not only were the studies of social characteristics grossly flawed by faulty categorizations of traits, but the statistical analyses of these traits provided only correlations. Even if a particular trait ran in a family, there was no way of knowing whether the trait resulted from family environment or family biology. This conflation of familial environment and familial biology did not, however, trouble the eugenicists. The public version of eugenic discourse was not and did not need to be strictly biological in its view of heredity (see also Pernick, 1996: 43ff.). Instead, much public eugenic discourse interpreted heredity as the product of a mix between familial environment and familial germ-plasm. An article by Mabel Potter Daggett, "Women: Building a Better Race" (1912), is illustrative. Potter Daggett reports on the activities of women in Iowa using the Iowa State Fair as a bully pulpit for eugenics. She extols the women's quest to help each other build the "perfect child," and she celebrates the $50 prize awarded for the "best baby of all." However, Potter Daggett's campaign included lectures during the week of the fair on "hygiene, dietetic care, and psychological training" by women physicians. It included laws to protect women from men who carried syphilis and to protect "Motherhood" by preventing women from becoming "sick and tired" through overwork, along with providing hospitals in rural areas. The same magazine article argued for a state law requiring the provision of plumbing and ventilation to clean up Indiana's slums. The discourse of heredity in this era thus merged genetics and environment.

The mixture of the hereditary influences of family environment and of germ-plasm was extremely convenient to the social goals of the eugenicists. Supporting the combination allowed them to argue that even if the new genetic research was wrong, the goals and means of eugenics were right. Whether bad children were created by the bad genes carried in feeble-minded parents or by the inability of the feeble-minded parent to provide a suitable home life was immaterial. In either case unfit parents produced unfit children, and so their breeding should be prevented. As "Our Drift toward Degeneracy" (1921) put it:

> At least three-quarters of all children are born to living conditions well below those of the average, as measured not by wealth, but by the quality of the parents—while a scant one-quarter have the advantage of homes above the average.
>
> Now, what enthusiast for the power of environment would deliberately raise most of his flowers and chickens under adverse conditions? Yet this is exactly what we are doing with the human species.
>
> So, from the view-point of either heredity or environment

> our method of perpetuating humankind is a complete reversal of nature's scheme for quality of species.

The predominance of this mixed view in the discourse of the era is indicated by the fact that 60 percent of the articles in the period attributed hereditary causality to the gene along with some other factors such as environment or family conditions, while only 34 percent of the articles assigned causality solely to the gene (see appendix 2, table 1).

Eugenics was thus a discourse about heredity, but it was not strictly a discourse about genetics. It incorporated both family environment and biological inheritance in its purview. Indeed, this was precisely the component of American eugenics that the biologists (especially the British geneticists) disdained as they sought to create a biological science of genetics. Interpreters working at the end of the century, looking back on eugenics, have persistently misread eugenic discourse by screening out these references to family environment because they have read eugenics reductionistically and anachronistically, as a discourse of genetics or protogenetics (e.g., Kay, 1993).[13]

The public discourse about heredity in the first three decades of the twentieth century clearly featured a vigorous set of arguments that called itself eugenics. The goal of eugenics was to perfect the society by using controlled breeding programs to increase the number of superior persons and to eliminate those deemed inferior. This was clearly a highly discriminatory discourse. The source of this discrimination, however, was not an unusually high degree of biological determinism. Eugenics was content to rest itself on an ambiguous mix of familial inheritance, understood as including both environmental and biological components. Instead, the discriminatory tenor of eugenic discourse was fanned by the particular model of biological process on which it rested. The model offered was a holistic one; the identity of the entire person was linked to the character of the microscopic part—the germ-plasm. Part and whole were tightly fused. In contrast with the expectations of later critics, who would argue for greater holism in genetic research, this fusion was not benign. Instead, the vision of an undifferentiated and transcendent germ-plasm constituted an entity that could be hierarchically graded in a dualistic manner—fit or unfit. The resulting discrimination was clearly linked to a perfectionism that urged the reproduction of only superior individuals. Moreover, superiority was judged as a uniform quality pervading a caste of persons based on their race, class, or ethnic heritage, rather than as a singular facet among a melange of personal traits.

The apparent feasibility of the eugenic project was clearly enhanced by the new science of genetics. The statistical measure of the transmission of hereditary traits and laboratory manipulation of heredity in animals authorized a vision of the control of heredity that human beings had not previously imagined. Eugenicists did not, however, simply follow the dictates of science; they selectively appropriated portions of biological research that served their interests. Public eugenicists refused the linguistic transition from *germ-plasm* to *genes* and ignored the early discoveries of the complexities of genetic processes because those discoveries delegitimized the eugenicists' polarized, reductionistic thinking and undermined the feasibility of the simplistic control levers they sought. The same selective vision was evident in the divided labors of those who were active geneticists but also avid eugenicists. While they might or might not apply more complex genetic models in their scientific work,[14] when they applied genetics to social policy, their own personal social position and political views guided their judgment as to which scientific formulations were relevant in the public sphere. The persuasive appeal of the stock-breeding metaphor, which glorified their status as scientists and promised to solve their problems as citizens, was compelling to them. Scientists who were eugenicists therefore tended to feature the simplified cross-breeding of Mendelian "single gene" traits as the model of genetics best fitted to public action.

Rigorous adherence to biological genetics of any form was not, however, necessary for the achievement of eugenic goals. Breeding programs could function to control human moral and intellectual traits whether the mechanism of transmission of the heritage passed on by families was conceived of as environmental or biological or both. Genetics did not then have even rudimentary mechanisms for separating familial biology from familial environment. Neither science nor the society at large had yet sharply divided the heritage one received from one's family into separate biological and environmental packets. This ambiguity allowed the eugenicists to interpret studies of familial heritage as genetic heritage when they found it convenient to do so, while also allowing them to deny the significance of the lack of a rigorous proof of genetic components when necessary. Eugenicists thus had it both ways. They gained the prestige and luster of science for their cause without being held accountable for scientifically rigorous proofs.

Eugenics was therefore a public appropriation of genetic discourse that was highly discriminatory, extremely perfectionistic, and substantially deterministic, albeit deterministic with regard to familial influences that were indeterminately biological and environmental. The widespread public dissemination of eugenic discourse is a testimony to the extent to which it promised to allay pressing concerns for its advocates and to some extent for the editors who printed it and the readers who endorsed it or acquiesced to it. But eugenic discourse was not a singularly triumphant discourse in the United States. It was not the only available account of human heredity. Consequently, to gain a fairer picture of the attitudes about heredity that circulated in the public arena during this era, the limitations of eugenics as a rhetoric and the challenges to eugenics should also be examined.

2 CHALLENGES TO EUGENICS

The previous chapter can be read as yet one more instantiation of the now-standard construction of eugenics as a frightening, overwhelmingly successful social movement that wrought horrible ends. Such a construction is useful for marketing ends. Books sell better if they are about frightening social phenomena of immense importance. More substantively, there is good reason to argue that eugenic thought is capable of authorizing truly evil outcomes. However, to let that portrait stand as it does, without contextualizing it within the competing discourses of the day and without noting the rhetorical and scientific weaknesses in the eugenicists' vision, would be to misrepresent the public attitudes about heredity. In spite of the frequent play it received in many public arenas, eugenic discourse in the United States was an odd curiosity, frequently laughed at by the populace whenever it became distinctively itself (that is, whenever it was more than simply the expression of racist and culturalist sentiments). Eugenics in the United States was also a failed rhetoric that had relatively small impact. This chapter will demarcate the limitations on the success of eugenics as a public rhetoric and outline the competing discourses that challenged it. The chapter will then trace the fall of eugenics to its implementation by the Nazis, to its undermining by the development of molecular biology, and to its own internal rhetorical limits.

The Lack of Implementation of Eugenic Programs

To recontextualize eugenics adequately first requires remembering that eugenics by no means went uncontested. Diane Paul (1995: 10) has touted the overwhelming popularity of eugenics by noting that it was one of the most frequently indexed topics in the *Reader's Guide to Periodical Literature* by 1910. About 20 percent of these articles, however, either opposed eugenics or included substantial arguments against the eugenic view (see appendix 2, table 7). An article title from *Current Opinion,* "The New Heredity as a Tissue of Absurdities" (1919), captures the sense of contest and even ridicule used against many biologically hereditarian perspectives. Moreover, most of the articles offering alternatives to eugenics were not indexed under that heading. This counterdiscourse was widely available in many forms. An instructive example is provided by the layout of the *New York Times* on 25 September 1921. On page 16, indexed under the heading of eugenics, is an article reporting a pro-eugenic speech by Major Darwin. Next to it lies an article on infantile paralysis that emphasizes environmental precautions parents must take to protect their children from disease. The latter article is not indexed in any fashion that would allow one to assess it as part of the nature/nurture debate of the era. Yet promotion of such environmentalist perspectives was widespread. This was, after all, the era when the behaviorism of John B. Watson attained the status of what Mark Haller has called a "fad" (Haller, 1984: 170), an era in which Horatio Alger continued to be a living myth (Paul, 1995: 11), and when religious revivalism, though past one of its cyclical peaks, continued to have substantial success.

Newspaper columns, magazine articles, and speeches offering many contrary views of the origins of human capacities were at least as prolific as eugenics. In addition, pro-eugenic rhetoric in the United States was not particularly successful at producing policy or programmatic victories, at least on its own. The two most significant achievements that have been attributed to eugenics were immigration restrictions and the legalization of compulsory sterilization. Immigration restrictions would probably have passed if Mendel had never grown a single pea. Although eugenicists were active proponents of immigration restrictions, the fundamental forces driving the restrictions were racism, cultural imperialism, and labor concerns. Eugenic discourse reorganized the racist and cultural biases of the broad coalition supporting immigration restriction, and it gave a patina of scientific legitimacy to these other forces, but the persuasiveness of that patina should not be overestimated. In previous eras, immigration restrictions against the Japanese and the Chinese had been successfully organized without "eugenic science," and it is probably more reasonable to see eugenicists as opportunists running alongside the anti-immigration wagon than as the horses powering it. This is not to deny that eugenics made some contribution to the victory of immigration restriction, but as the Pulitzer Prize–winning historian Edward Larson has demonstrated, it is a serious overstatement to attribute the Immigration Restriction Act of 1924 to eugenic discourse (Larson, 1995: 101–4). Racism and cultural imperialism were mixed into eugenics, but they were much larger, more powerful, and at that time more enduring rhetorics.

The power of eugenics with regard to compulsory sterilization laws was somewhat greater. In spite of the fact that some eugenicists repudiated sterilization laws, many others were active and successful in passing such laws and defending them in the legal system. Daniel Kevles is probably correct in concluding that "American eugenicists played a dominant role in bringing about the passage of state sterilization laws" (1995: 100; see also Reilly, 1991: 39, 42). Although American eugenicists were a predominant force in sterilization efforts, the program of sterilization had distinctively narrow success in the United States. While the plight of those who were sterilized under the eugenic laws should never be dismissed as insignificant, on a comparative scale sterilization laws were not heavily used in this country. Daniel Kevles himself notes that "from 1907 to 1928, fewer than nine thousand people had been eugenically sterilized in the United States, as against a 'feeble-minded menace' of—in Henry Goddard's estimate—three hundred housand to four hundred housand people" (1995: 106). Both Kenneth Ludmerer (1972: 95) and Kevles attribute this relatively low utilization of externally imposed sterilization to a general discomfort or outright opposition to sterilization that suffused the judiciary and the executive branches (see also Haller 1984: 124, 177; Larson, 1995: 64). Even Reilly (1991), who wishes to emphasize the significance of eugenics as a motivation in involuntary sterilization and to place it later in time, admits that sterilization was not so much a popular measure as a measure sequestered from public view, endorsed and administered by physicians and state superintendents of institutions for the mentally ill.

The eugenicists' vision of large-scale interventions in human reproduction was thus repudiated. Instead, they were successful only at gaining indirect sanction for use of the new technologies for involuntary sterilization upon those whose rights to individual autonomy had already been restricted.[1] To gain a sense of scale, the relative lack of popular endorsement and usage of sterilization can be contrasted with other tools of racist social control. Lynching, segregation, harassment, and white-induced race riots were popularly and enthusiastically implemented by American whites against American blacks throughout the decades surrounding the turn of the last century.

Arguments against Eugenics

The relative lack of public enthusiasm for eugenics is traceable to several arguments frequently and energetically voiced by opponents of eugenics.[2] The most important of these arguments was the defense of the rights of individuals to control their own reproduction. Though the term *reproductive rights* was not yet codified, there was a strong public sense that reproductive capacities were too "intimate" to be infringed by the government. Samuel George Smith (1912), in a rich, anti-eugenic article published in the *Atlantic Monthly,* noted: "Human freedom is not without difficulties and defeats, but it is the only hope of the race." Diane Paul has suggested that this era did not recognize parental rights (1995: 71), but her observation rests on the assumption that rights exist only when the U.S. Supreme Court explicitly endorses and protects them. Although the Supreme Court did not explicitly address reproductive rights until the decades of the sixties and seventies (in cases such as *Griswold* v. *Connecticut* [381 US 479 (1965)] and *Roe* v. *Wade* [410 US 113 (1973)], the Court itself argued in these cases that the right to privacy in reproductive matters had been a long-standing component of the American heritage. Eugenicists themselves also recognized the existence of a long-standing public presumption of reproductive freedom when they argued that reproductive privacy was outmoded and should be eliminated.[3] An article promoting eugenics in the *Literary Digest* identified the presumption in this fashion: "How have we come so far on the way to racial degeneracy without any attempt to check ourselves? Mainly, thinks Mr. Humphrey, because of a pious horror of interference with the right of parenthood. It is a hang-over sentiment from the ages of ignorance and superstition which we can not shake off" ("Our Drift toward Degeneracy," 1921). Eugenicists spent substantial time arguing that the right to parenthood needed to be overturned only because they generally believed their audiences took that right for granted.

The eugenicists were, however, largely unsuccessful in their quest; most people continued to take for granted the right to control their own reproduction. Many probably wished that the feeble-minded would not reproduce their kind. The majority may even have believed that in a few extreme cases individuals should be sterilized, either for their own good or for the good of their potential offspring, or even for the good of the society, which might have to care for the offspring in the face of the parents' incapacity to do so. These, however, were most generally understood as rare exceptions to be executed only in cases where the individual's ability to exercise free choice was demonstrably impaired. There was never broad support for governmental interference in private reproductive practices on a large scale.

Broad support for state control of reproduction was further stymied by the anti-eugenicists' argument that there was no agreement about which qualities should be bred for and who should control the eugenicists' knife. Opponents of eugenics correctly highlighted the lack of consensus on the qualities that should be bred into the population. Thus, *Current Opinion* quoted a Dr. Lowie as challenging Galton's criterion of "civic worth" because the phrase "remains undefined and indefinable" ("Risks Involved . . . ," 1920: 500). The article asked whether the creative thinker who challenged authority or the docile person was of greater value to the society. This counterrhetoric thereby emphasized that there was no single standard for human excellence but rather that humanity prospered because of its diversity.

Opponents of eugenics also asked who would control reproductive decisions if individuals did not do so for themselves. Dr. Lowie was further quoted to the effect that

> when men are mated with a conscious purpose to effect a definite end, there is bound to be a clash of ideals. The eugenist may have quite definite notions of what is desirable but whether his decrees would tally with the judgment of the wisest and best of mankind is another matter. Even a disinterested commission of experts would run the danger of yielding to professional bias and subjective preferences. ("Risks Involved . . . ," 1920: 500)

The fatal flaws in the stock-breeding model were precisely its inability to specify who among the human species should be the breeders and for what purposes human beings should be bred.

Finally, eugenics was challenged by arguments indicating that factors other than biological heredity were more important to individual human characteristics. A variety of different approaches were taken to this direct challenge to the genetic component of the eugenic creed. Two primary versions of this argument, though fairly widespread at the time, were historical losers to the genetic world view. The first was a defense of Lamarckianism in contrast with genetics. Not all Lamarckians were opposed to eugenic policies. However, direct public attacks on Mendelian genetics weakened the plausibility of the eugenic position because the authority of Mendelian genetics was the major source of the legitimacy of that position. The Lamarckian claim that characteristics acquired by parents could be transmitted to their offspring through the germ-plasm was so thoroughly routed after this era that it is easy to forget how strong were its advocates at the beginning of the century.[4] The confident tenor of the pro-Lamarckian argument in that era is, however, indicated in an article in the *Literary Digest* in 1919. The article abstracted and reprinted portions of an essay by C. L. Redfield that had been printed in the *Western Medical Times.* Redfield defended the Lamarckian idea that characteristics acquired through hard work by parents were transmitted directly to their children through the reproductive process, and he denied that genetic mutation was the cause of changes in any species. His demonstration of the scientific evidence in the case produced this conclusion: "Power is gained always by exertion and . . . the doctrine that the powers of living things can be altered by 'mutation' from generation to generation is 'scientifically unsound' " ("Gaining and Losing Human Power," 1919: 25; see also "Inheritance of Acquired Characters Proved at Last," 1919). The power of scientific authority was thus wielded as a rhetorical weapon on both sides of the eugenic controversy.

A second and probably more effective source of opposition to the biological assumptions in eugenic discourse was religious discourse. It has been widely noted that the Catholic church was a source of substantial opposition to eugenics (Kevles, 1995: 118–19). There was also, however, a nondenominational Protestant opposition to eugenics. The famous orator, politician, and religious advocate William Jennings Bryan argued against eugenics in his speech and essay "The Bible Is Good Enough for Me" (1925). The Great Commoner's trust in God as creator was joined to a preference for reliance on the Christian version of individual "free will." Like other religious opponents of eugenics, Bryan associated eugenic attitudes and genetics with biological determinism, the belief that one's status in life is determined by one's genes. In contrast, the religious perspective held that one's fate in life is determined by a combination of Original Sin, Christ's Redemption, and one's individual willingness to accept that redemption and act accordingly. This religious opposition to the familial fatalism of eugenics was joined by a few broad secular arguments against any biological castes based on genetic determinism ("Freakishness of Heredity," 1915).

Lamarckian and religious opposition to eugenics might have had some significant impact in their day, but they would not have much long-term staying power. Lamarckianism was soon thoroughly discredited, and, while religious discourse would continue to provide some opposition to particular applications of genetic technologies, most religious authorities would come to accept the basic functional processes of the genes. Even in this early era, however, there were some progressive scientific propositions that would form the basis of enduring arguments against classical eugenics throughout the century. In the early decades, the major element of this argument was that the genes indeed passed down the human inheritance but that what humans inherited, unlike other animals, was teachability. Hence, what humans learned was more important than what they arrived with at birth (Cooley, 1923). This argument did not deny that human beings are products of their genes; it simply set the product of the genes in such a fashion that culture is enormously important to the development of individuals. It thereby accepted biology but denied biological determinism. The use of scientific genetics against eugenics was rare in this early period and clearly not a major alternative view, but it foreshadowed important currents that would soon participate in the demise of classical eugenics.[5]

The Demise of Classical Eugenics

Eugenicists had always suffered some public ridicule for their claims ("Dean Inge Bids Church Heed Eugenics' Value: Says Advocates Are Not Fools or Cranks," 1930; Kellogg, 1924).[6] This ridicule became outright hostility and absolute rejection in the midthirties. Previous scholarship has identified three causes for this rejection: the Great Depression, the rise of the Nazis, and improvements in scientific knowledge of genetics. The last two are evident in the public discourse.

It has been argued that the Great Depression made it immediately obvious that eugenics was false (Pickens, 1968: 5, 205). Widespread unemployment, it has been suggested, cast doubt on the proposition that the unemployed were out of work because of their incapacities rather than because of factors in the social structure. Such a claim is not widely evident in the public debate about eugenics during the period. Before and after the crash of 1929, in spite of severe unemployment, eugenic arguments maintained their fervor because it was possible to view the Depression as a demonstration that the eugenic crisis was even more urgent than previously indicated. On this view, the multiplication of the unfit was far advanced, and drastic and immediate eugenic measures were therefore needed to rectify the crisis. Even absent such arguments, the meaning of the Depression was in great flux. Not only was it highly contested (as it is to some extent in our era), but it was also more inchoate in the process of receiving its first definitions (Sefcovic, 1997). Thus, the Depression was not immediately and widely interpreted as proof of the falsity of eugenics. That interpretation seems to have arisen through the lens of later reflection, which is more cognizant of the interdependencies of job supply, interest rates, and global markets.

This is not to say that the Depression did not have any long-term impact on the credibility of eugenic arguments but merely that its impact was not swift, direct, or dramatic. In contrast, the influence of the Nazis' applications of eugenics upon the plausibility of the eugenic doctrine in U.S. public discourse was both direct and immediately evident. The *New York Times* provides a clear case in point. Until

1933, articles in the *Times* about eugenics were somewhat mixed but preponderantly favorable, and eugenics received a fair amount of coverage. After 1933, eugenics was associated with the Nazis and was increasingly denigrated. The sharpness of this change in affect is illustrated by the treatment given experimental eugenic colonies. In 1931, in an article entitled "Success Has Come to Eugenic Village," the *Times* favorably described an experimental eugenic colony in Strasbourg as "growing and prospering." In 1934, the nation's premier newspaper reported on a similar colony without positive adjectives and with negative scare quotation marks around terms like *eugenically faultless* ("Houses to Go to Parents of 'Faultless' Children"). By 1935, the *Times* was openly derogating such efforts as a "national fetish" of the Nazis (" 'Perfect Man' Aim of Nazi Eugenists").

The attitude toward eugenics taken in the *New York Times* shifted from cautious approval to cautious disapproval to outright rejection. The trend in the national press as a whole was similar, though not as clear-cut and uniform. Just as there were anti-eugenic articles in the period when the rhetoric of classical eugenics was most widespread, so also did a few eugenic articles of the classical type continue to be published over the next 20 years. In the midthirties, however, American public discourse increasingly associated eugenics with the Nazis and their mass campaigns for sterilization (the death camps were not yet a reality and the full extent of segregation activities had not yet been exposed in the United States).[7] This association was well grounded, given that the Nazis had declared their debt to eugenicists in the United States and Britain and that Nazi eugenic programs were similar in their focus on the unfit and in their use of sterilization. The Nazi programs simply implemented the proposals of eugenics fully and rigorously. The Nazis thereby made vivid and incontrovertible the concerns that had been raised against eugenics in the previous three decades. To borrow a concept from argument theory, the Nazis gave an overwhelming "presence" to the anti-eugenic arguments so that these arguments now carried more force.[8] Before the war, the presence of the immigrant masses, the urban poor, the illiterates newly entering the job market, and the growing public institutions had instigated fears that were more immediate and therefore greater than fears of abuse. The Nazis redressed that balance. The concrete image of eugenics in action swamped the benign appearance of the stock-breeding metaphor, and eugenics was discredited as an enterprise whose human costs were too great.

Simultaneously, if less dramatically, genetic research was chipping away at the assumptions upon which rested the eugenic vision of simple and effective breeding of human populations (Haller, 1984: 160; Pickens, 1968: 207). The ambiguous notion of the germ-plasm, with its connotations of cohesion and homogeneity, had been replaced by the pluralistic notion of chromosomes and genes, which were now increasingly plotted to specific sites along those chromosomes. Although most scientists still erred in thinking that the genetic material was probably constituted of proteins, they had learned to induce mutations through the use of X rays (in 1927) and UV light (in 1930) (Portugal and Cohen, 1972: 132). Moreover, scientists on the cutting edge, such as George Beadle and Edward Tatum, were beginning work that would locate the first particular chemical substances produced by particular genes (which was achieved by 1940 with the identification of tryptophan and kynurenine as regulated products of genes in flies with particular eye-color variations; Portugal and Cohen, 1972: 165). The genetic material was therefore increasingly envisioned not as a unitary mass of germ-plasm but as specific, targeted subcomponents—particular genes—that had specific and narrow effects. Moreover, the polygenic vision of trait causation had been firmly established by numerous studies that made it clear that Mendelian ratios did not hold for many, if not most, biological characteristics. Genes and the characteristics of living entities were increasingly understood to be related not in a simple one-to-one fashion but in an assortment of complex pathways. The implications of sexual assortment of these multiple characteristics made it evident that selective breeding for all but a few characteristics would not be a simple matter. Exploration of population genetics also made it evident that recessive features could not be rapidly and easily eliminated from the population.[9]

This more precise and sophisticated science gradually began to be reflected in the popular discourse. The *New York Times* of 22 January 1928, for example, concluded: "There are thousands of possible combinations of genes, so that the coming together of complementary genes produces something not necessarily found in the parents. . . . It follows that for the present, at least, we must leave mankind to its own devices and permit our great men to appear as sporadically as they have in the past" ("Improving Mankind"). In addition, some advocates used developing scientific understandings of gene-environment interactions to replace the eugenicists' narrow vision of two castes—the superior and the unfit—with a more panoramic perspective on the diversity of the species. They argued that human beings came with multiple talents and that there was no way to prefer one set above another, especially since what constituted a talent in one situation might well be a disadvantage in another ("Knockout for the 'Perfect Man,' " 1932; Mohr, 1934).

There was a substantial time lag in the public incorporation of these new scientific views. Early work on the polygenic characteristics of heredity was completed in the first two decades of the century (Portugal and Cohen, 1977: 119). Yet these data did not appear widely in the popular media with their anti-eugenic implications until the middle thirties. Clearly, the scientific findings did not drive the public discourse to new anti-eugenic positions but merely supported reorientations urged by other forces (cf. Turney, 1998: 93). It is too much to imply, however, that the science played virtually no role (Paul, 1995: 114). Early genetics had provided to the would-be eugenicist the image of a control lever that conceptually distanced the means of manipulation from individual humans (toward animals or germ-plasm) and that offered the promise of rational and certain outcomes. But eventually the lever was no longer believed to be either attractive or effective. Its attractiveness was destroyed by Hitler's use of eugenics, which made it evident that individual humans were manipulated and harmed by eugenic policies. Belief in its effectiveness was also undermined by a variety of new advances in genetics, which made it evident that human genes are too complex to be perfected through a few simplistic breeding selections.

It has been suggested that the feasibility limitations revealed by the expanding understanding of the complexities of genetics did not have much impact on public attitudes. The proof of this claim has rested on arguments that many American geneticists were eugenicists first and scientists second or that American eugenicists, not being predominantly genetic scientists, either ignored these data or did not understand them (Paul, 1995: 69). But in public rhetoric, it matters little whether the true believers learn from new data and correct themselves, for the balance of public opinion is built not from the true believers who speak their contesting doctrines, but from members of the broader public who serve as audience and judge of these doctrines. While it is true that the first generation of eugenicists widely persevered in their errors, new research in genetics provided the ammunition for changing the mind-sets of a second generation of geneticists (Cravens, 1988: 175) as well as for delegitimating the earlier claims among audiences of nonspecialists.

The new account of biological complexity did not, however, definitively trounce its competitors. Throughout the century, a tension existed between a vision of genetics as simple Mendelian genes neatly sorting out cleanly categorized characteristics and a vision of genetics as a complex set of interactions yielding small and highly variegated effects. Both of these visions operate within genetic science itself, and each is appropriated for different purposes by different public advocates.[10] The dynamics of reduction and synthesis persist in science because they are productive there. The dynamics of reductive simplicity and full-blown complexity persist in the public discourse because each features substantial, though different, rhetorical advantages.

Some Rhetorical Features of Eugenics and Its Opposition

The respective rhetorical strengths of reductionistic and complex accounts of genetic processes are well illustrated in the iconographic representations of genetics that appeared in the classical era. The reductionistic view of genetics offered clear and colorful accounts of biological heredity. Pictures of cross-bred roosters and charts of cross-bred rabbits bedecked popular accounts that described genes as beads on a string and depicted simple manipulation through cartoon diagrams. These images appeared in two different types of articles. One set purported merely to teach genetics. These articles spent their pages describing sexual reproduction and Mendelian ratios. The other set of articles alluded to these research findings and then applied them to eugenics. The rhetorical advantage of both of these reductionistic presentations rested on the ease with which they could be communicated in popular media where space is highly restricted and visuals are highly effective.[11] In addition, the view of inheritance as a simple clocklike mechanism governed by mechanical genes in simple relationships offered in these articles was rhetorically appealing to scientists and the public alike because it was orderly, and such order is not only comforting but also allows the systematic development of scientific questions and scientific research.[12]

In spite of these rhetorical strengths, depictions of the diversity of the human species and the complexity of genetic interactions arose to challenge this account of genetics in the twenties and thirties. These challenging articles were not accompanied by vivid photographs or simple diagrams. Moreover, instead of careful explanation, the discourse often tended to sound celebratory with regard to human features, and therefore it probably appeared to be less scientific. Nonetheless, the popular view of biological complexity had rhetorical advantages of its own. The portrayal of biology as offering astonishing complexity and diversity, unpredictable and less controllable, is richer. Instead of providing a sense of control, it promises adventure and life forever new. If biology is a simple clocklike mechanism, subject to easily applied controls, then our conquering of it is too easy a matter. There is greater romance in the vision of a biological world constantly interacting to produce a diversity of outcomes (see, e.g., Mohr, 1934). Moreover, in a society where diversity is materially manifested in the public as audience, the model that emphasizes complexity is more likely to gain a sympathetic ear among a broader range of citizens.

Perhaps, one might argue, the complex account is also a truer account. If so, it is not surprising that later eugenic proposals became more modest when manifestations of complexity arose to complicate visions of order and control. Whether because of the social complexities of diverse human communities, or the biological complexities of molecular interactions, or both, even the one group in which the eugenic true believers wielded real power eventually bowed to the conditions of feasibility generated by innate complexity. When the Nazis' experience of selective breeding revealed its limits to them, they escalated their program of sterilization to the more rapid and comprehensive means of direct extermination.

Arguably, a rhetorical balance of strengths and weaknesses inhered in the underlying scientific models of eugenics and its opposition. Likewise, there were rhetorical weaknesses inherent in the stock-breeding metaphor that countered the rhetorical strengths described for that metaphor in the previous chapter. The most significant of these weaknesses was an unredeemable pessimism about human nature, which was linked to a lack of proportionate means of implementation.

"Humankind at present is a motley and mongrel lot," declared Samuel J. Holmes (1926: 350). For human beings, the stock metaphor suggested that bad inheritance naturally swamped good inheritance. The superior person or the genius (who bore the only wanted kinds of genes) was enormously rare, whereas the unfit were numerous and would proliferate. Eugenic articles equivocated on the number of persons who counted as unfit. Ten percent was a proportion often cited. However, even this substantial percentage understates in at least two ways the image of the masses of unfit created elsewhere in the same articles. First, those who cited the need to eliminate the bottom 10 percent often indicated the desirability of doing so in each generation (thus constituting a moving line of fitness, working up the pyramid each generation). Second, in many accounts the unfit also constituted the third or more of the population that was "alien," and thus the imagistic and psychological construction of "the unfit" was larger than the already unwieldy 10 percent explicitly listed as in need of eugenic elimination. "Our Drift toward Degeneracy" (1921), for example, claimed that fewer than a quarter of children were to be found even in "average" conditions.

The cataclysmically pessimistic judgment of the human race was necessary to motivate the drastic social changes urged by the eugenicists, but the same pessimism threatened to constitute a barrier so large as to be pragmatically insurmountable. The alluring promise of genetics as a control lever for improving the human breed was swamped by the overwhelming task set for it by eugenicists: not simply to improve the human breed, but also to eliminate the unfit and thus to eliminate crime, poverty, and cultural practices alien to the young nation's dominant tradition. In part, this difficulty was traceable to a fundamental contradiction in the stock-breeding metaphor. The metaphor said that human beings were animals but also that the truly animal in nature (the "feeble-minded") could not function in human civilization. Eugenics could not answer the question, If human beings are only animals, and animals cannot survive in civilization, how can civilization survive? On this account, the future of human civilization was dark indeed.

The pessimism about the human species implicit in the stock-breeding metaphor thus easily led to despair. As an anonymous biologist concluded in *Harper's Monthly:*

> If only we had the means of eliminating all inferior combinations, as the breeder of animals and plants does, we might expect to produce a very superior race in future America. Unfortunately we do not now have this power, and there is little probability that we shall ever be able to apply to our human stock the principles of good animal breeding.
>
> Under these circumstances it seems probable that history will repeat that of many ancient civilizations. ("The Future of America: A Biological Forecast," 1928: 539)

America, he concluded, would fall into decadence and extinction, as Greece and Rome had done. In the end, however, the belief that the available eugenic measures were insufficient to the cause and unacceptable to human nature led not to the downfall of Western civilization but rather to the downfall of classical eugenics.

The fatal difficulties had been evident throughout the eugenic movement; practical measures for achieving what the stock-breeding metaphor imagined were at every stage the sticking point. From the very beginning, Galton imagined practical eugenics as an effort that would be a very long time coming, and even the most enthusiastic articles on eugenics tended to become vague and indeterminate when the pragmatic issues arose. This was because all the available means had obvious and fatal flaws. Mass compulsory sterilization was widely seen as repugnant to human nature, that is, inherently at odds with our understanding of the fundamental character of human dignity (even if sterilization in "extreme cases" was adopted, as described below; see Larson, 1995: 64). Mass institutional segregation was not only repugnant to many—animals might be caged, but humans should not be—but also too expensive for anyone to contemplate. Extending existing limitations on marriage was understood as unlikely to prevent unwanted copulation. Thus, in the United States, the most frequent plan offered for implementing eugenics was not sterilization, or immigration controls, or institutionalization, but simply research and education in the principles of eugenics in the hopes that this would lead to voluntary adherence to eugenic prescriptions for better breeding.[13]

The method of education and voluntary reproductive self-control was manifested in the "fitter family" contests at state fairs. It resulted in widely disseminated popular eugenic handbooks that urged women to marry well and to attend carefully and lovingly to their children. It showed up in the rapid and sustained inclusion of genetics in high school biology textbooks (Selden, 1989). The method, however, was obviously incompatible with the task. One could not realistically expect unfit animals to act in a responsible manner and forgo their own procreation. From the very beginning of eugenics, therefore, most of its proponents understood the limits of their own difficult position to some extent. French Strother admitted these limitations to the audience of *World's Work,* saying that the eugenic scientist

> knows enough facts now to be very useful to people who are sufficiently far-sighted to use them—as a good many wise people are now doing. But he has not enough facts to claim the right to offer the world a panacea for divorce, unhappy marriage, or the ills of society. The most he hopes is that some day he will have so many facts, so clearly proven, that only the very ignorant will not know how nature deals with human heredity, and only the fools will be rash enough to try to beat nature. But that's a century or two ahead yet. (1924: 170–71)

The often-touted distinction between "positive" and "negative" eugenics thus hid a crucial problem for eugenicists. Negative eugenics, the prevention of breeding by the so-called unfit, could not be achieved by any popularly acceptable methods on a scale of any sufficiency.

Positive eugenics, the increased breeding of the superior, would not resolve the dark problems envisioned by the eugenic rhetoric, and it did not fit within the dominant metaphor of the stock breeder. The stock-breeding metaphor thus had substantial rhetorical weaknesses arising both from its oversimplification of human nature and heredity and from its lack of rhetorical fit for a diverse human audience.

As the twentieth century opened, early discoveries in genetics enabled human beings to imagine that the germ-plasm provided a rational control lever that would allow us to exert simple control over the characteristics of our species. Like stock breeders, we could sort and segregate the human germ-plasm to eliminate the unfit and to produce more superior members of the breed. This discriminatory and perfectionist vision was widely repeated in public venues, but it did not go uncontested. Religious opponents denied both the version of earthly perfection offered by the eugenicists and their replacement of free will with biological dictation. Other opponents to eugenics grounded their objections in scientific principles. Lamarckians denied the sequestering of the germ-plasm from human culture, while a new generation of geneticists denied oversimplified relationships among genes and outcomes, though they were eager to accommodate a role for more complex genetic influences in human characteristics. The strongest and broadest argument, however, was anti-discriminatory. Most opponents to eugenics agreed that stock-breeding programs for humans violated individual rights to reproductive autonomy. They emphasized that human beings were too diverse to be neatly sorted into two heaps—the fit and unfit—and that no one could be trusted to make such sortings. Eugenic doctrines were thus contested with religious beliefs, with political doctrines, with outdated scientific theories, and with brand new scientific research. Moreover, the rhetoric of eugenics featured its own internal limitations, especially a trenchant pessimism about humanity that cast doubt upon the feasibility of the program on its own terms.

Altogether, these arguments and limitations were sufficient to prevent substantial enactment of the eugenic program in the United States. This does not mean that various components of the public did not find various pieces of the eugenic arguments highly appealing. Involuntary sterilization of institutionalized mental patients gained de facto public sanction. Perhaps, given time, the eugenicists might even have overcome social inertia and gained sufficient public acceptance to implement their central goals of broad-scale reproductive control. However, when Hitler's implementation of the eugenic program in

Germany made the consequences of such views vividly obvious, whatever attraction the broad eugenic program had previously held for the public was effectively destroyed. With the costs of the eugenic projects vividly before the public's mind, the new vision of genetic complexity rapidly replaced the older vision of the transcendent germ-plasm. Eugenics was, temporarily at least, displaced from the public agenda.

PART TWO

FAMILY GENETICS

1940–1954

In the aftermath of World War II, the classical eugenic movement dissolved. Its advocates struggled to reconstitute the movement by reforming it—abandoning the overt racism and classism of classical eugenics while maintaining the explicit goal of improving the human species. This reform eugenics was not adopted by the American public as a framework for understanding human heredity. Instead, the public understanding of heredity was reframed through a vision of genetics that defined itself as being in opposition to eugenics. Public discourse in the era explicitly and energetically rejected governmental control of genetic choice and shifted the focus of attention about heredity from race and nation to the inheritance of nuclear families. To implement this shift, the private practice of heredity counseling was presented as a replacement for the public endeavors of institutional segregation and sterilization. The eugenicists' preoccupation with superior intelligence and behavioral morality was also repudiated in favor of a narrowed focus on mental and physical illness. Simultaneously, advocates of the new genetically based framework for heredity explored means by which new findings about the complexities of genetics could be meshed with a democratized vision of the reproductive potential of all human beings.

This transition from classical eugenics to family genetics was achieved through transformative rhetorics that drew from the scientific credibility of genetics as well as from an external source—the changes in global politics of the postwar era. Chapter 3 describes the rhetorical transformation of eugenics to family genetics through its association with the splitting of the atom, international freedom of inquiry, and positive postwar appraisals of science. Chapter 4 analyzes the contents of the new discourse of family genetics.

3 THE TRANSFORMATION TO GENETICS

Genetics and eugenics had come into Western consciousness bundled together. Primitive genetic research was used by eugenicists to argue their case to the public, and genetic research attracted scholars interested in eugenic projects. Nobel Prize winner Hermann Muller, for example, declared, "It was basically because of my belief in the future artificially controlled biological evolution of the human species that I started studying genetics myself in 1908 at 18" ("Speeding up Evolution," 1946). This linkage between eugenics and genetics was challenged by the rise and fall of the Third Reich.

The central uses to which the Nazis put eugenic and racist policies delegitimated both overt racism and overt eugenic programs in Western public consciousness. Scholars have generally assumed that the delegitimization of eugenics meant a similar (if temporary) delegitimization of genetics in favor of accounts of human behavior based exclusively in various versions of "the environment." This assumption has rested largely on the apparent substantial decline in explicit public discussion about eugenics and genetics in the United States that occurred with the onset of World War II (Kevles, 1995: 164, 169; Ludmerer, 1972: 2, 127; Paul, 1995: 120; Pickens, 1968: 4).[1] In the years from 1940 to 1970, there was a noticeable drop in the number of articles indexed under headings such as eugenics and genetics in the *Reader's Guide to Periodical Literature* and a similar drop in newspaper coverage. The apparent drop in numbers of articles, however, belies the central significance of this period, for it was in these years that the crucial struggle to redefine the public role of genetics occurred, and in these years its general direction was largely formed.

Equally important, the drop in the number of articles might reasonably be interpreted as only apparent, for two reasons. First, the attention to the world war shifted the relative percentage of articles attending to international versus domestic concerns in almost every topic area. As international news absorbed more of the available page space, coverage of many domestic issues experienced a relative decline, but it would be a mistake to assume that all these issues lost adherents in the nation. Second, and more important, one major change in the eugenic movement in this era was a refocusing on the population control movement (Pickens, 1968: 69–85). In the immediate postwar era hundreds of articles about birth control flooded the press. Although the birth control movement had many strands and motivations and was not reducible simply to eugenics, a substantial number of the birth control articles simply transferred the discourse of eugenics from an internal, intranational frame to an external, international frame. In these articles, the "unfit" breeding excessively were not recent immigrants to the United States or lower-class Americans, but rather the alien hordes on other continents. The world overpopulation problem became the focus of anxiety, and "birth control" instead of "eugenics" became its name. The birth control movement thus absorbed much of the energies of eugenics in the postwar era, and if articles indexed under its headings were included in the eugenic article count, they would actually constitute a substantial increase in attention to eugenic themes. This new movement, however, was not directly identified with genetics, so efforts to count articles as an indicator of public attitudes become quite subject to issues of category selection. Given such conditions, it seems most accurate to identify the era not so much as a period of declining attention to issues of heredity as an era in which thinking about the constellation of issues surrounding human heredity was radically realigned.

The postwar change took the form of a realignment rather than an extinction because powerful forces had a sustained interested in the continuation of the study of heredity. Had genetic research been supported solely by popular and eugenic imperatives, it might have died with the eugenic movement. But genetic research was part of a network of scientific investigations, and it was driven in part by the intellectual and social force of that network, as well as by its more direct connections to the public.[2] In addition, those supporting eugenic programs sought nonpublic ways of promoting eugenics, feeling that eventually public opinion would accept a new, more scientifically advanced eugenics. Genetic research was seen by these persons as a method for promoting eugenics without incurring its immediate public stigma (Kay, 1993: 39–45). Commonsense observations of the role of biological heredity in human characteristics may perhaps also have helped to maintain discourse about genetics. There were, therefore, strong forces functioning to ensure that research into genetics and a public discourse about genetics did not die with classical eugenics.

The schism between classical eugenics and genetics did, however, constitute a potent rhetorical problem: Without the framework provided by eugenics, what social meaning could genetics hold? What discursive places would talk of genetics fill? What impact would "genetics" have in U.S. practices and policy? Creating new answers for these questions required first that the linkages between genetics and the tarnished vision of eugenics be decisively broken. The field of genetics had to transform the strong negative affect it had earned from its association with Nazi eugenics into a positive, or at least neutral, public affect.

José van Dijck (1998: 61) has argued that this transformation occurred during the late fifties through the early seventies, and she has identified factors including the space program and James Watson's book *The Double Helix* as key components of this transformation. These other factors were surely important in building the ethos of genetic research, but the original transformation took place a decade earlier. On the scientific side, this transformation occurred through a displacement of human genetics in favor of plant and animal genetics and through the medicalization of human genetics. On the public side, the shift accompanied two developments in international politics—the rise of the Soviet Union as the new enemy of the United States and the moral and psychic trauma of the atomic bomb. After realigning genetics with the new world order, advocates of the new genetics could then deploy the credibility and utility of science to focus the attention of the American nuclear family on their reproductive practices.

World War to Cold War

After World War II, the public vision of the United States had expanded to take in the world scene. The nation was now a global actor, and domestic consciousness was heavily influenced by the global scene. Discourse about genetics simply rode the coattails of the major shifts in international alliances that occurred rapidly at the end of the war. As the balance of enmity for the United States shifted from alliance with the Russians and opposition to the Nazis to strident opposition to the Soviet Union, the affective alliances surrounding attention to human heredity changed from the negatively connoted, Nazi-associated eugenics to a positive valence arising from Soviet opposition to genetic research.

During the war, eugenics had been associated with the Nazis and thus with the fount of evil. The emotional strength of this association is suggested by a *New York Times* article that reported on a presentation at the National Academy of Sciences claiming that chromosomes in cancerous plants took the form of a swastika ("Says Cancerous Plants Show Sign of Swastika," 1940). Reform eugenicists made an effort to breech this towering wall of disapproval by offering a new version of eugenics and by recasting the meaning of the Nazis' use of eugenics. The reform eugenicists tried to locate the evil of the Nazis in their racism rather than in their state-sponsored manipulation of human breeding. Simultaneously, they dropped the assumption that "good" qualities were linked to specific races or social classes in their own rhetoric and programs. They argued that degeneracy could be found in anyone's heritage, so that everyone would be a target of eugenic supervision. Reform eugenicists even argued that progressive degeneration (rather than eugenics) was the cause of the rise of evils such as Hitler, for an increase in morons would lead to such bad government. The reform advocates, however, retained support for the use of reproductive controls to improve the species. Earnest Hooton (1943), for example, asked the readers of the *Woman's Home Companion* a rhetorical question about the human species: "Does anybody know how to mold him into a decent animal?" The answer of this Harvard professor of anthropology was "yes," and he advocated a new "Department of the Population" to include a Bureau of Marriage and Genetics. Like classical eugenics, reform eugenics was pessimistic; Hooton argued that half of the American population was degenerate. Also, like classical eugenics, reform eugenicists argued that the conditions of civilization destroyed the processes of natural selection and produced progressive degeneration of the human race.

The reform eugenic view constituted a substantial minority discourse in this era. Major news magazines reported neutrally, if not favorably, on state sterilization plans and the British eugenic movement. Reform eugenics came under hostile critique from other sources, however. Opponents of reform eugenics located the monstrosity of the Nazi actions in their use of reproductive controls, focusing on the methods they had employed, not simply on the specific targets the Nazis had attacked. These articles did not present the victims of Nazi terror as members of subordinated racial classes, but rather as consubstantial with the reading audience, and by extension all of humankind. The *Nation,* for example, reported on sterilization by the Nazis, saying:

> Its mixture of madness, science, orderliness, and obscenity is characteristic of the special blend of poison with which Nazism infected mankind. . . . Of course, we have long known about Nazi methods of biological warfare, among which sterilization was one of the favored weapons. But there is a difference

> between knowing of their existence and hearing the actual details. (Gumpert, 1946: 597–98)

The *Nation* concluded that the report on sterilization "shows more clearly than it could be imagined the miasma of torture and perversion in which Nazi-dominated Europe lived." Therefore, it was government-directed reproductive control, not merely racism, that was stigmatized by the postwar opponents of eugenics. Reform eugenics was, from this perspective, little better than classical eugenics. In both cases, outside influence over individual reproduction was the required tool, and improving the species or race as a whole was the ideal. In the aftermath of the defeat of the Nazis, these arguments of the opponents of eugenics were persuasive; in general the American public concluded that both the means and the ends of the Nazis were unconscionable.

Reform eugenicists were unable to surmount the association of eugenics with the Nazis. Shortly after the war, however, the nation's enmities shifted. Although the Nazis were not forgotten, the Soviets soon came to dominate the national consciousness as the preeminent enemy and visage of evil. Thanks to Trofim Lysenko, the Soviets could be identified as resolute foes of genetic research, and by way of the emotional postulate that the enemy of my enemy is my friend, that made the citizens of the United States into the allies and supporters of free research into genetics.

Coverage of the career of Comrade Trofim D. Lysenko constituted one of the three most popular story types under the headings of genetics, eugenics, and heredity in this era. Early in the period, immediately after the war, magazines and newspapers reported Lysenko's experiments and his arguments that acquired characteristics could be inherited. These reports were generally not enthusiastic and sometimes contained a barbed challenge, but there remained some ambiguity in their tone. Lysenko's claims, perhaps because they were initially perceived as anti-eugenic, had at least interest and some grain of plausibility in the popular press. With the rapid acceleration of the cold war, however, the nation's press corps sharply slanted their stories against Lysenkoism. Before the end of the forties, reporting on Lysenko came to focus on how he had risen to power in the Soviet Union as a matter not of science but of ideology. The mass media identified Lysenko's rise as "the destruction of science in the USSR" in an article of that title (Muller, 1948). The press reported that Lysenko falsely claimed to be able to change major characteristics of species merely by exposing them to different environments and that Soviet geneticists who disagreed with Lysenko were sent to Siberia or forced to follow the party line.[3]

The ideological heart of these articles was the strong contrast between the freedom of inquiry, which purportedly characterized Western science, and the Soviet model, in which the party determined what was true. The *New Republic,* for example, quoted *Pravda* as denouncing a Russian geneticist on the grounds that "Zhebrak, as a Soviet scientist, ought to have unmasked the class meaning of the struggle taking place in connection with the problem of genetics. But, blinded by bourgeois prejudices, by a contemptible subservience to bourgeois science, he adopted the views of the enemy camp" (Lash, 1949). Through such articles, genetics came to stand as a powerful icon of free scientific inquiry.[4] This association was reaffirmed in the next two decades as well. During 1955 and 1956, the U.S. press hailed the downfall of Lysenko when his school lost favor with the passing of Stalin. Then again, with the fall of Kruschev in the midsixties, the American press celebrated the decline of Lysenko and the return of classical genetics based on the Western model to the Soviet Union. Across three decades, the storyline of the Russian resistance to genetic research normalized genetic research as a part of the virtuous Western tradition.

Atoms and Genes

The difficult transformation from a public value structure in which genetics was associated with eugenics and the Nazis to one in which genetics was identified as good because it was opposed by the Soviets was bolstered by a second major international focus of the postwar press. The two major scientific undertakings of World War II that also played major roles in the moral accountings of the war were eugenics and nuclear research. Whereas eugenics had attained a simple emotional valence equivalent to the evil of the Nazis, the moral associations given to the atom bomb were more complicated. The bomb was alternately a savior technology in the war against evil and a destructive force so powerful that its moral worth was questioned. Atomic energy and genetics became entwined in new and peculiar ways in the postwar period.

The gene had long been compared to and imagined as the "atom" of genetics (Haskins, 1945: 98). Weismann described Darwin's theory of pangenesis by noting that cells have "the power of giving off invisible granules or atoms, which, at a later period and under certain conditions, can develop again into cells similar to those from which they originated" (1893: 3). This comparison was made overtly in multiple ways in multiple articles in the period. For example, the atom was taken as a basic model and akin to the gene in Joseph Bernstein's comment: "To Lysenko's scoffing observation that no one has yet seen a gene, the obvious answer is that no one has ever seen an atom" (1946: 524). Such quick comparisons revealed basic assumptions about the nature of the gene. The gene was almost always described as a particle, and both the press and the scientific community long had assumed that when the gene was identified, it would be a discrete entity, probably round, imagined in much the same way that the atom had been imagined. This expectation is visible in the publication in *Newsweek* ("Genes: Sliced and Pictured," 1949) of a purported picture of a gene as a round, discrete dot, a photo-image reprinted from *Science* (Pease and Baker, 1949: 9). This is a picture of the transcription unit from a fly salivary gland. Such transcription units are visible in these unusual chromosomes, although they are not discrete particles but rather continuous portions of the DNA strand. Moreover, these physical segments do not map directly to the concept of the "gene," and these are not typical of genes.[5] However, both scientists and the public found what they were looking for in this initial (somewhat mistaken) identification of the gene as a discrete entity, similar to the imaginary atom. These visual analogies captured the deeper sense in which the public thought of genes as the fundamental particles of biology in the same way that they had been taught to think of atoms as the fundamental particles of physics.

The figurative analogy between genes and atoms deeply influenced the project of genetics and its public understanding, but the functional role of atomics in making genetics important to the public came from a different merger. In 1946, Hermann J. Muller was awarded the Nobel Prize for a series of experiments in which he used radiation to induce mutations in flies. The press reported this as an artificial speeding up of the rate of evolution, and they portrayed it as a model of human control of genetic processes. However, these experiments were also taken as indicators of a threat that radiation posed to the human genetic heritage. Several newspaper and magazine articles in the period reported that the spin-offs from atomic research posed a serious danger to the human gene pool. Both fallout from atomic bomb tests and exposure from medical radiation were indicted as threats to the long-term survival of the species. These articles transformed the older, discredited evolutionary worries of eugenicists into newly legitimate and urgent grounds for the pursuit of genetic research. Eugenicists had argued that the human gene pool was threatened with gradual degeneration because of the constraints that civilization placed on natural selection. The discrediting of eugenics and the inherent pessimism of it made that argument largely unpresentable. However, with the coming of atomic energy, the argument could be raised in a new and even more pressing form. Not merely was natural selection being impeded, but also through radiation human civilization was now actively inducing mutations that would degenerate the fitness of humankind. Humans were not merely preventing natural selection from weeding out degenerates; human action was also accelerating the rate of natural (biological) degeneration.

This negative perspective on induced mutation took a decade to permeate the national consciousness. *Mutation,* a neutral word for genetic change, became progressively associated with bad outcomes.[6] This shift can be seen between the two major sets of articles that linked genes and radiation. The first set of articles was published in 1946, when Hermann J. Muller won the Nobel Prize for his 20-year-old research indicating that X rays cause mutations. The press dubbed Muller a "gloomy nobelman" for his predications that "most mutations are bad" and that, "in fact, good ones are so rare that we can consider them all bad" ("Gloomy Nobelman," 1946). In contrast, the reporter accepted that we were "slated for an evolutionary speed-up," but asserted, "Whether this [radiation-induced mutation] will be good or bad is not at all clear." Instead, *Time* held out hope that "the few superior mutations would survive to improve the race." Thus, in 1946, Muller was a gloomy prophet facing an optimistic audience.

By 1956, when the second slate of articles on genes and radiation came out, the popular voice had adopted Muller's negative view. This set of articles covered British and American reports on the hazards involved in the use of radioactive technologies. The press now portrayed the intersection of radiation and genes purely in a negative light. The result of radiation was genes that were "permanently damaged" (Waddington, 1956: 137). The press now concurred with Muller, saying that "practically all radiation-induced mutations with effects large enough to be detected are harmful" (Thompson, 1956). In the popular imagination, all genetic change came to be viewed in a negative way. *Mutant* became a science-fiction label for monstrous beings who plagued human civilization. The intersection of atomic and genetic research gave rise, therefore, to one of the most fundamental mythic expressions of the fears generated by advancing technology. Movies and books portrayed our species as menaced by genes degenerated by human actions, especially radiation (Turney, 1998: 127).

In spite of its gloomy prognostication, this intersection between genetics and atomic research provided a rationale for continued public attention to the human gene pool.[7] This resulted in institutional measures that gave genetic research long-term stability. The investigation of the effects of radiation on genes established the Department of Energy as the eventual (if odd) home of much genetic research and contributed in the long term to the development of the Human Genome Project.

Physicians and Fertility

While the developing formation of international alliances and other fallout of World War II generated structural support for genetic research and transformed the valuative connotations of genetics, the application of genetic knowledge continued to suffer from a lack of structural mechanisms. The vision of state-controlled breeding that had dominated the classical eugenic vision had been discredited; thus there was a need for developing a popular substitute. Reform eugenics identified this substitute structural authority clearly: it was to be the medical profession. Earnest Hooton articulated this turn to medicine as a clear response to the disasters of Nazi eugenics and governmental control:

> Who, you ask, can undertake a program for the restoration of the American people to physical and mental health and to social and economic efficiency? Who can improve human quality of future generations? Certainly not the politicians, nor the conventional educators. . . . Definitely not the lawyers.
>
> The only large group that is at all qualified to undertake this gigantic task is the medical profession. Doctors understand human nature better than any other class of persons. (Hooton, 1943: 96)

Although reform eugenics was eventually unsuccessful, its suggestion that the medical profession constitute the social institution that would provide eugenic guidance was adopted. At the time, however, there was limited linkage between control of reproduction and the medical profession. That linkage was strengthened by the normalization of medicalized birth control, as well as by the direct and substantial number of articles published in the era addressing the issue of infertility.

Between 1945 and 1950 a dozen articles appeared in the mass media regarding infertility. These articles seem to have only the most tangential relationship to genetics, but they asserted the need for couples to come to physicians before pregnancy. In tandem with the enormous growth in articles about birth control, the attention to infertility directly asserted the authority of the medical profession over the process of procreation. The press counseled married couples: "Don't wait too long for your baby; don't take measures suggested by your friends, and don't go to a quack. Your family physician will refer you to the proper specialist or clinic" (M. Davis, 1947: 43). Moreover, the articles made infertility appear to be a widespread phenomenon, thus asserting the widest possible scope for the involvement of medical physicians in the procreative process: "The childless marriage is often a sad one. Yet there are all too many childless homes. About 17 percent of marriages in this country are barren, and another 5 to 8 percent suffer from 'one-child sterility' " (M. Davis, 1947: 42; cf. Condit, 1996b). Identifying a quarter of all marriages as medically deficient and encouraging all women to worry about their potential infertility effectively generalized medical concern about procreation. This move built a discursive, and eventually structural, bridge between the medical profession (which was initially none too eager to play the role of genetic arbiter assigned it by reform eugenics) and prospective patients. With the combination of birth control and infertility treatment, a clearly defined role for physicians in reproductive processes before pregnancy was established. This normalization of the contact between physician and couple desiring pregnancy assisted in the translation of physicians into hereditary counselors.

The next step would be a series of "baby prediction" articles which counseled the couple to address the family physician about any of their hereditary concerns. Together, the advice of the baby prediction articles and the insertion of physicians into birth control and fertility enhancement created the structural space for genetics to be enacted in everyday life. Pregnancy had already been medicalized in the previous century. Now the process of medicalizing conception began to be established. The practical role of the new genetics was thus located in a medicalization of genetics and the featuring of pregnancy (an obvious site for hereditary concerns).

Many elements thus came into play in the placement of the new genetics as separate from the old eugenics. Reform eugenicists sought to maintain classical eugenics minus the overt racial ideology. They did not lose the struggle completely, for mandatory sterilization of a small, mentally atypical segment of the population continued to have at least moderate public acceptance (Reilly, 1991), and articles in the era conveyed these attitudes. Throughout the era, the advocates of reform eugenics continued to receive some neutral or even favorable attention in the press. However, the reform eugenicists' vision of broad governmental control or even influence on reproduction was defeated, and as the period closed, control was beginning, very slowly, to pass to the medical profession. This required, however, major reconfigurations in the public meaning of genetics, and these revisions were accomplished by linking opposition to genetics to Russian dogmatism and by merging atomics and genetics as a threat that could not be ignored. Out of this struggle and process of transformation, there gradually evolved a vision of genetics as a vital research arena and as serving a concern for couples planning their families. Two types of articles introduced this new genetics to American families.

Introducing Family Genetics

If reporters were to present a favorable image of the new genetics to the public, they had to do so in a fashion that could overcome the skepticism generated by the association of genetics with eugenics. Eugenics had claimed to have scientific roots, but it had been discredited as a group of "cranks" ("Genetics and Medicine," 1941). In reaction to the stigma constituted by its association with eugenics, articles introducing genetics in this era worked intensively to emphasize its scientific foundation and its exclusive authority to produce knowledge about heredity. This was accomplished by a foundational series of articles in the press that reported new discoveries in genetic research, such as the identification of mRNA, genetic experiments with mice, and research on crops. In general, these discoveries were presented purely for their interest value, and though the presentations were uniformly positive, they were less ecstatically celebratory than similar articles would be later in the century. These reports of new discoveries served to legitimate genetic research and to familiarize the public with the assumptions behind genetic studies and genetics' most spectacular findings.

The scientific objectivity and technical expertise of genetic treatments of heredity were also directly defended in many articles in the era. Thus, in reporting on the first large-scale and relatively systematic hereditary studies of twins, Morton Hunt presented meticulousness in research as a refutation of the bad image of genetics created by eugenics. He argued: "Kallman has often been accused of promoting a backward, medieval, racist and even fascistic theory. By the dogged accumulation of facts, however, he has succeeded in gaining a measure of recognition for his unpopular concepts" (1954: 21). Norman Carlisle similarly instructed the readers of *Coronet* in the superiority of genetics to other methods of hereditary analysis:

> All too often, such questions are still answered by ancient superstitions and old wives' tales. . . . Yet the vital field of heredity need no longer be a matter of guesswork. By peering at the tiniest blocks of life with the electron microscope, by countless experiments with plants, insects, and animals, and by patiently studying thousands of human family records, scientists are at last solving many baffling biological mysteries. (1951: 140)

Thus, those who might offer competing accounts of heredity were dismissed as uninformed—"old wives"—and geneticists were assigned the role of the "heredity experts." The grounds of expertise were defined by technical equipment (microscopes), experimental manipulations, and patience of study. In a similar vein, James V. Neel, M.D., warned off competitors of genetic medicine:

> Probably the most widespread and serious misconception is the idea that people without scientific training can make predictions about the offspring from casual observations of the parents alone. Scientists can now make predictions about some conditions, but for most of them this requires *scientific* study of the whole family. . . . (1951: 56)

The new genetics was therefore first a science. Genes were complex entities, decipherable only by the expert, and experts could produce powerful information not available to lay persons.

The popular articles that presented new scientific discoveries indicated that genetics had broad general interest, and they granted it legitimacy based on special expertise, but they provided little public sense of the utility of genetics in everyday life. These discovery-focused articles may have made it easier for people to conceive of biology as an important factor in human heredity, but by and large these stories relegated genetics to the distant realm of Nobel Prize winners, animal experiments, and science enthusiasts. A second type of press coverage brought informational genetics more directly into daily life. These articles promised to use the knowledge of genetics to predict what a couple's baby would be like. They declared that "no science more directly concerns everyone" (Carlisle, 1951: 140).

Often presented in women's or family magazines such as *Parents Magazine*, *Coronet*, and *Woman's Home Companion*, stories about the prediction of the characteristics of one's future children introduced the average American reader to "heredity counselors" and offered the information of genetics as an issue of interest for family life. Such articles told readers how genetics could help them know in advance "important" things about their babies. With a few insignificant exceptions, these articles were scientifically sound, given the science of the day. They introduced readers to the concept of "hidden" (i.e., recessive) genes and dominant genes, to sex-linked inheritance, and to the basics of Mendelian ratios.

While these articles introduced the practice of scientific prediction of characteristics of future children, they also identified the appropriate decision makers as a combination of nuclear families in consultation with medical personnel. Like classical eugenics, reform eugenics had urged that outside agents should make decisions about human reproduction. Although reform eugenics was not as overtly classist and racist as classical eugenics, it still imagined families in an extended sense. Because the reform eugenicists' goal remained improvement of the human population as a whole, the extended lines of families were their relevant objects. Nuclear families were simply convenient points of focus for exerting influence on the longer-term characteristics of extended families. The new focus on family genetics, however, repudiated the interest in the extended family and specifically identified the good of the nuclear family as its goal. In keeping with the democratic spirit of the era, the rhetoric of family genetics fused the belief in individual rights with the nuclear family and, more particularly, with the reproducing couple.[8]

Reform eugenics had also designated the medical profession as the appropriate replacement for the government as the agent of eugenic control. Family genetics accepted medical involvement but generally rejected unilateral medical control in favor of a combined decision-making process, whose precise boundaries remained unsettled. For most writers in this era, the revulsion in regard to Nazi eugenics in combination with the democratic attitudes of the era resulted in the conclusion that couples should make final decisions about their own reproduction. A central, declared public value of genetic counseling became client autonomy. *Newsweek* reported that the genetic counseling clinic "lets the couple decide whether or not to take the risk" ("Clinic for Ancestors," 1946: 55). *Today's Health* similarly relayed the concept that genetic counseling clinics gave information about heredity "not to tell people whether they should or shouldn't have children, but what the risks are, if any, so they can make their own decisions" ("Bad Blood," 1951).

In some cases, this respect for individual choice was expressed in a relatively full sense, even accounting for disagreement between the medical professional and the client: "Any couple of reasonable intelligence will make a decision which is right for them. I don't always agree with them. . . . But you don't need any restrictive laws for people of normal intelligence" ("The Blood Line," 1955). This vision of individual choice was, however, in tension with an expressed medical opinion that people with particular conditions, especially Huntington's chorea and mental illnesses, *should* choose against reproducing their genes. Morton Hunt expressed hope for an improvement of the human species in the future, at a time when "individuals begin to use genetic knowledge constructively, and voluntarily limit their own reproduction if they carry major susceptibilities to mental illness" (Hunt, 1954: 82). Other sources implied an even stronger role for medical prescription, arguing that parents might be persuaded or induced by medical personnel to limit their reproduction. *Today's Health,* for example, suggested that early testing for Huntington's chorea was desirable because, "if they could then be induced to forego having children of their own, it would greatly reduce the chance of the gene being passed along" ("Heredity and Mental Illness," 1956: 54). Couples thus would be allowed to choose, but it was presumed that there were better and worse choices and that medical personnel would play a direct role in helping couples to understand the factors behind such choices.

Such putatively medical prescriptions were not, however, based on any strictly medical criteria for reproductive choices. One article cast 1 in 8 as "poor betting odds" ("The Blood Line," 1955), and another saw 1 in 25 as "good odds" (Neel and Bloom, 1953: 14), but there was no basis provided for such a bracketing of decisions. Indeed, as time went on, geneticists would increasingly raise the criteria for "good odds" (Kolker and Burke, 1994: 100–101). Technical data alone could not ground these decisions, because evaluative criteria were needed to decide which diseases were so bad as to warrant eradication through reproductive abstention and how large a chance should be taken on reproducing any given condition. Not surprisingly, the tension between the medical profession's values (the production of healthy children) and parental choices (which engage multiple, culturally variable values) would prove to be a long-standing difficulty while genetic counselors debated, throughout the century, the values of directive as opposed to nondirective counseling (Bernhardt, 1997; Michie et al., 1997; Wolff and Jung, 1995). The basic grounds for that debate, however, were established in this early era.

Whomever the ultimate decision maker might be, the baby prediction articles of this era faced a central difficulty in introducing family genetics as a useful practice for the average American family. To claim significance, the articles had to promise far more than genetics could then offer. The articles opened by noting that all parents wondered what their children would be like—their appearance and their talents. The authors then promised that genetic counselors could tell the readers "things you've always wanted to know" ("What Will a Baby Look Like?" 1943), because they could "predict quite accurately various traits of an expected baby" (Scheinfeld, 1953: 40). As the stories progressed, however, it became clear that the most that could be done was to "state the odds for and against the appearance of specific traits." Genetics could predict averages and probabilities but could not predict what traits particular children would have or even answer the oft-repeated question, Will your baby be normal? Instead the articles recited the ways hair and eye color are inherited and how sex is determined. They then dealt with some of the illnesses and deficiencies that can be inherited genetically. In the end, the articles had to concede that genetic variation did not allow prediction in individual cases. They noted that "the fact that every child represents a new combination of genes helps to explain some of the surprises" (Scheinfeld, 1953: 78). Moreover, they sometimes conceded that items such as intelligence and behavior were more heavily influenced by environment, and thus counseled: "If you worry at all about bringing a child into the world it should be less about his inheritance than about the kind of home and world awaiting him [babies were then generically male]. For while you have no control over the genes you pass on to your children, you can do a tremendous lot about the kind of place in which he lives and grows" (Scheinfeld, 1953: 80). The new family genetics thus foundered on the same difficulty as the old eugenics—implementation. Baby prediction articles touted the benefits of genetic counselors—"a successful case of heredity counseling has the far reaching and multiple effect that it guides parents from personal tragedy and exchanges children doomed to suffering for sound, healthy human beings" (Fromer, 1956: 131). However, such "exchanges" were based on absent or unlikely mechanisms. A very few articles suggested that readers check out prospective mates' family trees before marrying or that specific people avoid having children or adopt. But abortion was illegal and genetic tests nonexistent in this period, so all that genetics could offer was statements about odds that were of dubious use or functional meaning.

These articles overcame the lack of power genetics then had to offer parents real answers and real technical solutions by casting the genetic counselors primarily as sources of reassurance. Unlike eugenics, where scrutiny of heredity was a source of pessimism, these articles featured optimism, reassuring parents that they had "every reason to expect" that their baby would be "normal" (Scheinfeld, 1953: 80). The appeal of predictive genetics in this era thus rested primarily on its putative ability to reassure people that they should do what they were going to do anyway—have children.

Genetics as a science for predicting or preventing hereditary problems in one's children did not have much to offer in the forties and fifties. Yet this vision of genetics, combined with the "discovery" articles, successfully outcompeted reform eugenics, as well as Lamarckian or other culture-based alternative accounts of human development in a substantial segment of the public consciousness. The triumph of family genetics is evident in a numerical shift in the major topics of articles. In the era of classical eugenics, the largest single group of news articles had focused on the issue of reproductive control from outside the individual (42 percent; appendix 2, table 3). Additionally, in the pre–World War II era, the major ethical value expressed had been preservation or improvement of the quality of the gene pool (57 percent of all articles; appendix 2, table 4). After that war, there was a statistically significant shift in the topics of mass media articles, to a focus predominately on individual reproductive goals and health (31 percent vs. only 13 percent on general reproductive control). Moreover, the focus on the quality of the gene pool as a central social concern dropped dramatically, from 57 percent in the era of classical eugenics to 16 percent in the later era. At the same time, there was a statistically significant increase in the number of articles that emphasized individual rights (appendix 2, table 3). These transformations can be characterized as a change from eugenics to family genetics, and this shift virtually eliminated any overt public opposition to genetics. While the discourse of the early century had been relatively strongly polarized, with strong supporters and opponents of the eugenic vision, the discourse of family genetics at midcentury was largely uncontested. Only three articles in our magazine sample presented dominantly negative responses to genetic accounts in this era, making the mean favorableness rating 2.89 on a 3.00 scale, the highest during the century (see appendix 2, table 7).

The rhetorical transformation process that moved U.S. public understanding away from a debate over eugenics and toward a more uniformly accepted and narrowed vision of family genetics relied on both a repositioning of eugenics according to international alliances and a reassertion of the scientific basis of genetic understandings. It required casting off explicit racist and classist discourse and replacing the goal of the improvement of the human species with enhancing the choices of couples and the well-being of nuclear families. It involved a further medicalization of human reproduction, involving physicians in decisions about family procreational decisions. These transformational processes would have long-term consequences. Efforts to regulate genetics would later come up against its status as an emblem of free inquiry, and the strengthening association between the gene and atomic radiation may have further encouraged a particulate, noninteractive model of the gene. In the immediate term, however, the transformation from eugenics to family genetics included new conceptions of the gene itself and a new effort to redefine genetic choice within the bounds of normalcy and equality.

4 THE FAMILY GENE

The understanding of heredity that was developed in the rhetorical formation of family genetics between 1940 and 1955 featured distinctively new perspectives on human heredity, including more precise images of the gene and more egalitarian efforts to portray all human genes as equal. These new images of the gene involved the crafting of a sharp distinction between biology and environment. This distinction utilized models of gene-environment interactions that gave primacy to the gene and made possible a vision of genes as powerful and confining forces shaping human characteristics. These portrayals of a powerful gene did not, however, lead to social policies based on genetic determinism, because genes were understood to be too complex for human control and because genetic influences were attributed primarily to physical characteristics rather than to moral and intellectual qualities.

The development of the framework of "family genetics" also involved repudiating the significance of broad-scale patterns of genetic inheritance; the questions of interest about heredity were relocated from the nation and race to the nuclear family. Thus, in concert with the growing egalitarianism of the era, some advocates began a conscious effort to provide formulations of genetics that were egalitarian with regard to both race and gender. These overt formulations did not erase latent racism and sexism in much treatment of genetics, but they represented a new direction and new possibilities. In this respect, as in most others, the era was a relatively cautious one, mostly lacking in fanfare or debate. The formulations of genetics it offered, however, opened up distinctively new avenues, many of which would prove to establish the directions in which the public discourse would develop across the rest of the century.

Genes and Environments

One of the sharpest breaks from the era of eugenics was the drawing of a clear and even oppositional distinction between the relative contributions of genetics and social environments to human outcomes. Genetics became associated with heredity, while the role of family nurturance in shaping children began to be subsumed in the larger, less active, and less temporally sensitive term *environment*. This process of redefinition included both the depiction of an autonomous gene and the offering of a range of models for the relationships between gene and environment.

THE GENE IN CONTROL

No metaphor dominated this period in the way that the stock-breeding metaphor had dominated classical eugenics. However, the figurative description of genetics distinctive to this period was the controlling gene. The portrait of the gene as in control of human beings increased to a statistically significant degree from the previous period and would decline fairly sharply thereafter (appendix 2, table 2). This vision of the controlling gene and the discomfort it caused was reflected in the *Saturday Evening Post*'s comment that "it offends the average man to think he is not captain of his own fate" (Hunt, 1954: 80). This shift in vision was a profound one. In the eugenic era, man (the era had been sexist) had seen himself as in control of his genes. Smart people could reshape human reproduction by determining who would breed prolificly and who would be prevented from reproducing. Since one's "germ-plasm" was simply and directly related to who one was, such control was possible.

The attitude of the new era was far less self-confident. Conrad Berens mocked those with the older attitudes: "So you think we can control the genes! But first we must understand them, and science is still unraveling their mysterious mechanisms" (1955: 34). Faced with increasingly complex models of genetic action, people now felt themselves to be subject to the whims of their genes. These genes, moreover, were tricksters: they played a complicated lottery game in which a distinct combination of hereditary units was delivered to each person to set the path of his or her fate. The complexities of the genetic game were beyond human manipulation, and genes were more powerful than humans. We might have begun to understand them, but we could not control them. They controlled us, and all we could do was work within the boundaries that they set.

This vision of genetic complexity drew from research genetics, which had advanced rapidly beyond the simplistic picture of Mendelian ratios and "unit factors" discretely controlling each namable aspect of human characteristics in a one-to-one fashion. These advances were almost uniformly integrated into the popular media's discussion of genetics by the end of the era. For example, James Neel provided a relatively detailed explanation of dominant and recessive genes and then noted:

> These two types of heredity are the very simplest. There are many more complex types. For instance, some conditions are dependent upon the simultaneous presence of several different genes. Other conditions, such as color blindness and hemophilia, the bleeding disease, are what we call sex-linked characteristics, appearing largely in males. Some other conditions tend to run in families, but the exact mode of inheritance is obscure and has not yet been worked out. (1951: 56)

Benjamin Gruenberg and Joel Berg provided a similarly complicated notion of inheritance to the readers of *Parents Magazine:*

> The processes of our minds are very complex; they are influenced by a great many dominant and recessive genes. We all carry both "smart" genes and "dumb" genes. Their interaction may result in widely varying intelligence in children, superior or inferior to either of their parents in some particulars. (1956: 43)

No longer, therefore, was the public led to think that there was a gene for intelligence and a gene for musical ability. Instead, the media noted, each characteristic "depends on the combination" of genes ("Heredity," 1952). The addition of complexity to the portrayal of genetics, however, did not reduce the power of the gene but rather increased it.

MODELS OF GENE-ENVIRONMENT INTERACTIONS

The vision of active and controlling genes was in part dependent on the creation of a clear separation between the forces of environment and genetics.[1] For the public, this separation was articulated through a boundary-setting model. This model held that "Nature draws a set of outside limits for any given person, and Nurture selects a specific site within those limits" (Hunt, 1954: 80). Within this model, there was a limited role for environmental manipulation, but the gene acted first, and its actions could not be trumped. Environmental manipulations (or individual will) could not change basic tendencies or structural relationships: "Environment worked upon a built-in tendency and either minimized or accentuated it." Gruenberg and Berg emphasized the role of environment more heavily but repeated the same statement that limits are drawn by nature, using a different but consonant metaphor: "heredity loaded the gun" and environment "pulled the trigger" (1956: 116).

This vision of the relationship between the gene and its environment posited a fully insulated and stable gene. Writers claimed that each person's heredity was "fixed at conception" (Gruenberg and Berg, 1956: 116) or even "long before conception" (Carlisle, 1951: 140). Once a set of genes was delivered to a fetus, their program was treated as irrevocable.[2] This tendency to assume a static genetic configuration was manifested in its most extreme form in articles about radioactivity, which emphasized that radioactive exposure hurt the germ cells and thereby harmed future generations, but which did not note that radioactivity could injure the persons immediately exposed ("Gloomy Nobleman," 1946). Direct damage from radiation to exposed persons was known to be significant by this time, but there was a sustained ambivalent attitude toward radiation among the public (Turney, 1998: 125). Emphasizing the transgenerational harm of radiation while deemphasizing the direct harm to the recipient probably helped to make sense of the apparently contradictory nature of radioactivity, which appeared simultaneously to be violent and threatening and yet also useful for life-saving medical treatments, energy generation, and national defense. Such an interpretation of radiation damage also was consonant with the assumption of a stable gene, impervious to external influences, but it conflated genetics with heredity. It falsely assumed that since one's set of genes was determined at birth (heredity), those genes would work in the same fashion throughout the lifespan rather than change through time.

In spite of the impetus constituted by this territorial, boundary-based model of the stable gene and its dependent environment, the mean level of genetic determinism in this era was similar to that of all the other eras. Sixty-one percent of the articles listed both genes and environment as playing a role in human outcomes (see appendix 2, table 1), and a variety of levels of genetic determinism existed in various articles. This variation was in part attributable to the fact that the range of territory carved out by the gene could be understood as relatively broad or relatively narrow, and thus either the gene or the environment could be emphasized. On the most extreme end of the spectrum was Anne Fromer, who declared that genes "decide everything basic that the new person is going to be, physically and mentally" (1956: 133). Nearby was *Newsweek*'s endorsement of a psychologist's opinion that "human nature is nine-tenths inborn and one-tenth acquired" ("The Power of Inheritance," 1947).

Further toward the middle of the scale rested a large group of articles that began to integrate the new findings of genetics through a process of differentiation. That is, they conceded that different traits had different balances between gene and environment. This allowed them to offer strong portraits of powerful genes while still admitting the significance of environmental action or individual will. Thus, Arthur Kornhauser presented a poll of "experts" to the readers of the *American Magazine.* The experts had been asked about a range of human traits, and they had been able to indicate whether they thought that differences among persons were determined (1) almost completely by heredity, (2) largely by heredity, (3) somewhat by heredity, or (4) to no significant extent by heredity. According to his experts, intelligence was best described as determined: "In large measure by heredity, but also affected considerably by normal differences in environment" (1946: 52). Kornhauser interpreted the results of his poll as indicating that intelligence was primarily determined by heredity, but he contrasted this with race prejudice, criminality, and alcoholism, which he indicated were determined by environment. Disposition, he concluded, was determined by both heredity and environment. This tendency to differentiate the degree of genetic influence among various conditions was significantly more widespread than in the previous era, and this shift represented the beginning of a long-term trend toward increasing differentiation of degrees of heritability among various types of traits (appendix 2, table 8).

Further tempering the portrait of the powerful gene was a greater recognition of the probabilistic nature of the relationships between genetic configurations and particular outcomes. In contrast with the eugenic era, where heredity was portrayed as uniform and predictable, heredity in this era was portrayed as fickle, a "lottery," in which some outcomes might be more likely than others, but fate could always strike in unpredictable ways. Thus, Laird (1952: 23) declared that "somewhere around 80 to 90 per cent of [a human's] brain power is inborn," but he conceded that it "does not run uniformly" in families and that children from the same family had as much as a 15 percent difference in their IQs because the heredity of intelligence was not controlled by a single gene. He emphasized that heredity was "mixed up from hundreds of ancestors" and described the effects of regression toward the mean.

The shift from the national scale to the nuclear family accounts in part for this change in the public interpretation of the meaning of the statistical relationships underlying genetic processes. On a social scale, the existence of a probabilistic relationship produces apparently rational and predictable outcomes for the aggregate. Mendelian statistics thus appeared to produce the possibility of control over the national germ-plasm. On an individual scale, however, the outcome of the probabilistic relationship is experienced as relatively random. When families have only a few children, the underlying statistical patterns are not evident in the bodies of their children. Hence, probabilistic relationships at the level of the nuclear family seem less like rational prediction and control and are experienced more like a lottery—where fate delivers its odds to particular individuals seemingly at random.

Increasingly sophisticated notions of statistics in research genetics also enabled the exploration of even more progressive visions, which began to depict less unilaterally powerful genes. In these new accounts, genes participated in human outcomes without uniformly and autonomously determining those outcomes. Authors at the less genetically absolutist end of the scale portrayed genes as creating "susceptibilities" rather than determinate paths. Thus *Today's Health* not only narrowed "mental defectiveness" to "mental disease"; it also explained that "in some conditions—mostly the rare ones—heredity plays a direct and major role; in others it produces tendencies or susceptibilities, which will give way to insanity only if there is some strong, adverse environmental push. In any case, in each type of mental disease heredity may work in a different way and in a different degree" ("Heredity and Mental Illness," 1956: 22). This vision portrays the environment as a potentially powerful factor in human outcomes. Readers of *Coronet* were similarly reminded that "just because you inherit a tendency toward a disease does not mean that you will get the disease," and they were told that heredity and environment were "both important in determining a life history" (Carlisle, 1951: 144). In contrast with Kornhauser (1946), Carlisle also concluded that intelligence "is subject to environmental influence probably more than any other human trait" (144). This description of genetics as conferring susceptibilities that required interaction with the environment to produce particular (especially unwanted) characteristics was also emphasized with regard to a variety of physical diseases, including cancer.

Accounts of genes as conferring "susceptibility" rather than causing a disease state allowed alternatives to the vision of the all-controlling gene. However, the discovery of such scientific facts did not automatically produce more liberal views. The social meaning of these facts was still struggled over. Some authors, for example, emphasized that one might have a susceptibility to tuberculosis but that susceptibility would be relevant only in some environments. This made environment of central importance in the avoidance of TB. However, it was also possible to subsume the discourse of susceptibility within the model of genetic dominance simply by stigmatizing or devaluing those who were susceptible. Thus, where a progressive article might assure the reader that tuberculosis was not caused by genes, a more eugenically minded author could assert that "a case of t.b. may require the presence of a germ and a host for it to lodge in, but whether that host gets sick or not evidently rests, to a great extent, upon his hereditary resistance" (Hunt, 1954: 80). In this account, the person is just a host and the germ is just a presence; the active agent is the heredity which offers "resistance." Thus, when gene and environment were both admitted to play a substantial role, placement of one or the other as the active and central agent became largely a matter of preferred emphasis.

Given this context, the move going on in the period to relabel the dichotomy from a relationship between nature and nurture to a contest between the gene and the environment was probably significant. *Nurture* was clearly an active term, while *environment* was more passive. In a contest between nature and nurture, nurture might have had the connotative advantage in gaining the upper hand for the position of active agent, because nature was culturally understood as either active agent ("mother nature") or passive background. In the contest between environment and gene, however, the connotative advantage went the other way. *Environment* was surely used as a term with more passive, scenic overtones, and the gene, envisioned as a discrete particle, was more readily made into an actor. Consequently, when the warm, active term *nurture* was replaced by the passive background of *environment,* and when *nature* was replaced by the active, complicated *gene,* it became easier to imagine the gene as the dominant partner in the relationship.

By the close of the twentieth century the image underlying the boundary model—the image of a stable gene—would be replaced. Genes would increasingly be thought of not as the specific particles delivered, "like beads on a string," to the individual at conception, but as information—subject to corruption or augmentation—that was contained in the "genetic code" that the individual received. With this different model, the genes, which were distributed throughout the body and constantly reproduced, could be understood as changing throughout life, and their actions could therefore increasingly be understood as dependent on developmental interactions and environmental influences in more particular ways. In the meantime, the atomistic, mechanistic model would be a scientifically fruitful one. Scientists would have increasing success in locating the problems the model told them to look for; single-gene disorders such as sickle-cell anemia and PKU became visible and increasingly treatable. The effects of the model on public behavior were somewhat more ambiguous.

SOCIAL POLICIES

Though genes might be powerful and constrict the operation of environment, the public discourse of the era suggested that there was little that could be done about their random interactions, so that, in practice, environmental control and action deserved more attention than any applications of genetic information.[3] Some of the suggestions for pragmatic actions made at the time seem odd or offensive by the standards of the end of the twentieth century. For example, *Today's Health* counseled its readers that "the man who wants to found a dynasty should swallow his masculine pride and marry the smartest woman he can find" (Laird, 1952: 31). In a more contemporary vein, Gruenberg and Berg (1956) suggested that knowledge of genetics would allow the identification of inherited conditions in advance so that compensatory actions could be taken. Most typical, however, was the advice offered by *Coronet.* After a broad treatment of the role of genes in human characteristics, the popular magazine counseled its family-oriented readership that although genetics clearly played an important role in intelligence, "in this, as in countless other factors of personality and physique, heredity may make us what we are, but the geneticists agree that what really counts is 'What you do with what you've got' " (Carlisle, 1951: 146; see also "Heredity," 1952). The vision of the gene as enormously powerful was thus compatible with social prescriptions that emphasized manipulating the environment instead of genes.

The vision of a powerful but isolated gene also functioned, in conjunction with other aspects of the new genetics, to narrow the focus of discussions about genetics increasingly to medical applications. Facets of human beings, such as intellectual aptitudes, personality, and antisocial behaviors, were now understood as too complex for current human decipherment. Research genetics itself was focused on a few conditions that were clearly inherited biologically. Consequently, public discourse about genetics focused increasingly on physical disease and much less on behavior and mental qualities. In the era of classical eugenics, 43 percent of the articles had asserted genetic links to behavioral conditions, and 66 percent had asserted genetic links to mental qualities. In the new era, only 19 percent of the articles asserted such links to behavior and 38 percent to mental qualities, and these shifts were statistically significant (appendix 2, table 9). In contrast, the percentage of articles dealing with genetic links to physical characteristics remained about the same (63 percent vs. 64 percent). Thus, where earlier articles focused on "mental defectives," articles in this period increasingly focused on a narrower and rarer class, mental diseases. Where earlier articles emphasized alcoholism, prostitution, crime, and poverty, these articles featured discussions about tuberculosis, diabetes, and cancer.

The associations between cancer and genetics would become highly important to the funding of the genetic community by the end of the twentieth century. The role of genetics in cancer began to be studied in this period through experiments in which animals were exposed to known carcinogens and genetic changes were observed. One of the significant aspects of this research was that it opened up an important alternative path, one wherein genes were viewed not primarily in their hereditary role but rather in their functional role within an existing human body. As the *New York Times* put it, cancers "should be controlled not by eugenics (controlled matings), but by genetics, the identification within the body of that mechanism which determines both susceptibility and resistance to cancer" (Kaempffert, 1947). This hope for gene-based therapies was eventually to become the much-desired alternative to prenatal genetics. This was, however, an era in which the news media still predicted that within five years cancer could be controlled ("Cancer and the Brave Hope," 1947).

Although the dominant trend of the era was toward medicalization of genetics, there remained some articles that focused on potential genetic links to behaviors (such as alcoholism or homosexuality). Where the old eugenics sought to eliminate the individual's putatively bad offspring, however, the new genetics sought to destigmatize the behavior. Rather than an alcoholic being a bad individual, for example, the *New York Times* (Laurence, 1949) reported that alcoholism resulted from an inherited inability to utilize nutrients. On this account, genetic inability could be compensated for by diet. Alcoholics were thus not "bad seed" or weak willed but simply persons in need of a better diet. Such progressive reinterpretations of the implications of genetic processes were also applied to broad social concerns such as race and gender.

Defining Equal Genes

The elimination of a vision of genetics in which human beings were easily able to control genes had a distinctive equalizing effect. There were no more visions of superior humans, acting as breeders, designing and controlling the rest of the human stock. The classical era's focus on "superior" children disappeared. In its place, a variety of formulations occurred, the most important of which was the preference for "normal" children. Over the next decades, the opposition between normal and abnormal would become the defining structure of genetic judgment. The trend began in this era (though the shift is not yet statistically significant, see appendix 2, table 6). This new normal/abnormal polarity was more inclusive than the old dichotomy of superior/unfit. The earlier dichotomy had judged very few people to be desirable and large numbers to be undesirable, and the great middle had generally been ignored. The new dichotomy envisioned the vast majority as normal and a small minority as abnormal. The use of the prefix *ab* instead of the prefix *sub* reflected an effort to avoid demeaning some persons. This was obviously not, however, a completely egalitarian vision. Fear that one's children would be abnormal was portrayed as a serious and pressing concern, even though few children would eventually fall into that category. By narrowing the range of persons who would be described as having undesirable heredity to persons with enormously debilitating conditions, the new vocabulary may have stigmatized unfavorable genetic configurations more intensively, at the same time that it granted "good heredity" to a much larger group—almost everyone. Thus, Down syndrome was portrayed as a "hideous condition" and given the label *mongolian idiocy* (Neel and Bloom, 1953: 14). This rhetorical formation reinforced the institutionalization and sterilization of those who were not perceived to be normal by increasing the perceived distance between the abnormal and everyone else. Moreover, the dialectic of normal versus abnormal revised but replayed the assumption of the previous era that difference was bad. In spite of these limitations, the new formulation was a more inclusive vision of hereditary potential than the old.

This incomplete egalitarian shift was fueled not only by the new vision of genetics but also by a general democratizing of American public opinion after World War II. In part, this was a reaction against Hitlerian hierarchism, but it was extended far beyond the conflict between Aryan Christians and Jews. Many scholars have observed the equalizing effects of the war, during which a large number of women were placed in good-paying jobs in factories and black soldiers performed in highly visible positions defending the nation's safety (Page and Shapiro, 1992: 63, 77, 100–101). The attitude shifts so frequently attributed to the sixties had their roots in the immediate postwar era.

Whatever its source, this democratization of attitudes was directly integrated with the new findings about genes. Public writers on genetics thus emphasized that the genetic inheritance of most persons was roughly equal. As Neel and Bloom reminded the readers of *Collier's*, it was not the case that some people were genetically undesirable and others desirable; rather, any pair of parents might produce a child with genetic problems because "most of us carry at least one undesirable recessive gene" (1953: 12). Whether this was a cause for pessimism or not was still contested. As Neel and Bloom saw it, "We all have good heredity as well as bad, and it's often more likely that a child will inherit superior ability than a minor or even major defect" (1953: 11). In cases of cousin marriage, they therefore recommended that individuals decide whether or not they were willing to risk the one in eight chance of passing on the hereditary defects in their families. *Newsweek,* however, said that these were "poor betting odds for child bearing" ("The Blood Line," 1955).

Whether or not individuals would want to take such a chance, the democratizing implications of the generality of disease and the complexities of inheritance were also widely used as refutation for eugenic arguments in favor of eliminating bad genes from the human pool. One of the fathers of genetic counseling advised the *Newsweek* audience: "If we eliminated all the defectives in the human race at once, not many people would be left. There isn't just one skeleton in every family closet—there are a dozen" ("Clinic for Ancestors," 1946: 55). The readers of the *Saturday Review* received a similar refutation of the concern over the disproportionate reproduction of the less-than-superior:

> High intelligence undoubtedly is based not on single varieties of genes but on the cooperation of many genes. The valuable varieties must be present, singly or in partial combinations, even in the great mass of individuals who score low in intelligence. From there they can in the course of a single generation reconstitute an appreciable number of the "best" combinations. In other words, the population at large constitutes a great reservoir, and the possible loss of valuable genic varieties possessed by the small upper layers of the population tells only a part of the story. ("The Future of Our Genes," 1953: 25)

After the war, therefore, new findings in genetics were fairly successfully wedded with increasingly democratic popular attitudes. The personal characteristics of individuals were separated from their reproductive potential. Any person could produce a child of genius, and any genius could, through natural causes or human-induced radiation, produce a child who was abnormal (which was understood as subnormal). Everyone's potential inheritance and reproductive potential were therefore roughly equalized. There was an attempt to extend this vision across the races, but this effort was incomplete.

RACIAL STRUGGLES

There was a conscious effort to use the new genetics to make arguments about racial equality. As noted above, "race prejudice" was attributed to environment, not heredity, by much of the media. Similarly, the *New York Times* (Kaempffert, 1944) reported the comments of a speaker at a scientific conference: "There is no evidence that one 'race' is mentally superior to another—an assumption that Professor Bowden calls 'ethnocentrism.' " In a more extended vein, *Life* magazine ran a substantial article that attempted to portray races not only as equal but as genetically intertwined with one another:

> Though races differ genetically, UNESCO's scientists feel that major factors—intelligence and aptitude—show little or no variation. Environmental factors such as education and opportunity, they feel, are far more important in producing the differences among people today. But as the economy of backward areas advances environmental differences will diminish. And as races mingle more and more freely, most genetic differences may disappear. ("How the Races of Man Developed," 1953: 101)

There are clearly limitations in this effort by *Life* to provide an egalitarian vision; for example, cultures and economies that are organized differently from the West are portrayed as "backward" in the article, and it presumes that they will eventually conform to the Western model. Moreover, the article uses the terms *Negroes* and *Mongolians* to designate nonwhite groups and describes Mongolians as having "slant eyes." Some today will even object to its endorsement of intermarriage. However, given that in this era a primary argument of racists was that races were unequal because they were different, the article's effort to assert commonality was clearly progressive. It defined human beings as sharing a common ancestor—an Adam, who was dark-skinned, not white. It emphasized that "the people of Europe and America show traces of all three primary races" (105).

In contrast with these conscious efforts to use genetic research in support of a vision of racial equality, the popular articles on genetics constantly revealed the racial hierarchies still built into the national consciousness. Especially early in the period, blacks were used as models to illustrate genetic abnormality ("Polydactylism in Georgia," 1945), and nonwhite ancestries were treated as genetic problems. Thus, in "The Lottery of the Genes," *Newsweek* reported on a new book, *Our Living World,* which purported to use genetics to provide reassurance about the chances that two white parents could produce a child with black racial characteristics. The magazine assured its readership that the book buried the "ghosts of the old scare stories about black babies born to seemingly white parents" and provided an account of the inheritance of skin color. The magazine provided these genetics as an explanation of "how it has been possible for upwards of 20,000 persons with Negro blood in their veins to 'pass' into the white race every year in this country" (1943: 86). The assumptions of the dominant racial hierarchy, that white/Caucasian children were desirable and others not, were similarly evident with regard to Asian children. In "Brown-eyed Parents, Blue-eyed Child, WHY?" Neel and Bloom were asked this question: "I married a Japanese girl. If our children marry white Americans, is there any way of telling how many generations it will be before the children of those marriages look like ordinary Americans?" (1953: 14). Not only did the questioner clearly prefer white children, but he also equated white with "ordinary." In their answer, Neel and Bloom scientized the racial traits, probably in an effort to deny the stigma in the question, referring to "the assemblage of hair form and color, facial structure and other features we associate with Mongolian ancestry" (14), but they did not challenge the racist assumptions in the question, perhaps thereby indirectly legitimating the hierarchy that it reflected.

GENDER AND GENETICS

Research in the new genetics was thus consciously used to democratize our understanding of people's hereditary potential. There were difficulties, however, in stretching this egalitarianism across racial lines, for racial hierarchies were firmly entrenched. Not surprisingly, the same tension between progressive efforts and regressive residuals manifested itself with regard to gender. On the progressive side, magazine articles about infertility worked to dispel the myth that women were always the party responsible for a couple's inability to bear children. *Good Housekeeping* led its article on infertility with a bold banner: "Gentleman, let's face it. Women aren't the only ones responsible for childless marriages" (M. Davis, 1947: 42–43). On the regressive side, writers of both genders attributed a couple's infertility to women's emotionality (Kleegman and Gilman, 1947: 31) or their unwillingness to "abandon the masculine role" (Wassersug, 1947: 834). Additionally, questions about sex selection were frequently asked, reflecting a strong desire for control over the gender of children.

The new genetics could thus be made compatible with an egalitarian ethos, but it also could serve the questions and interests of those with anti-egalitarian values. In this period, the struggle between these two interpretations of the social implications of genetics replaced the simple dominance of anti-egalitarian perspectives that had predominated in the classical eugenic era.

End of the Era

The "new genetics" defeated reform eugenics as a public discourse that would account for heredity through biologically based mechanisms including genes. To achieve public legitimacy, the new genetics defined itself as being opposed to the chief values of eugenics, and it overcame much of the stigma of eugenics through a valuative transformation based on competition with the Soviets and interactions with discoveries about radiation. To support the new vision of familial genetics, new social structures were established, involving the medical profession in private reproductive processes, especially through a focus on birth control, infertility, and family medical histories, but also through limited genetic counseling.

Like the old eugenics, the new genetics was concerned with producing people with particular qualities. Unlike classical eugenics, the new genetics focused on familial rather than national or racial outcomes, and it was increasingly concerned with disease conditions rather than mental superiority. The new discourse thus scaled back the perfectionism of the old. In substantial ways, however, the new discourse was at least as deterministic, if not more so. Although the growing appreciation of the complexities of genetics radically reduced confidence that humans could control genes to any significant degree, the new model of the gene depicted it as powerful, autonomous, and impervious to external influences.

These understandings of genes were imbricated in a struggle over the respective role of genes and environment. The dominant model placed the genetic particles as determining the boundaries of each individual's potentials and constrained environmental effects within those boundaries. Genes were thus understood as more powerful players, though the relative power of environment ranged across a fairly broad scale depending on the particular conditions at issue and the particular authors who were writing about them. In spite of the gene's power, the environmental realm was emphasized as the region in which practical effects could be achieved by human actions. Genetic determinism did not, therefore, lead to fatalism or a genetically based social program.

The boundary model represented a significant dichotomization of heredity and environment. In the previous era, heredity had been understood as that which was passed down by the family. One's inheritance thus included family values, property, and biological characteristics in an unsorted mix. In the new era, genetics and heredity were defined as equivalent to each other, and other familial influences were described as opposite forms of influence. Many articles of the time thus firmly defined genes as the "determiners of heredity" and indicated that "everything a child inherits is determined by the genes" (Scheinfeld, 1953: 40). Family actions provided the realm for active control, but these were not understood to be a matter of "heredity." This change was perhaps facilitated by the ongoing demise of the extended family, which made less obvious the ways in which family practices were passed on through the generations. It was probably also encouraged by a growing academic emphasis on social structure, which demoted the significance of familial history in favor of social institutions and economic systems. Whatever the multiple causes, the commingled actions of nature and nurture were redefined as a struggle between biological heredity and immediate environment.

A substantial, conscious effort to integrate egalitarian attitudes with genetics was also new. The rhetorical formation of the era replaced the eugenicists' sorting of persons into a small superior class and a large unfit class with a large middle class of normals and a very small class of abnormals. All people were portrayed as having equal reproductive potential, but this egalitarianism did not successfully quell racist assumptions about genes in the culture, nor did it resolve sexist assumptions about the genetics of men and women. The unsettled nature of the paradigm was reflected in the uncertainty about the proper decision-making locus for genetics. Egalitarianism demanded that individuals make their own choices, but assumptions of medical expertise urged directiveness on the part of the medical profession.

Thus were laid the basic foundations of a view of humans as beings whose character was strongly shaped by the actions of submicroscopic chemicals, a vision of heredity and genetics focused on individual outcomes and individual choices rather than the interests of the nation and the control of the state. No major wars or dramatic abuses would terminate this era. Instead, with the discovery of the double-helix structure of DNA and the rapid unraveling of its mechanisms of operation, modifications of this view of hereditary action would develop. The double-stranded DNA would come to dominate the collective vision, replacing the "particle gene." With it, a vision of genes as code, rather than as mechanistic particle, would rapidly come into the public media. These scientific visions would interact with the excitement of the sixties to produce revisions and extensions of the new public account of human heredity as a genetic phenomenon.

PART THREE

GENETIC EXPERIMENTATION

1956–1976

The discovery of the double-helix structure of DNA in 1953 gave rise to a rush of molecular experimentation in genetics that continued through the end of the century. This research was broadly conceptualized through a coding metaphor, which crossed effectively into public discourse about heredity. The energetic new programs of scientific research were introduced to the public with wide-ranging speculation about their implications for the future reproduction of the human species. Thus, the era was dominated both by a fresh and increasingly vigorous wave of scientific experimentation and by fresh and vigorous public experimentation with the social meanings of molecular genetics. Initially, these public thought experiments focused on radical modifications of humanity and its reproductive practices. Chapter 5 will explore the ways in which the coding metaphor encouraged these explorations of the possibility of "rewriting humanity." Genetic counseling however, became the dominant social practice described in the public media for genetics. Chapter 6 will describe the ways in which genetic counseling served as a means for employing the research findings and technologies of molecular genetics. As chapter 7 will explain, this era of experimentation included increased controversy over issues of discrimination. That chapter will also indicate how this wide-reaching rhetorical formation ended when the fear of "biohazards" symbolically redirected public attention away from radical experimentation with the human species toward microorganisms and individual therapies.

5 RECODING HUMANITY

Watson and Crick discovered the structure of DNA at the end of February 1953. They published their results in the premier British scientific journal, *Nature,* on 25 April 1953. It was not until 13 June that the *New York Times* gave the discovery attention, and then only on the last column of page 17. The mass magazines provided no announcement of the discovery whatsoever.

The lack of publicity in the United States for this scientifically and historically important breakthrough may have been partly attributable to the initial hesitance of the preeminent American authority and competitor in the area, Linus Pauling, to confirm immediately the correctness of the discovery in the British lab.[1] A larger cause, however, was the tremendous gap between the public vision of heredity and the new chemical model of genetic action that the Watson-Crick discovery congealed. The new vision, which integrated microscopic biochemistry and large-scale evolutionary processes, was difficult enough for the scientific community to grasp, but at least they had a basic understanding of chemistry and its world view. For the public, however, the *New York Times*'s announcement that the chemical structure of a nucleic acid had been deduced no doubt seemed unremarkable.

Within six years, this public attitude was reversed. Molecular genetics soon captured the eager attention of the mass media in an intense fashion that continued through the close of the century. This reversal occurred when scientists were able to elaborate a rich and compelling new metaphor about genes and heredity—the metaphor of the genetic code. Instead of the old view of genes as particles similar to atoms—beads on a string—the new metaphor saw genetics as a language system and the gene as a "message" written in the chemically based code of DNA. Public attention to genetics grew dramatically when news reports began to focus on the broadly comprehensible effort to break the code.

The Coding Metaphor

The shortcomings of the coding metaphor for hereditary processes have been extensively critiqued by Evelyn Fox Keller

(1995), Lily Kay (1995), and others (Doyle, 1997; Lippman, 1992; Rosner and Johnson, 1995). José van Dijck provides an especially useful review of these analyses (1998: 149–62). These critiques are correct in noting the way in which the usage of code metaphors focused too much attention on the gene and not enough on the cell and in protesting the way in which the metaphor idealized (and hence dematerialized) genetic processes (Condit, 1999b). However, such critiques have not explored the productive side of the metaphor—the way in which it transcended the limitations of the atomistic, particulate models it replaced and encouraged a more fluid sense of genetic processes.[2]

Placing "the genetic code" as the source of human heredity provided a scientifically and socially powerful image indeed. It literally allowed the rewriting of our basic understandings of what heredity is like. Because of its chemical basis and the focus on DNA, the new research and the metaphor that conveyed it enabled more flexible models of gene-environment interactions, and thus less simplistic and deterministic views of biological heredity. All the new potentials of the metaphor were not, however, seized upon and explored. Instead of elaborating upon gene-culture interactions, the public attention was absorbed by a different interaction, that between human consciousness and the gene. The first wave of public reactions to the new research on the genetic code fixated not upon the relations of the genetic code to its environment but rather on the technical possibilities for self-conscious reconstruction of humanity that the new genetics might enable. The reactions to these early explorations were simultaneously awestruck, fearful, and giddily enthusiastic, a set of reactions quite appropriate to the major reconstruction of thought about heredity entailed in the new effort to imagine a "genetic code."

THE INTRODUCTION OF THE CODING METAPHOR

When Watson and Crick described the structure of DNA as a double helix, they immediately recognized the way in which the structure allowed DNA to replicate and thus allowed for hereditary characteristics to be passed on. The molecule could simply be unzipped, and each of its halves could serve as a template for rebuilding a complementary half. Thus, two DNA molecules could easily come to be where only one had been before. When the *New York Times* tried to describe this process in metaphoric terms in 1953, it said that "what nobody understood before the Cavendish Laboratory men considered the problem was how the molecules were grooved into each other like the strands of a wire hawser so they were able to pull inherited characters over from one generation to another" ("Clues to Chemistry of Heredity Found," 1953). Small wonder the public was not interested or impressed; the description was not likely to be clear to the public, nor was it heuristically rich.

Although the mechanism for self-replication was immediately obvious from the structure of the DNA molecule, it took substantial additional imagination and experimentation to figure out how this same structure could also interact with the other portions of the cell to manifest characteristics that might be simultaneously hereditary and essential to the functioning of the cell. Across the next four years, the necessary reimagination and experimentation took place. In 1958 the "central dogma"—that DNA was transcribed by RNA to produce proteins—was formulated; by 1961 Jacques Monod and others had discovered messenger RNA, and it had been established that the DNA code consisted of triplets (see Portugal and Cohen, 1977: chap. 12). These studies relied on and developed a much clearer and more intellectually powerful metaphor than the wire hawser to communicate the potentials of the new research. The key metaphor successfully used for accounting for the process was that of the dawning information age. It was a metaphor about communication that was influenced by the information technologies developed in World War II, by the burgeoning global communications revolution, and indirectly by the philosophical revolution called semiology (Fox Keller, 1995; Hayes, 1998). The metaphor employed by the scientists was that of codes.

The coding metaphor was beautifully adapted for describing genetic processes. A coded message could simultaneously be passed on to others intact and provide guidance and direction for action. Like life itself, codes were orderly and yet complex, real and yet intangible. Codes also had the unique ability to feature both relative stability and openness to change. A message could convey a meaning with relative reliability, but it could also be altered or corrupted, and this combination of qualities seemed to be important for genes in the evolutionary context. It was also important that the vehicle of this metaphor was widely comprehensible, at least for the reading public.

Scientists themselves were rapidly coming to rely heavily on these code metaphors, but the shift from the old atom-particle vision to the new communication vision took until the mideighties to manifest itself fully in textbooks.[3] The new public vision was similarly uneven. A few articles continued to employ the old beads-on-a-string metaphor on into the midseventies. However, in 1960, in popular magazines such as *Time, Newsweek,* and *Commonweal,* the breakthroughs were introduced through heavy reliance on the coding metaphor. *Time* announced the fervent scientific quest to "break" the genetic code and presented DNA as carrying the "genetic 'instruction code' " that told an animal how to develop and function. The code was said to consist of four chemical bases that acted like letters, "in roughly the same sense that the Morse code of telegraphy has three letters, dot dash and space" ("Genetic Rosetta Stone," 1960). Like archaeologists, scientists were searching for "a Rosetta Stone for the language of life." They would then know what sequences of bases were responsible for what characteristics. What scientists faced, *Fortune* declared, was a "problem in cryptography" (Lessing, 1966: 151). The business magazine marveled that "the genetic code is the oldest of all languages, running back through evolution to the lowest bacteria" (152). It was "the universal language."

Over time, the code metaphor became increasingly elaborated. Bases were "letters," but their groupings spelled out "words." "Messenger" RNA was discovered to carry the instructions of the DNA to the rest of the cell. Mutations were simply changes in spelling, and also important, messages could be "rewritten" or "edited." These metaphors carried some distinctive World War II baggage. "Breaking a code" was a patriotic and heroic act, a key to victory. Further, DNA, as bearer of the code, was not a mere messenger but a "director" or commander of the cell at large. Most significant, because it carried a code, DNA carried the *meaning* of the organism, and hence it provided the "secret of life." Since the secret of life was written, there was hope that we could learn it.

To some extent, this new metaphor meant that "the gene" took a back seat to the molecule of DNA. The new vision was founded in chemistry rather than atomic physics, and the chemical of heredity—the molecule—is DNA. The atomistic gene, in an important sense, did not exist. There was no autonomous particle directing human life courses. Instead, there were complex strands of chemicals interacting with one another. Although it took genetic scientists themselves some time to come to this understanding (and not all yet understand it fully even today), genes were enormously difficult to define.[4] Two different ways to define them evolved in the scientific community. The most common was to define the gene as a strip of DNA that codes for a particular enzyme. This definition best fit the "one gene, one enzyme" model of the geneticists' self-endorsed central dogma, which stipulated that a stretch of DNA was transcribed to RNA, which then assembled a specific protein (these principles were described for the public several times in various major magazine articles). However, that definition was not really physically accurate, for on the chromosome itself, the code which can be mapped to the specific amino acids in a finished molecule is not a precisely delineated, uninterrupted set of atoms. Coding sequences are usually interrupted by "noncoding" sequences, and the particular translation given a sequence is often dependent on noncontiguous sequences and on non-DNA molecules, which determine whether a particular sequence will be read from one point or another or in one direction or another. DNA also comes wrapped in nucleosomes, and the form in which these are assembled and packed also plays significant and complex roles in what the "code" can mean. In other words, although DNA clearly exists as a specific and definable physical entity, genes are concepts—to be defined functionally rather than in terms of their physical boundaries. Scientists in this era thus defined genes either in terms of an imprecise, truncated physical account (a coding sequence of DNA) or in terms of a more abstract functional account (a coding unit responsible for the production of a particular enzyme). The two components—concept and physical entity—were overlapping rather than identical.

The public definitions of genes gradually came to reflect that dual reality. Article after article emphasized that heredity was a chemical matter, and several journalists defined genes as "long strips of DNA" ("The Gene Machine," 1976) or noted that "genes are basically sections of an extremely complex molecule called deoxyribonucleic acid (DNA)" ("The Secrets of the Cell," 1970: 43). As early as 1958, *Time* magazine appeared to sound the death knell of the previous view of genetics, saying: "No longer were genes considered abstract units of heredity. They became actual things, not entirely understood but known to be concerned with definite chemical actions" ("The Secret of Life," 1958: 53). This new vision of the gene as a subset of DNA was also represented iconographically. Replacing the particles and dots of the previous era, the new images accompanying the magazines featured drawings of the DNA double helix instead of the gene or represented the gene through a photomicrograph of a section of DNA (e.g., Gelman, DeFrank, and Copeland, 1976). The new chemical model, however, merely added to the old conceptual model; it did not replace it. A gene was still a "unit" of hereditary information: "Genes are the fundamental units of heredity, each one carrying the commandment for a specific biological function" ("The Gene Makers," 1970).

Therefore, a tension existed between a material account of genetics, which would focus on the chemistry of DNA, and a functional account, which would focus on the concept of the gene. The choice of emphasis was potentially consequential. Focusing on DNA could open up more interactive perspectives that saw the molecule of DNA as part of a complex process of cell machinery; focusing on the gene could encourage more traditional perspectives that saw the gene more reductionistically, as a center of meaning for the cell. As a rhetorical device, DNA had the advantage in visibility and materiality. Accordingly, it was depicted in 14 different magazine articles, whereas the gene was depicted in only 4. But the gene had other advantages. The tradition had long taken genes as the core concept. Moreover, in the public realm, the chemistry was much too difficult to convey (and more technical errors crept into popular reports during this era). As a consequence, though the image of DNA was pervasive, perhaps even dominant, the simplified concept of the gene continued to guide both public and scientific thought in important ways.

The treatment of sickle-cell anemia provides a good example of the consequences of the focus on the gene over the complexities of the DNA as a whole. The discovery that sickle-cell anemia is caused by the mutation of one "letter" in a particular gene (i.e., stretch of coding DNA) was widely publicized, and the scientific establishment as a whole took this single-gene discovery as a basic model for future research. But as a clinical fact, the misspelling of the hemoglobin gene does not reliably result in a particular set of physical symptoms. Some people with this misspelling develop severe and early sickle-cell disease and die young; others develop relatively slight symptoms. Clearly, it is the DNA sequence as a whole, in the context of the cell as a whole, in the context of the organism as a whole, in the context of specific environments, that produces varieties of manifestations of sickle-cell disease. But because the search for specific genes came to dominate the scientific project of the era, these crucial interactions received proportionately little attention. Even worse perhaps, in the public realm people were misjudged by being labeled in terms of the specific status of that one gene rather than in terms of their entire genetic configuration. Arguably, this was not bad science—finding simple cases may have been a necessary precursor to tracing complex interactions. But when reductionist scientific protocols were transferred into the public sphere, there were major consequences. The public discourse continued to focus on isolated "words" (genes) of the DNA code rather than on the paragraphs and texts that contributed to the wholeness of lives.

DETERMINISM AND THE GENETIC CODE

The coding metaphor, especially to the extent that it retained a focus on individual genes as the texts to be read, limited the full potentials available in the new chemical models. Nonetheless, the new code metaphor also had distinctive advantages over the older view of the gene as a self-contained particle. Most important, the new view allowed explicitly for interaction with the environment. Both chemicals and codes were already understood as sensitive to contexts and interactions. As in the previous era, most authors emphasized that genes are not autonomous and that environment is important in determining who we are. The relative degree of this balance did not change to any statistically significant degree (see appendix 2, table 1). Moreover, the range of opinions about gene-environment interactions was still wildly variable. Some articles still featured a relatively strong dose of genetic determinism; others emphasized that genes only confer "susceptibilities"; and others prioritized the environment. There was, also, however, a proliferation of the models through which these relationships were envisioned. Three important alternatives to the older boundary model of the relationship between genes and environment were explored in this era.

The older, boundary model had specified that genes set the limit and environment determines where in those limits we fall. The first new model instead envisioned genes as merely a starting point for the individual. Following the lead of geneticist Theodosius Dobzhansky, for example, the readers of *Parents Magazine* and *Catholic World* were told that "man as a species has been selected for educability" (Puner, 1959: 96; Dobzhansky, 1966). The genetic code of humans, in other words, was to be infinitely recodable by culture. *Time* magazine also reflected the view that individual initiative and appropriate social structures could rewrite the individual from the starting point provided by the gene, allowing one to move toward unpredetermined ends:

> The human is born with a hereditary capacity to be bright, stupid or anything in between. His starting position on the intelligence scale is predetermined—a biological sentence, like the one that orders tigers to give birth to tiger cubs and the human female to produce human babies. But nothing prevents a normal man from enriching his intellectual birthright, if it is allowed to mature in a hospitable environment. The obverse is equally true. ("Intelligence: Is There a Racial Difference?" 1969)

In a similar vein, the readers of *Saturday Review* were counseled by an expert biochemist that "every inherited factor can be influenced by an appropriate adjustment of the environment" (Williams, 1971: 61). This image of genetics as a plastic starting point had significantly different implications from the image of the gene as an immutable particle that set rigid boundaries. The starting-point metaphor treated genetic influences as real and conceded that they should be taken into account, but it also assumed that they could be overwritten. This new model was probably encouraged by the coding metaphor because codes in general were not thought of as offering much tangible material resistance; it was therefore easy for readers to imagine that environments could rewrite or overcome codes.

A second alternative to the older model accepted the image of DNA as a setter of boundaries but, instead of seeing those boundaries as restrictive and limiting, envisioned them as expansive and nonrestrictive. *Seventeen* magazine counseled its young readers that "heredity may set limits on your achievement, but these limits are broad indeed and, in fact, often never reached" ("Where Did You Get Those Eyes?" 1968: 165). Prefiguring a trend toward assumptions of human malleability that would be a hallmark of the last third of the century, the magazine noted that looks could be altered by orthodontia and plastic surgery; astigmatism could be hidden with contacts, and even eye color could be changed by plastic lenses. Moreover, it was noted, young people routinely changed their own genetic heritage by using LSD. Within this vision, genetics set a range of options; however, the heritage bequethed was not narrowly confining and defining, but rather so generous that, in practice, environment rather than genetics proved to be the limiting factor. Similarly, the readers of *Mademoiselle* were provided a dynamic account of genetics; although the article argued that there was a genetic component to mental illness, it qualified the significance of the biological component heavily, noting that "the room for environmental influence is large enough that parents may have children with much higher intelligence than their own. This is because the genes for the parents' intelligence may never have had a full chance to express themselves, because of a poor environment" (Rensberger, 1973: 119). In these articles directed toward young readers, the older relationship between genes and environment was radically altered by the simple assumption of the relative generosity of genetic endowments.

The third alternative model of gene-environment interactions employed the dynamic potential built into the coding metaphor to incorporate a developmental perspective into public discussions of genetics, perhaps for the first time in the public discourse. Whereas the old view emphasized that genes were imparted to a person at conception and could not be changed, the new focus on codes highlighted the changeability of messages. Genes were expected to be rewritten and edited by the processes of the cell. Articles exploring this model noted that the specific events of the developmental process in the organism had a significant impact on what characteristics that organism took on. *Parents Magazine* counseled its readers that, although inherited traits and characteristics were "determined" by genes, it was "of course" true that "some characteristics are modified by environment. A person's basic body structure and potential are inherited, but a good diet or a poor one, illness or accident suffered prenatally or after birth, help to determine body shape and functioning" (Wohlner, 1973: 45). Some attention was also given to prenatal environments as key contributors to individual characteristics. Additionally, the public discourse corrected its earlier errant tendency to depict radiation as affecting only the genes in germ cells and to neglect its impact on somatic cells. The developmental perspective had at least gained a brief hearing in this era.

The new model of the gene as a chemical coding unit made it easier to imagine the gene and the environment in more dynamic interactions than had been possible with the atomistic particle model. This was not, however, the opening most exploited in this period. Instead, the majority of articles chose to emphasize the opportunities for dynamic interaction between conscious, individual human minds and the gene. Perhaps important in this shift was a decreasing emphasis on animals and an increasing focus on human genetics. In the previous eras, photographs of various kinds of animals had constituted a major part of the visual images accompanying popular articles about research genetics. In this period, only six articles featured images of animals (animals were replaced by pictures of scientists). The use of the coding metaphor reinforced that change in focus because humans were understood to be the coding animals par excellence. Whereas animals might be fatally precoded by their genes, human beings, by learning the language of genes, could become coders rather than merely coded.

This attention to the linkage between code-writing abilities and genetics seems also to have contributed to an important shift in the types of characteristics for which genes were seen to code. There was a statistically significant increase in the number of articles asserting a link between genes and mental characteristics. In the previous era only 38 percent of articles had asserted such a link, whereas in this period 51 percent of articles asserted that genes code for mental characteristics (see appendix 2, table 9). These articles included discussions of both intelligence levels and mental illnesses, such as manic depression and schizophrenia. A similar increase did not occur with regard to behavior (often more directly linked to animals), lending support to the speculation that the symbolic associations between "codes" and the human mind may have encouraged such exploration and assumptions.

Critics have worried that the new genetics offered images that were excessively "deterministic." This analysis suggests, however, that if anything, the coding metaphor could be faulted for imparting a false sense that it is easy to intrude upon the microscopic world of the gene. In this era, at least, codes were understood as easily rewritten. One simply erased a word on a page and entered a new word. What could be simpler? This was no doubt a psychologically welcome contrast to the image that had dominated the previous era—the closed-off autonomous gene that imperiously dictated our fate. But the image went too far in the opposite direction. The metaphor of a "code" hid the material difficulties of intruding on a complex series of material interactions that operate at the submicroscopic level. It offered a vision of easy answers and too facile manipulation of genomes that ultimately would prove to be quite false. It also may have encouraged assumptions about intelligence and genetics that were based more on symbolic associations than on rigorous research (more on that below).

The coding metaphor was not the only metaphor offered in the era. Some of the many other options have been preferred by critics. Katz Rothman (1998: 23) and Hubbard and Wald (1993: 11–12), for example, offer the metaphor of a cookbook or recipe as a superior metaphor for the gene. The recipe metaphor received occasional use in the era, as did the metaphor of a musical tape and building blocks. More common still was the metaphor of the chromosome as the thread of life. Each of these metaphors had its own limitations. The recipe metaphor might have been particularly unlikely to be endorsed by scientists, since it made all too obvious the relationships between chemists and cooks, thereby feminizing the occupation. But both the recipe metaphor and the musical tape metaphor might have had the same failings as that of the code, for they were simply particular versions of coding.[5] The building blocks metaphor might have failed more dismally by reinforcing the atomistic qualities of the previous era. The weaving metaphor implicit in "the thread of life" had attractive potentialities, but it ultimately focused on the chromosome as a separate autonomous entity as well. Thus, for all its faults, the coding metaphor had important and unique advantages, and it was as widely and enthusiastically seized upon as the stock-breeding metaphor had been in the eugenic era (see appendix 2, table 2). That focus had a major consequence in returning the public discourse to the preoccupation with improving humankind that had been prominent in the eugenic era, but in markedly different ways.

Rewriting Humanity

The first popular article to explore the possibilities for remaking humankind through biotechnology appeared in 1943. It was an isolated foray that received little response and no emulation. As the fifties closed and the sixties began, however, a torrent of such articles was loosed. Their titles captured their central focus: "On Re-doing Man" (Hirschhorn, 1968); "Heredity Control: Dream or Nightmare?" (Lasagna, 1962); "Genetic Engineering: Controlling Man's Building Blocks" (Lederberg, 1969); "The Brave New World of the Unborn" (Rorvik, 1969a). Showing varying degrees of enthusiasm, awe, and worry, these articles began to explore the new possibilities for human reproduction that had been opened by the combination of new scientific discoveries and new metaphors about them.

Several scientists spurred this exploration with their speculations and pronouncements—Hermann Muller, J. B. S. Haldane, Robert Sinsheimer, and Jacques Monod, among others. However, several of these figures had long been agitating for similar world views with only the occasional paragraph of press attention. Their musings, which had previously been dismissed as science fiction, were now heard as scientific fact. Part of the difference in public reception rested in the snowballing scientific discoveries about molecular genetics, but the snowball at this point was quite small; major changes included only the Watson-Crick structure of DNA and a gradual increase in the understanding of how DNA (and then RNA) functions. It was only near the end of the period that technological breakthroughs such as recombinant DNA techniques had accumulated to the point where the technology actually permitted substantial new manipulations. Moreover, most of the techniques that received central mention in the articles were not yet realities—cloning, ectogenesis, in vitro fertilization, surrogacy, and genetic engineering. For example, articles that explored the social implications of cloning appeared before the momentous cloning of a frog in 1968. The few and minor techniques that existed early on made a major difference in the scope of public speculation precisely because the metaphor of the genetic code made the discoveries and techniques both easily graspable by the lay public (albeit in simplified terms) and because that metaphor so readily accommodated the image of humans rewriting nature's codes about themselves.

The image of humans rewriting nature's codes was rapidly absorbed into the public consciousness as a replacement for the depressing image of the gene as controller of human destiny, which had underlain the previous era. Thus, the new discourse assured us that "a gene is not just some rigid, never-changing monster inside us that we can't do anything about" and insisted that "human beings are not merely passive subjects to the rule of the gene. The gene is very adaptable, and its activity is ultimately under our control" (Goodman, 1965: 95). Relieved celebrations of the new vistas of "human control" were everywhere. As early as 1959, *Coronet* forwarded the prediction that "the time draws near when man, to an increasing extent, can control his own biological evolution" (Rosenfeld, 1959: 126, quoting Tage Kemp). In 1969, the *New York Times Magazine* expressed hope that "man may be able to transcend his hereditary limitations, to turn back the inevitable, to control the lottery" of the genes (Stock, 1969b: 25). Similarly, *Time* magazine opined that "hopefully, in a generation or so, it may lead to means of controlling genetic processes in humans" ("Three Men and a Messenger," 1965). *Newsweek* saw the grandest potential in this control, that the "knowledge could also lead to the controlled evolution of the human race—a power as awesome as harnessing the atom" ("Exploring the Secrets of Life," 1963: 63).

In the previous era, 12 articles had explicitly presented the gene as in control of heredity, and only 4 had presented human beings in control. In this later era, defined by the breaking of the genetic code, 24 articles explicitly presented human beings in control, while only 7 assigned that control to the genes (see appendix 2, table 2). These images of human control were fully compatible with the metaphor of the genetic code. Journalists routinely referred to "rewriting" genetic blueprints full of "garbled instructions" and "gobbledygook messages" (Hicks, 1969: 44) or concluded that "there is much in life that needs editing" ("Exploring the Secrets of Life," 1963: 65).

At some level, the journalists were aware that their essays constituted a major discontinuity in thought from previous times, and they struggled to justify the radical shift. They noted that "fantasy has turned into near-reality with astonishing rapidity" (Rosenfeld, 1975: 44). Direct references to particular pieces of science fiction were common; *Newsweek* noted, "Page by titillating page, the mind-boggling inspirations of H. G. Wells and Aldous Huxley seem to be springing into full-blown reality" ("Genetic Engineering," 1972). Fourteen articles in the period made direct reference to Aldous Huxley's *Brave New World;* three to the science fiction movie *The Andromeda Strain;* and several more to Frankenstein's monster and other fictional beasts. Many scientists bemoaned such comparisons as misrepresentations, because they interpreted such references as hostile to science or frivolously embarrassing. But these references were not generally hostile. *Brave new world,* for example, became a phrase with an ambivalent valence rather than one with a uniformly negative cast.[6] Nor did the resort to science fiction represent frivolity. Writers in the period turned to science fiction because human society as yet lacked any other concrete representations from which to make sense of the new possibilities and the new view of human heredity. Moreover, any approach to these subjects inevitably presented substantial challenges. The new research programs were powerful and directed at fundamental human issues.

One way in which the popular writers limited the symbolic challenges constituted by the new discoveries was to focus heavily on methods and techniques. Even the most sober of the articles could not avoid sounding just a bit like a "gee whiz" catalogue of new technological possibilities. Ectogenesis, cloning, artificial insemination by donor (including sperm banks from "superior" men), human-animal hybrids, embryo surrogacy, tissue regeneration, cyborgs—all these possibilities were explained and introduced to the reading public. Several articles went further; if superpeople were no longer in vogue as concepts, one could meet the spirit of the space-age by imaging new "specialized" humans who would be designed and cloned to fit the new conditions of rapidly technologized worlds—astronauts without legs for space exploration, humans with gills for exploring and living under the sea, four-legged people for colonizing heavier worlds.

Though the catalogue of fantastic potential tools was titillating, frightening, exciting, and more, it served primarily as background to introduce the central technological concept of genetic engineering. Scientists and the public both imagined the possibility of "erasing" misspellings in the genes with tiny lasers and replacing them through viral vectors carrying new and better pieces of genetic code. These imagined technologies soon took on increased specificity when the use of viral vectors and plasmids to import foreign DNA into bacteria became routine, and then when the discovery of restriction enzymes revolutionized the process, making genetic engineering, according to several magazine articles, easy enough for a high school science project. Magazines like *Fortune* and *Time* conveyed to their readership basic models of the ways in which all these new genetic engineering techniques functioned, along with a surprising amount of detail about repressors, inducers, histones, and controlling enzymes that framed and contextualized the workings of genetics.

One might explain this focus on technique by noticing that this was precisely what was missing from both the old eugenic visions and the revised ideologies of the new family genetics. A lack of adequate means of implementation had always been the stumbling block of eugenic programs.[7] Thus, presenting effective techniques was simply the logical last step for completing the eugenic puzzle. However, the discoveries and metaphors also superseded the eugenicists' goal. The eugenicists' longed-for "superior race" may have been built from "superpeople" but these persons were basically still human. The new visions stretched further. The scientific dreamers sought self-consciously to "draw out of ourselves a being that will excel us" (Rostand, 1959: 98). The project of building humans who were quadrapedal or had gills or who were intended to be immortal cyborgs was recognized as crossing the borders from the species of humankind to something else. Even the project of adding an additional developmental doubling of the brain cells of the human fetus was expected to bring about a person that was not only smarter but also thought in different ways. This set of discourse was thus not a mere throwback to classical eugenics but rather proposed a "breaking" of the human code in a decisive way.

The focus of this strain of hereditarian thought was not race or family, but rather something beyond the human race as it had known itself. It is not surprising that the status of this discourse remained unsettled and inconclusive. Ultimately, conclusions would be made, in that policy directions and social structures would be established to accommodate the new knowledge and technologies within particular moral, philosophical, and discursive frames, but these were neither "rational" conclusions, nor simply "consensual" conclusions, nor merely what the technology would allow (though each of those factors played a role). The project would remain open because there was no consensus sufficient to close it off, and no consensus was necessary to sustain it, since smaller projects such as basic research and clinical research on human disease effectively functioned to continue to develop the technical and social possibilities for the larger project.

The lack of consensus about the appropriate boundaries for human manipulation of genes is evident in the wide range of positions taken in the debates of the era. Aging scientists like Muller and Haldane sat in their ivory tower, displeased with the human masses below them, and dreamed of creating a new humankind, better than its current self. They proclaimed that an objective review "shows plainly that the idea of a creature superior to man emerges quite naturally from the facts" (Rostand, 1959: 98). These scientists were well-supported by conservative outlets like *Fortune* and *Horizon,* which sponsored heavy doses of miscellaneous fragments of classical eugenic discourse. Such sources routinely suggested that the human gene pool was losing its quality because of the end of natural selection, that governments might require genetic testing before marriage licenses could be obtained, that genetic diseases could be wiped out in one generation if carriers would simply not mate, and that people who chose not to abort fetuses with deformities should not receive public support.

On the other end of the spectrum sat purist environmentalists, who insisted that genes were meaningless and consideration of them was motivated only by racism. In between rested a myriad of positions, including Catholic press outlets that approved wholeheartedly, uncertainly, or not at all of "manipulating man," and progressive voices that reveled in the new ideas and new vistas and potentials simply because they were new and interesting and gave great food for thought. Some voices championed going ahead cautiously, but toward what was not always clear, since various procedures had been discussed, criticized, and praised, with no conclusions drawn.

In most public debates in most eras, there have been a similarly wide range of positions. Public policy has usually evolved from shared ground, that is, when certain themes are repeated more often and integrated more successfully with other themes (Hasian, Condit, and Lucaites, 1996). For public discourse about human genetic self-modification in the sixties and early seventies, however, it is difficult to define much widely shared ground. Most of the mainstream journals took no clear position on the ethical desirability of any of the new techniques. Instead, they produced articles that read like a hodgepodge of positive and negative considerations. The theme of "dream or nightmare," "good or evil," was a common one, with articles alternating between positive potentials and negative problems with the new genetics. The articles were difficult to read, for decisive statements in favor were rapidly countered by decisive problems against, followed by more decisive advantages, or a switch to a new batch of techniques, problems, or benefits. The journalist's voice was usually unclear, and when the journalist did reach the grand ending, one could always be surprised at what one would find there, for no clear argument or path had led to the conclusion. The tone within a single article might range from awe-struck contemplation, to gee-whiz sci-fi enthusiasm, to fearful regret.

The public was clearly struggling to come to grips with the daunting range of implications, possibilities, opportunities, and dangers that the new world view and new techniques made possible. Since each of the considerations was itself uncertain, it was impossible to add them up to any kind of cost-benefit projections. The result was a kind of radical undecidability, which was reflected in the generally vague conclusions of the articles, such as that by Peter Medawar, whose "Future of Man," an article weighing the positive and negative potentials in the new science and its technologies, summed up his consideration this way: "The bells which toll for mankind are—most of them, anyway—

like the bells on Alpine cattle; they are attached to our own necks, and it must be our fault if they do not make a cheerful and harmonious sound" (1960: 93).

Indeed, it would be our own fault if we could not choose how to use the technologies wisely, but that hardly told us what wise choices might be. Not only did chaotic mosaics constitute most of the articles addressed to the issue of modifying man, but also the articles did not develop in a clear direction over time. Early in the period, warning articles and comments were more frequent than in the middle period, but at the close of the era a strong spate of warnings appeared again. Arguments in less mainstream outlets such as *Christian Century, Psychology Today,* and the *Reporter* might be picked up later by more mainstream outlets such as *Time, Harper's,* and *Saturday Review,* but the new arguments were simply incorporated into the heap. They added new dimensions, new excitement, but they did not seem to transform the undecidability, to settle it in any way, or to move it in any coherent direction.

This undecidability might be dismissed as simply a consequence of the formal characteristics of the new journalism of this age. The new journalism sought to remain "open" in order to maximize its readership. In keeping with the journalistic mandate to present both sides of all issues, coverage of genetics could be expected to include both positive and negative potentials of the new discoveries in genetics. However, on other subjects, the new journalism tended to contain one or the other of the sides of an issue in an effective manner that generally supported a dominant consensus or at least delineated a stable battle between two positions (Condit, 1994). This is much less clearly the case with genetic discourse about transcending humankind. Moreover, the success of the "both sides" formula was itself economically useful only because the public itself had grown increasingly heterogeneous. This growth in the variability of the audiences addressed by the mass media meant that multiple values would come into play on many topics, challenging one-sided moral foundations and formulations. The topic of genetics provided the most extreme exemplifications of these postmodern conditions because of the fundamental nature of the project: to expend extraordinary resources to remake our species into something beyond what we are implies that there is something that is lacking in the species, but that lack can be determined only by the value criteria used currently by the species, criteria which must themselves be put into question by the claim that the species needs to be remade. A radical undecidability was thus built into the nature of the subject and the project. The debate also suffered, however, from basic uncertainties about what the project's advantages and disadvantages might be and from a lack of a direct clash between those who supported the new visions and those who opposed them.

The uncertainties in the new project arose both from its potential scope and breadth and from the fact that the idea of transcending humanity was radically new to most of the public. It was, of course, this very scope and breadth that inspired excitement about it. Altering humanity was clearly the ultimate manifestation of the scientific dream of rational control, but as a pinnacle of achievement, the project extended even beyond enlightenment science. The pull of the image of a better humanity has had very strong attractions throughout the ages. To make us better than we are has always been a compelling goal of both religious and political reform groups. It is almost a moral tautology: How could we not want to make ourselves better? This was no less true in the idealistic sixties, when visions of humankind prospering in a peaceful supercommunity of "drugs, sex, and rock and roll" abounded. Thus, in 1972 the *New Catholic World* indicated that "it is a fundamental law of human nature that man is a maker of himself" and that genetic engineering has the potential to "transform man from a social trouble-maker into a compassionate, community-oriented citizen" (Sheerin, 1972: 212). Similarly, Jean Rostand (1959: 97) offered the probability that "there would arise individuals whose spiritual caliber would surpass anything that humanity has so far known" and declared that "nobody has the right to assign an upper limit to man." In an era when humankind seemed poise to launch itself into space, when all social institutions were being questioned and many restructured, it seemed plausible that human biology might well be altered to fit the new conditions. Humankind could be adapted to fit its rapidly expanding universe by diversifying itself in biological, as well as social, ways.

What could we create? What could we be? What enormous questions. And, of course, the enormity of this powerful vision made it not only attractive but also threatening. Repeatedly, the manipulation of the human genome was linked to the recent experiences in physics. Articles suggested both directly and indirectly that the power of controlling the gene could lead to "some vast and terrible mistake" (Sheerin, 1972: 212) similar to that of the use of the atomic bomb. Not only was power feared for power's sake, but also many commentators recognized that remaking humanity biologically threatened to destabilize all that human beings knew. Hence, concluded the science writer Horace Judson, "I think we are afraid of the plasticity of man" (1975: 41). It is no surprise, therefore, that many articles, while simultaneously celebrating the new discoveries and the potential power they brought also worried that we were "not yet wise enough to make man" (Rorvik, 1969a: 83).

The new potentials imaginable with the breaking of the genetic code were breathlessly presented to the public. The problems with these potentials received substantial coverage as well. For the most part, these problems were framed in parallel with the earlier arguments against eugenics. Such a framing did not constitute a straw argument designed simply to discredit the new possibilities through guilt by association, nor was it a simple artifact of history. There were substantial connections between the discourses of eugenics and the new genetics. First, those who touted the new advances often did so alongside fragments of the old eugenics. Further, the two discourse sets shared important assumptions—that humanity was not biologically good enough and that genetic controls could provide a means of improvement. Moreover, short of governmental planning it is difficult to imagine how persons designed for particular niches would develop (what parent would want to design a legless child?). Thus, one of the major objections to the project was that which had first been raised by Aldous Huxley's *Brave New World*—that big government would take over and control breeding, fashioning humans for use in particular niches, thereby violating human freedom.

As with eugenics, therefore, two major types of critiques were launched against the project of "participatory evolution" or "modifying man" (the era continued to harbor substantial linguistic sexism). The first set of critiques was based on the biology of genetics; and the second, on social structural considerations.

Most of the biologically based arguments against genetic manipulation sought to deny that such manipulations could really provide a means of human improvement. One such line of argument held that the species could not be improved because manipulating a set of genes to achieve desired qualities often led to alternative genetic problems. Genetic selection was thus presented as a zero-sum game. One could improve certain features of a biological entity only by selecting against other features. Sickle-cell anemia became the primary example. The mass media pointed out that, while sickle-cell anemia was a highly undesirable disease, eliminating the genes for the condition would simultaneously reduce resistance to malaria. Through such examples, the public received a science lesson in what the press called "heterotic vigor": genes that appeared in two doses might be detrimental, while one dose of the gene produced "supernormal" characteristics. One could thus not eliminate an undesired condition without simultaneously eliminating a desired one. Furthermore, the opponents of genetic manipulation indicated, the dynamics of constant mutation and recessive conditions meant that new (undesirable) forms would constantly reassert themselves, so that humanity could not be perfected.

The descriptions of heterotic vigor were part of a larger argument for the value of genetic diversity. Concerned writers warned the public that the "total genetic pool for the human race is very small. . . . If we destroy or alter any of these genes, their original blueprints will be lost to mankind" (Hicks, 1969: 47). The diversity argument was also easily applied as an attack on cloning, which was said to lead to dangerous homogeneity in the human gene pool. The specter of extinction due to missteps in manipulation was also raised; making new forms of humans, some writers suggested, risked destroying old forms.

A substantial amount of the biologically based refutation also attacked the premise that the human gene pool needed any improvement. Several sources took substantial space to refute the eugenic claim that the gene pool was becoming worse or that it could be improved. The readers of the *Saturday Review* (Scott and Fuller, 1965) were told that parents who had one child with a serious defect tended to avoid additional pregnancies because they were overwhelmed by the problems of caring for the first child. Further, they argued that it was not true that "feeble-minded" individuals had more children than average. Instead, they reported statistics showing that the mentally disadvantaged tended to have fewer offspring than average and that their relatives did so as well. *Commonweal* extended the argument, citing statistical computations by the geneticist Lionel Penrose indicating that if persons of normal intelligence generally limited their offspring to two while persons of reduced intelligence had no children, the total intelligence in the population would decline (Hirschhorn, 1968: 260). This would occur, they explained, because superior intellects were generally not born in large numbers from the upper strata of the population but raised themselves from other strata. Thus, they concluded, "neither positive nor negative eugenics can ever significantly improve the gene pool and simultaneously allow for adequate evolutionary improvement of the human race" (Hirschhorn, 1968: 260).

These attacks on the perfectibility of the human species were buttressed by an optimistic portrayal of the human condition that stood in obvious contrast to the dominant story of the previous eras, which had portrayed civilization as disrupting the processes of natural selection. The optimistic assessments of humankind included inventive accounts based on new information and new understandings in genetics that portrayed human civilization as a neutral, or even positive, development within human evolution. The *Saturday Review* in 1965 offered an exploration of the evolution of dogs that carried a moral lesson for human beings. The article concluded: "The super-man is not to be found as an individual but as a well-developed human society" (Scott and Fuller, 1965: 49). In the wild, the article counseled, nature selected against brilliant specialists in favor of "tough all rounders," whereas in civilization, geniuses of various sorts could be supported by the social structure, in spite of the inevitable personal shortcomings they might have in other areas. The article referred directly to the Hardy-Weinberg law to suggest that the loss of natural selection would mean, not a decrease of fitness, but a stability in the gene pool. Eventually, the authors claimed, human heredity would come into a state of balance. That balance would include human diversity, which was good: "In a human society a diversity of individuals is even more useful [than in nonhuman creatures]" (Scott and Fuller, 1965: 51).

These biologically based refutations of the need for intensive intervention to create better human beings were not "anti-genetic." They did not deny the importance of genetics, but rather found ways to construct genetics within progressive value structures. Even an article in the *Nation,* one of the most leftist popular magazines, suggested that the problem was not in the study of genetics, for science had "become an inseparable part of the intellectual adventure of man" (Luria, 1969: 408). Instead, the problem was to attend to social implications. The *Nation* thought that instead of repudiating genetic research, what was needed was a reapplication of the fruits of research through an effort that would "awaken the public and their elected representatives from the complacency that lies behind the distorted priorities of present-day society" (Luria, 1969: 409).

In contrast with these biologically based arguments against genetic manipulation of humans, most of the social arguments against genetic engineering sought to challenge the importance of genetics and to refocus attention away from genetic interventions toward social interventions. Several articles argued that cultural evolution had taken the place of biological evolution. Authors of nonmainstream journals asked, in various ways, why we needed to evolve wings when we could simply get into an airplane to fly. A more mainstream argument was built on the new vision of the liberality of the human biological endowment. The geneticist George Beadle asked, "Why should we try to engineer a more intelligent human being when we cannot yet realize our existing intellectual potential?" (Hicks, 1969: 44). *Commonweal* similarly put environmental intervention ahead of genetic manipulation, citing Theodosius Dobzhansky's claim that "it is useless to plan for any type of genetic improvement if we do not provide an environment within which an individual can best use his strong qualities and get support for his weak ones" (Hirschhorn, 1968: 260). Combining this focus with a biological perspective, the Catholic magazine further suggested that once an egalitarian environment was established, the forces of natural selection would provide the types of individuals best suited for such an environment.

The magazines (and some newspapers) also repeated the concerns and criticisms of earlier eras: Who can decide what is a better species? How can this be done without violating individual rights? Some saw any attempt to alter human genetic lineages as a violation of the basic dignity of humanity, and others asked the searching question, When do we cross the line so that what we have created is no longer "man"? ("The Body," 1971: 39). There was a fairly sustained concern that our basic humanity was at stake when we started altering ourselves genetically. This concern was heightened by a belief that such tinkering would be "irreversible."[8]

Arguments against cloning were particularly sustained. Critics suggested that genius is unique—a product of a particular interaction of a particular genetic configuration with a particular historical configuration—so that cloning would reproduce not genius but merely persons destined to try to live in the past. The cutting argument used against sperm banks—that its proponents were "always having to take great names out of the bank"—applied also to cloning, which the *Los Angeles Times* referred to as producing "carbon copies of humans considered in some way extraordinary" ("Scientists Urged to Go Slow," 1972). In *Harper's,* Lucy Heisenberg reported a widely repeated story that Hermann Muller, a prominent geneticist and advocate of eugenics, had been heavily involved in communism and had early nominated Lenin and Darwin as potential members of his proposed genius sperm bank (Eisenberg, 1966: 57). When Muller had fled Stalinist Russia, he reportedly changed his list of putative supermen. Similarly, it was frequently noted that candidates for the list of superior gametes or clones, such as Abraham Lincoln, Emily Brontë, and Ludwig von Beethoven, might have had to be removed from the shelves when it was discovered that they had Marfan's syndrome, or low resistance to tuberculosis, or a hereditary predisposition to deafness. Looking back on the candidates listed for cloning by the advocates of the era reinforces our sense that perhaps the cloning choices of one era would not be so favorably viewed with the substantial added perspective provided by time. Everyone's lists were different, but names on the lists included J. Edgar Hoover, Marcello Mastroianni, Jean-Claude Killy, Joe Namath, Rudolf Nureyev, Raquel Welch, Lew Alcindor (now known as Kareem Abdul-Jabbar), Mark Spitz, Arnold Toynbee, and Charlie Chaplin. Visual images of clones included a Barbra Streisand genetically modified to serve as an entire chorus line of Rockettes. They also included Adolf Hitler, reference to whom served often as a cautionary tale, the need for which was anticipated by advocates of cloning with the repeated statement that "of course, precautions will have to be taken to keep fanatics and madmen from cloning themselves" (Davidson, 1969: 7).

The last set of social critiques of genetic manipulation offered in this era was aimed at protecting an institution heavily and directly affected by the new proposals and techniques for "modifying man"—the traditional nuclear family. Even *Ms.* magazine, hardly a staunch proponent of the traditional family structure, worried that negative consequences for families might be significant (Rivers, 1976). The heart of these concerns was that parents would begin to see children as a consumer good and to think of procreation as yet another arena for "shopping." Parents might then treat their children as another arena for status competition (Weitzen, 1966). Asked *Harper's,* "Do we really want to encourage parents to play one-upmanship with their children's bodies?" (Etzioni, 1973). Moreover, they speculated that parents might feel free to reject children who were "not what we ordered" (Etzioni, 1973).

The ability to program children biologically might thus fundamentally alter the relationships between parents and children. The *New Republic* asked its readers to think about one troubling type of situation likely to arise: "The father, eager to sire a professional basketball player, may prescribe the gene or genes for a seven-footer. But suppose the rest of the genetic complement makes the son better suited to be a chessplayer?" (Gilman, 1966: 13). Although the incompatibility between parental desires for children and the children's own goals is an age-old story, the shopping metaphor might encourage parents not only to hold false expectations but also to hold them with a sense of right (as a buyer expects particular performance from any other good purchased). Those expectations might be further exacerbated by the separation of the genetic linkage between child and parent. *Mademoiselle* asked its young audience, the people most likely to partake of the revolution in genetic technologies, "You may be able to choose the sex of your baby and prefabricate its character, but will it be *your* child?" (Weitzen, 1966: 71). Given that the desire to perpetuate one's own identity is a distinctive part (though only a part) of the drive toward parenthood, the magazine noted that genetic selection might increase alienation between the child and the parent. Unlike adoption, generally seen as a last resort, parents would have chosen these forms of alienation and control, with what psychological consequences no one knew.

The exciting novelty of the new technologies and the potential power they offered were tempered by these critiques. However, there was ultimately no direct clash between the two sets of arguments. Those promoting the new technologies expected not only new family structures but also new worlds. Those opposing the new technologies argued that there were other ways to achieve the improvement of the human species, but the pro-genetics forces could simply reply, "And why not these too?" Both sides agreed that personal liberties would need to be protected in the process and that big government should keep its hands off control mechanisms (though not funding). The debate therefore had no clear outcome.

Although some proposals were made and committees established, there were no comprehensive legislative proposals forbidding genetic engineering of human beings or regulating it, and few articles came even close to the kind of coherent vision necessary for such legislative proposals. Of course, such a deferral of decision was made temporarily irrelevant by the fact that most of these genetic proposals were still not technically feasible. The basic research on "the genetic code" had made it possible to dream in more concrete ways than in the eugenic era about making a better species, but the dream still could not be implemented.

The lack of technical feasibility did not mean that extended public deliberation was not appropriate.[9] Most of the time estimates for the technical achievements made by the prognosticators of the era were for dates that were near at hand, and many of these predictions turned out to be surprisingly close to the mark. *Esquire* provided a table that summarized the Rand Corporation estimated dates for various achievements in genetic techniques. That table predicted that artificial inovulation in humans would occur in 1972; the first test-tube baby was born in 1978. It predicted genetic surgery by 1995; experimental surgery, now called gene therapy, produced some short-term results in the case of adenosine deaminase deficiency in the early nineties, though it was not yet clinically successful by the end of that decade. The report predicted routine animal cloning by 2005, a mark that at this writing still seems possible. It predicted widespread human cloning by 2020, a date that remains at least feasible, and it predicted routine breeding of hybrids and specialized mutants by 2025, a date that now seems unlikely, though the project itself seems ever more likely unless social mores and controls are radically changed (Rorvik, 1969b: 115).

There was, of course, a range of predictions, including, "Don't hold your breath" (Stock, 1969: 8). However, most sources predicted genetic engineering would be a reality before the close of the century.[10] The unfolding of the historical journey seems to be following a course somewhere between those who expected immediate results and those who expected many decades before the new knowledge and technologies would have much impact, with the closer estimates being strikingly more accurate. Given these estimates, it is significant that social action was stymied by a general lack of consensus.

The lack of consensus on the radical issues of remaking the forms of humanity or substantially altering the means of human procreation (through cloning or other means) tended to obscure the development of one major point of agreement. The use of genetic manipulation for the treatment of genetic disease was widely identified as acceptable. The magazines and newspapers of the era made the public newly familiar with rare diseases such as Tay-Sachs and Huntington's chorea, and they held out the promise of genetic intervention for curing these tragic conditions. They were told the story of family after family who faced staggering trials from a daunting roster of debilitating genetic diseases. In general, the conditions selected as illustrative of genetic disease were so severe that almost all authors agreed that if one could "cure" genetic disease, this might be a good thing. Thus, a minimal consensus formed that genetic technologies could be researched, developed, and applied toward the goal of "producing healthy children" (Streshinsky, 1966: 61). As *Newsweek* put it, "Despite the controversial background of eugenics, it would be undoubtedly merciful to remove the instruction for some of the 250,000 genetic defects in babies born each year" ("Exploring the Secrets of Life," 1963: 66). The magazine argued that such genetic surgery would clearly be more acceptable than controlled breeding or sterilization of "genetic misfits." The *New York Times Magazine* concurred, declaring that "the least controversial area of genetic control would be the restitution of some congenitally abnormal individuals to relative normality" (Lasagna, 1962: 58). The reporter suggested that genetic alteration be confined to cases that were voluntary and that involved gross defects, and these stipulations became generally accepted criteria for programs of research that would pursue human genetic manipulation. Such research programs offered a safe zone of consensus, and they therefore became the central product of the combination of the powers of the new scientific research into human genetics and its mixed public reception.

The second half of the twentieth century would rely heavily on the coding metaphor to understand the way in which biological heredity influences human life. This metaphor opened many new pathways to the human imagination, some of which were explored energetically and some of which were neglected. In its early years, these explorations featured an extreme perfectionism, extending to dreams of a new and better species to transcend humanity. Because the coding metaphor had replaced the atomistic-particulate metaphor, these discussions were also, however, able to employ models of genetic action that were less uniformly deterministic, without denying substantial roles to genes.

Ultimately, the broad and compelling problems of genetic intervention with which the era wrestled were able to find resolution only in the arena of the health and well-being of seriously ill individuals. Even this limited project, however, harbored substantial difficulties that demanded public address. The idealized vision for genetic selection was gene therapy for those who were ill, but the practical mechanisms that were available for genetic manipulation increasingly featured prenatal testing and abortion. Discussions about the growing practice of genetic counseling became the place where these difficulties were allayed.

6 GENETIC COUNSELING

Serious genetic diseases are, by nature, relatively rare. The mechanics of evolution mean that any condition which is substantially and uniformly debilitating is unlikely to be transmitted in the population through time with any frequency. Few could, however, object to treating those who experienced such debilitating diseases if genetic medicine could offer those treatments. In the era of our current focus, the third quarter of the century, the goal of scientific research into genetics was therefore most generally identified as producing medical treatments that restored health-enhancing genetic sequences or gene products to those who lacked them. Because no such treatments existed in this era, the actual practice of genetic medicine meant something quite different. It meant the use of genetic technologies to identify illnesses with chromosomal or single-gene origins and in some cases to select against potential persons who were more likely than average to be seriously ill. In order to avoid a self-evidently eugenic framework, in which some outside entity controlled the reproduction of individuals, some means of involving individuals in this selection process was needed. Genetic counseling provided that mechanism. It is puzzling, therefore, that previous studies of popular discourse about genetics have paid virtually no attention to this pivotal discourse.

"Heredity counseling" had been introduced to the public in the previous era. In the second half of the century, however, the relatively rare and specialized practice of heredity counseling was transformed into the increasingly general practice of "genetic counseling" through the careful negotiation of several ideological tensions. The basis for genetic counseling was a strong, negative attitude toward genetic disease. Given the increasingly egalitarian enthusiasms of the world views of the sixties and early seventies, however, it was difficult to structure and place a practice that was so intrinsically hierarchical and intensively technical. In order to portray genetics as a socially appealing practice, its supporters had to deal with the problems presented by the characterization of "genetic defects" and to face a struggle over who would wield decision authority and how. The widespread use of genetic technologies was also, however, dependent upon a change in law and medical practice with regard to abortion.

Abortion Stories

Genetic selection and genetic counseling might conceivably have developed in a different direction if abortion had remained unspeakable and illegal throughout the end of the twentieth century. In such a world, genetic counseling might have continued to be a minor practice whose major outcomes were reassuring couples about their reproductive prospects or urging a relative few to adopt or use artificial selection until such time as fetal genetic reengineering became feasible. That was certainly what the popular imagination originally foresaw, and the reengineering of embryos conceived by human parents remained a mainstay of science fiction concerned with genetic selection through the close of the century. But abortion was legalized in this era. Reform laws were enacted in the sixties, permitting therapeutic abortions in several states, and in 1971 New York legalized abortion on request, after which the annual legal abortion rate for the nation escalated rapidly to nearly a half million. The transformation was completed in 1973, when *Roe* v. *Wade* legitimated legal abortion throughout the nation, regularizing an abortion rate around 1.5 million per year and creating a nation where almost one in three women would have an abortion sometime in their lifetime. These changes were spurred by physicians involved in seeking therapeutic abortions for their clients, sometimes on genetic grounds (Condit, 1990).

The change in the legal status of abortion shifted the public rationales for genetic counseling rapidly and significantly. In the early sixties, before abortion was widely legalized and routinized, the press had described three applications for genetic counseling. First, and most typically in real practice, genetic counselors in that era helped couples who had a child with some birth disorder determine whether the causes of the condition were hereditary and whether they might face a high risk of recurrence in any additional pregnancies. Second, counseling had been presented as a means to discourage young people whose genetic risks were high from marrying or planning on having children. As the Catholic journal *America* put it, "Ideally, of course, counseling should be received by the unmarried, so that they may choose a mate with whom they will have normal healthy children" (Baumiller, 1964: 221). Some sources even envisioned genetic counseling as a mandatory practice for getting a marriage license, though there was little practical effect to such proposals. Third, genetic counselors were described as helpful in alerting families to a susceptibility in the family that required preventive treatment such as telling a family in which rheumatic fever was common that they needed "to see that the baby receives especially good care whenever he has a cold, or when he might be less resistant to infection" (Streshinsky, 1966: 167).

These rationales for genetic counseling potentially carried strong negative affect, which might have deterred individuals from involving themselves in genetic counseling. Reporters, however, often accepted the suggestions of geneticists for recasting negative prescriptions and compensations in a positive light. One means of giving these stories a more positive affect was to emphasize that even though the couple was saddened by their choice not to bear further children, they were relieved of personal guilt by learning about the genetic causes of the disease, and they were able to adopt "healthy children" (M. Thompson, 1967: 13). Another typical balancing narrative in these early articles recounted the experiences of a frightened couple who came to the genetic counselor certain that they should not reproduce, only to find that their risks were low. Happily they went off to conceive a "normal" child (e.g., Crane Davis, 1968: 32).

The stories that could be told about genetic counseling in the early sixties were surely dramatic, but they were nonetheless limited in their scope. As the seventies opened with liberal abortion laws, these stories began a rapid expansion and metamorphosis. After 1969, amniocentesis with abortion began to make a regular appearance in the mass media as the central technique of genetic selection, and the press portrayed the new technologies as "giving the counselor new, precise facts and options" (Stock, 1969: 25). In 1971 several stories presented dramatic examples that identified the social value of the new practice. Genetic counseling had a new approach, the press reported, "devised to ensure that [couples] had only normal children" (Steinmann, 1971: 20). *Good Housekeeping* and *Life* battled for the most heartwarming story. *Life* told about the trying experiences of the Storm family in "fighting the genetic odds." The Storms had a child with Hurler's disease, a disorder that results not only in mental retardation but death in childhood. The Storms wanted a larger family but felt unable to handle multiple children with Hurler's disease. Fortunately, the reporter revealed, medical genetics had learned to detect Hurler's disease in utero, and this (with the option of legal abortion) gave Paula Storm "the courage to try a pregnancy" (Steinmann, 1971: 20). Tests showed that the new baby would be "normal."

The full depth of the happy ending made possible by genetic testing was even more thoroughly portrayed in the story offered by *Good Housekeeping* that same year (Block and Stein, 1971). *GH*'s article, "The Miracle Baby of Carolyn Sinclair," recounted the life trials of Carolyn Sinclair. Ms. Sinclair was widowed during pregnancy and bore a child who was mentally and physically retarded by a hereditary disease. When she became pregnant after remarrying, genetic counselors could tell her only that there was a 50–50 chance of the next child having the same disease as her previous child: "In anguish, she underwent an abortion" (87). A year later, she was pregnant again, but this time, through the "miracle" of amniocentesis and genetic testing, she was able to learn that the baby would be "normal." No abortion, the story implied, was therefore necessary.

The clear lesson of these stories, and others (Berg, 1971; Spencer, 1971), was that genetic counseling was a productive practice. It allowed women to go about "producing normal children" (Sarto, 1970: 14). The stories suggested that without counseling and amniocentesis, women would use abortion to avoid having any further children. Genetic selection was therefore presented to the public not as a practice of preventing deformed children, but rather as a process for producing healthy children. Rather than increasing the number of abortions, the articles indicated, genetic counseling and amniocentesis allowed abortion to be turned to the end of producing healthy children. The stories argued that these procedures actually saved healthy normal fetuses. Without precise genetic testing, for example, parents with sex-linked hereditary diseases might abort all male embryos because they could not know whether the fetus was "affected" or "normal." Precise genetic testing allowed parents to abort only "abnormals" (Sarto, 1970: 21).

Those who advocated genetic selection through abortion recognized that not all of its audience would be happy with this framing. Some people would choose not to abort any fetuses. The media argued, however, that nature was on the geneticist's side in this moral debate. "Nature is a great abortionist," *Fortune* reported (Bylinsky, 1974: 152). *Parents Magazine* made the lesson quite detailed: "In many instances nature itself interrupts the pregnancy if the fetus has a severe chromosomal imbalance or a very harmful genetic defect. . . . This is nature's way of ensuring that most children born will be healthy and vigorous" (Rusk, Swinyard, and Swift, 1969: 146). Genetic counseling was therefore portrayed as completing nature's work.

Such normalization of selective abortion became crucial as physicians and geneticists began to expand the usage of genetic testing beyond those families who had experience with birth defects to the more general population. Early in the seventies genetic selection techniques were brought to wider potential user groups in two ways. Soon after abortion was legal and available in New York, communitywide testing began within Ashkenazi Jewish groups for Tay-Sachs disease, a relatively severe condition with relatively high prevalence in that ethnic group. Second, amniocentesis began to be advocated for all pregnant women over 40 (and later, those over 35), on the grounds that they were more likely than younger women to produce children with Down syndrome (Wohlner, 1973: 45). A variety of articles encouraged greater use of genetic counseling in general, some in highly enthusiastic fashion, exclaiming that "the opportunities for genetic counseling are endless" (Rusk, Swinyard, and Swift, 1969: 62).

Genetic selection through amniocentesis and then the "triple screen" test would expand across the next decades to virtually all pregnancies. Because of its expense and because of the incompatibility of the rules and procedures of good counseling with routine medical practices, as this expansion occurred, genetic counseling would become increasingly truncated. The offering of these tests would too often be accompanied by a pamphlet or a 10-minute description from an undertrained nurse in an obstetrics office (see chapter 9). These developments were not foreseen in the articles about genetic counseling of this era and were not part of the practice of genetic selection as it was publicly depicted. However, the articles did keep one eye on a better future. They insisted that genetic selection through amniocentesis was only a stopgap measure. Someday, true genetic engineering would enable medicine to recode faulty genes rather than discard fetuses (Wohlner, 1973: 118).

The press in the era established a public consensus that abortion was to be used for genetic selection against fetuses carrying genetic configurations that conferred serious hereditary diseases. They did so by giving their audiences both good reasons (parents had a right to choose to avoid what they perceived of as emotional, economic, and personal burdens or tragedies with regard to their own reproduction) and rationalizations (genetic selection didn't increase abortions, it really saved normal fetuses). The question remained, however, What counts as a "normal" fetus?

Genetic "Defects"

At the base of any application of genetic technologies is a hierarchical judgment: some genetic configurations are better than others. This is an inescapable component of a world view that would seek to apply genetic technologies to change human genes. Otherwise, the costs of intervention are too great. For some people, any such hierarchy is unacceptable. For others, however, it makes a great deal of difference whether the hierarchical judgment is seen as applying to an essential or to a secondary characteristic of a human being. While we might be willing to make value judgments about secondary or "accidental" qualities of human beings (e.g., it is better to be healthy than to have pneumonia), we might not be willing to differentiate about the essential worth of human beings (instead, deciding that all persons have equal dignity and value).

The vision of heredity brought about by research and genetics, especially as understood through the coding metaphor, made it possible to imagine even very severe and sustained characteristics of human beings as secondary or accidental rather than essential. The contrast with the era of eugenics is instructive. When the germ-plasm was a mysterious whole, people with various nonstandard genetic configurations were written off as "feeble-minded," and children with Down syndrome were dismissed as "idiots" or even monsters. The germ-plasm was an indissoluble unity and was taken as isomorphic with the identity of the individual. In contrast, the focus on genes as pieces of a larger code made it possible to imagine undesirable features as "erasable," that is, as separable from the fundamental being of the individual. One could, for example, imagine erasing the extra chromosome 21 from a child with Down syndrome, and thereby imagine *the same child* without the mental characteristics, heart problems, and/or facial features that the extra dosage of the chromosome seemed to confer. Similarly, one could imagine replacing the errant sickle-cell gene without changing the whole aspect of a person's underlying identity. Whether this was good philosophy or not is a question that deserves far more space than is available here, but the imaginative possibilities opened up by the metaphor made it tenable to stop talking of "defective children" and start talking of "children with disabilities."

This new way of looking at things allowed negative judgments of particular hereditary features without thereby judging the whole person. This separation of the identity of the child from particular characteristics also encouraged the imaginative emphasis on genetic engineering throughout the second half of the century. In 1966, as "reform" laws that permitted abortion in isolated cases were just beginning to appear in a few states, *Fortune* magazine explained to its readers that "it is no longer difficult to imagine the day when chromosome analysis will be a routine part of prenatal care, and DNA abnormalities will be corrected in the foetus before birth" (Lessing, 1966: 176). In contrast, the use of amniocentesis and selective abortion to eliminate a whole being, a fetus or unborn child, had been absent from the public vision. It appeared in none of the early predictive time-lines as a technology that would be available in the future. Genetic engineering was thus originally imagined as a process done in utero or postnatally to correct misconfigured genes in a child. This vision, however, was soon swamped by the technological fact that amniocentesis and abortion were available and prenatal manipulation of genes was not. The method of "preventing" genetic configurations that caused undesirable conditions therefore became not the preferred vision of genetic reengineering of parts of a fetal genetic code, but rather abortion of that code in its entirety in favor of another entity with a different code.

The shift from the imagined vision of genetic engineering to the real practice of prenatal genetic selection was consequential. The use of genetic selection made it much harder to separate negative judgments of particular health and mental conditions from negative judgments of the persons who had those conditions.[1] This made the new genetics more like the old eugenics than anyone wished it to be. A separation between the selection against people and the selection against diseases was still philosophically plausible for those who believed that a fetus was only a potential, rather than a full, human person. However, this separation was far more difficult discursively than genetic engineering would have been, and it muddied the lines in troubling ways. These difficulties were reflected in the changing range of value judgments placed on children with particular "defects."

As the period progressed, the movement away from treating children with disabilities as outcasts from the human social fabric was slow and incomplete. There was a fairly clear shift from the term *mongoloid idiot* to *Down syndrome,* though that transition was not completed in the period. Also, many articles worked relatively new discursive ground in emphasizing the worth of people with disabilities. The *New York Times* reminded its readers that "even a deaf, blind child can appreciate the sweet morning fragrance of a pine forest or the smell of honey-suckle on a summer evening," and suggested that "the careers of Van Gogh, Chopin and Helen Keller remind us that serious handicaps need not make existence pointless" (Lasagna, 1962: 60). Similarly, instead of characterizing children with severe mental retardation as monsters or idiots, several sources encouraged seeing these children as happy beings: "These children are so precious," *Life* magazine quoted one parent as noting, "so much more sweet and adoring than a normal child" (Steinmann, 1971: 25). In an era where self-fulfillment was a new, highly valued cultural priority, the happiness of the child and the people around him or her constituted sufficient validation of the worth of the life of any individual. This was an important contrast to the prior era, which had judged such children to be of no value because they could not be sufficiently productive.

As important as these new characterizations of children with disabilities might be, they were still pioneering efforts. There clearly remained not only a strong cultural bias against genetic disabilities, no matter how minor, but a recurring tendency of those supporting the use of genetic technologies to totalize those diseases and the persons who had them in discriminatory ways. The term *defectives* was still sometimes used to describe persons who had particular disabilities, and their genetic configuration was described as "tainted," or as a lurking serpent, or as a "special menace" (Lederberg, 1970: 49; Stock, 1969b: 26). Most distressingly, this sentiment occasionally extended beyond the prenatal period; when *Parents Magazine* discussed adoption, child-seeking parents were told that one "baby wasn't a good risk" based on its genetic history (Hammons, 1961: 103).

The depth of the tension between egalitarianism and selectiveness was evident in the tendencies of single articles to mix more and less progressive attitudes. An article in *Fortune* reported a vision of "a new morality" in which avoiding the birth of children with genetic defects was understood as a moral commandment (based largely in the financial savings to the community). Later in the article, however, it also gave favorable space to consideration of the view that abortion of "mongoloid" children would not be desired by all families, noting that "these children have sunny dispositions, they are very responsive, and their demands are not excessive" (Bylinsky, 1974: 152). These articles did not resolve the tension between valuing persons with genetically based diseases and disvaluing the diseases. One way in which the triumph of selectiveness over egalitarianism was rationalized, however, was the increasingly frequent conclusion that it was kinder to the children themselves to avoid the birth of those who would suffer greatly (Alvarez, 1972).

This conclusion was most effectively justified through an extended portrait of the suffering experienced by the children who had genetically based diseases. A newspaper article entitled "Tay-Sachs Victims" provides the best example. The *Times-Picayune* (of New Orleans) gave an extended account, which served to acquaint the public with the difficulties presented by the disease, as only a lengthy quotation can effectively illustrate:

> Picture a small boy, about two or three years old, lying paralyzed and blind in his hospital bed. A nurse feeds him by tube and turns his body regularly to prevent him from getting bed sores.
>
> This underdeveloped child was normal once. . . . But after a

> few months, the glow of his eyes disappeared along with his muscular strength, sight and appetite. These were the first signs of Tay-Sachs disease. . . . Victims of Tay-Sachs disease don't live over five years of age.
>
> This once beautiful, happy child smiles no longer. Things that used to interest him are now of the past. His mind has stopped functioning and he has hourly seizure [*sic*]. . . . And the most terrifying thing is—nothing can be done. . . .
>
> So he lies now, waiting for his early death. (Haynes, 1974)

In such portraits, the child is portrayed sympathetically, but the disease overwhelms the life of the child. The newspaper presents a tragic situation, and tragedies are to be avoided.

Genetic diseases were often identified in this era with the most extreme possible conditions, since that gave the best justification for the use of genetic selection. Extreme cases, though they are relatively rare, helped best to maintain humane sympathy for the person who had the disease while also portraying the disease as of such a hideous nature as to be worth avoiding.

Whether children with disabilities were portrayed sympathetically or ostracized, however, the general conclusion was still that it was better to have a "normal" or "healthy" child than an "abnormal" or "defective" one. In the period, six different articles used some version of the title "Will Your Baby Be Normal?"[2] Overall, the most preferred term for describing the characteristics of children who were desired was *healthy* (see appendix 2, table 6). This health-focused term may have masked some of the negative connotations of the normal-abnormal pair, partly by focusing on what could be taken to be a transitory quality of a person (his or her health) rather than an intrinsic quality (his or her normality). However the terms *(ab)normal, defect,* and *defectives* were still frequently used to label children in ways that submerged the identity of the child in his or her problems and difficulties.

The incompleteness of the movement toward more egalitarian vocabularies was partly a result of the relative newness of an ideology that recognized the rights and worth of persons with disabilities. However, the slow progress also was traceable to the need to justify genetic selection. If atypical genetic configurations were not seriously problematic for those who bore them, for the family, or for the society, there would be no justification for the expensive and emotionally painful process of genetic selection. Therefore, though the genetic code made it possible to de-essentialize the human person, making that person the product of multiple parts, none of which was related to the fundamental worth of the person, that potential could not be fully activated in a setting where the primary offering of genetic medicine was selective abortion.[3] The introduction of genetic technologies was therefore the introduction of selection among human beings. The unavoidable question was, Who would make these decisions? Although the answer on most every tongue was "the parents," deciding exactly what that meant in the context of the highly technical information provided by genetics and the highly technical procedures that necessitated the involvement of medical personnel was not a question squarely faced.

Nondirectiveness and the Paradoxes of Information

Given the context of the "do your own thing" sixties and the emphasis on freedom of choice in the seventies, it is not surprising that legal control over genetic testing was delegated to individuals and families. Because there was no societal consensus on which human qualities should be promoted, there came to be an explicitly articulated and inexplicitly consensual decision that parents would decide about genetic manipulation on the basis of "personifications of their own ideals—the generally admired primary virtues of high character, keen all-around intelligence, and sound physique," along with characteristics such as "a joyful disposition, musical proclivities, aptness at repartee, rapid calculation, courage or endurance" ("Frozen Fatherhood," 1961).

The potential problems with vesting such control in parents did not go unnoticed. Both *Esquire* and *Harper's* worried that this could well result in homogeneous sets of children, all with the most current popular qualities, that is, children by "fads and follies" (Etzioni, 1973; Rorvik, 1969b). The *New Republic* noted that this meant that decisions might be left with "private profit and ignorance" (Gilman, 1966: 13). Ignorance was, indeed, the central problem. The average parent did not have the technical knowledge to understand the options offered by genetic testing without some substantial assistance. Therefore, the choice to vest decision authority in parents was really a choice to vest decision authority in parents and medical genetic professionals.

The incorporation of genetic professionals into the choice-making process might seem somewhat ominous in the historical context offered by eugenics. However, this was not the context in which genetic professionals were framed in this era. In spite of the fears raised by the atomic bomb, scientists in general held positive public esteem, and the scientists making genetic discoveries were hailed as heroes and celebrities. Pictures in magazine articles about genetics in this period featured scientists substantially more often than they pictured double helixes, pedigrees, or genes. No fewer than 41 pictures of scientists enlivened our sample of magazine articles (out of 134 articles). The iconic celebration of the scientist was supported in verbal form as well; George Beadle, for example, was introduced as "one of the great living scientists in an age of science" (Hicks, 1969: 44).

There are many reasons for this hero worship. It was an age increasingly unlikely to treat military leaders, business persons, or politicians as heroes. Scientists seemed a relatively safe alternative, one perhaps more laudable than the next era's exclusive worship of movie stars and athletes. Moreover, what these scientists were doing was psychologically gratifying. They were delivering us from the iron grasp of genetic determinism. They were promising to eliminate the power of the genetic lottery and putting control of our fates into our own hands. In so doing, they were also exemplifying a broader cultural ideal for humans—the use of our brains to improve our lives and our selves. In the exuberant words of one science writer, they offered to make us "free at last from the tyranny of matter" (Davidson, 1969: 10).

This hero worship made scientists an attractive source for vesting the power to make decisions about genetics. The previous era's distrust of politicians and government was widely reiterated in this period. *Time* magazine, for example, sneered at the city council in Cambridge, Massachusetts, when it tried to regulate potentially hazardous biological experiments within its borders ("Genetic Moratorium," 1976).[4] As an alternative, some sources were fairly overt in their statements that the experts on the subject should guide policy initiatives: an article in the *New York Times Magazine* suggested, for example, that because genetic counselors understood both the scientific and human dimensions of the issues of genetic selection, they would be central to national policy discussions (Stock, 1969: 26).

Not everyone saw geneticists in this favorable light. Some sources recognized that, like everyone else, geneticists had vested interests and partial perspectives. They thus concluded: "Just as war is too important to be left to the generals, so genetics is too vital to be left to the monopoly of scientists and physicians" (Etzioni, 1973). Such concerns constituted the basis for efforts to regulate genetics on the social level (see below). However, these concerns were certainly not strong enough to rule geneticists out of the decision process. Given the historical precedents and the accepted combination of strengths and limitations recognized for individuals and for genetic experts, the cultural consensus easily settled on the idea that decisions about genetics were to be vested in the collaborative process named genetic counseling.

The precise character of the collaborative relationship between parents and genetic professionals was carefully sculpted in the mass media. The counselor would provide information in a "nondirective" fashion so that parents could then decide. This norm was repeated in every article about genetic counseling in the period: "The counselor calculates the odds for or against the appearance of the suspected defect, and explains them to the prospective parents. The choice—to have a child or not—is then their own" (Spencer, 1971: 162). The media described genetic counselors as "translators" (e.g., Stock, 1969: 25) who would communicate the information needed to allow parents to make their own decisions: "In every instance of genetic counseling it is the parents who decide what to do. The geneticist merely presents the facts—or the probabilities—so that the couple is able to make an informed decision" (Wohlner, 1973: 45). Although a few articles outside the genetic counseling genre challenged that assumption with statements from classical eugenics that were highly directive, the general consensus was that parents should make the decisions but that professionals would provide the information necessary for the decision and execute the procedures.

Repeated studies have shown that this translation process is a difficult one, fraught with problems (Brunger and Lippman, 1995; Michie et al., 1997). There is always more relevant information than can be conveyed or absorbed, so counselors must make value-laden decisions about both what to present and how to present it. Such choices will inevitably influence decisions in some ways. Moreover, it is difficult to draw clear lines between the informational components and the valuative components of a decision. Including genetic professionals in these decision processes therefore meant giving them some substantial room for influencing the decisions. This inclusion was justified by portraying parents as ignorant and emotional.

The difficulty parents had in grasping the concepts involved in genetics was signified through a recurrent story about the need to reexplain to them the implications of statistical probabilities. Several different sources explained that when parents were told that their children had a 50–50 chance of inheriting a dominantly transmitted genetic disease, they misinterpreted this numerical estimate. If they already had one child with the disease, the parents reportedly interpreted the one-out-of-two chance as meaning that their next child would not have the disease. The counselors corrected this misunderstanding by saying: "We had to explain that this was not the case. The odds that any child will be affected are 50–50, independent of what has happened before" (Rusk, Swinyard, and Swift, 1969: 62; see also Newton, 1976: 62). The repetition of this example of misunderstanding highlighted the need for genetic counselors to "translate" knowledge of genetics to parents.

The nature of patients as ill-equipped decision makers was also conveyed by portraying their emotionality. One counselor relayed the story of a husband who insisted that his wife had "failed" him with her first pregnancy and who used the counseling session to try to confirm that the guilt rested with his wife. The reporter described the husband's feelings as an example of "the emotional tangle through which the genetic counselor must find his way" (Stock, 1969b: 26). Another article, written by a physician, described the client-patient as "pale," "shrill and frightened," while the doctor described himself as speaking slowly "to make sure that Marian understood" (Newton, 1976: 18). Thus, parents were portrayed as irrational agents whom counselors had to set straight.

Sometimes, the decision-making deficiencies of genetic counseling "patients" were conveyed simply by condemning their decisions as errant ones. The *New York Times Magazine* recounted the instance of a woman who had given birth to a child with a cleft lip. She was told that the chances were 1 in 20 that a second child would have the same feature. In spite of the fact that "most people find [those odds] reassuring," she decided against having any more children and chose an illegal abortion when she became pregnant. The counseling doctor concluded that the medical personnel had "failed to convey a realistic view of the risk in this case" (Stock, 1969b: 27). Although the counselor explicitly accepted the blame in this instance, it is clear that the woman's decision was judged to be a faulty one, and the difficulties of "successful" counseling were highlighted.

The media thus presented the contradictory messages underlying the practice of genetic counseling: Parents should make decisions about genetic selection among their own progeny, but parents were ill-equipped to make such decisions; genetic counselors should convey neutral and necessary information, but there were right and wrong answers, and clients who did not agree with the counselors were in error. These difficulties have not been fabricated by the media; they stem from a practical paradox inherent in mass utilization of highly technical equipment. The usual solution to the paradox is to try to simplify the decisions involved so that the users of the technology do not have to understand the mechanisms but merely operate them with simple switches. In the case of genetics, some steps would be made in that direction in the next era through the use of mass-produced informational pamphlets and video training of parents, but ultimately decisions such as these cannot be reduced to the simplicity of a remote control box.

In this era, this informational paradox gave support to those who wished to use persuasive means to influence the decisions of parents. While they allowed that parents should certainly make final decisions, various sources advocated and modeled various degrees of suasion to be exerted on parents. For example, two sources argued that parents who were hesitant to abort fetuses with Down syndrome should be taken to institutions where children with this disorder were kept, in the belief that seeing the children would dissuade the parents from continuing such pregnancies (Rorvik, 1969a: 80). Others suggested that parents be "encouraged" by society through some unspecified mechanism to prevent the birth of children with Down syndrome (Bylinsky 1974: 152). The central source of suasion, however, was understood to be physicians; the press indicated that, with a severe disease such as Tay-Sachs, "doctors may advise that the pregnancy be terminated to prevent the birth of the doomed infant" (Rusk, Swinyard, and Swift, 1969: 146). Similarly, *Look* reported that a prominent physician at Columbia "invariably trie[d] to persuade the mother" to abort a fetus if she already had a child with a deformity or retardation (Rorvik, 1969a: 80). Thus, though parents were to have a final say, physicians were presumed to have some sort of expertise about which babies ought to be born and which fetuses terminated. Some of these judgments included covert class issues; echoing eugenic discourse, one physician opined, "It is to be hoped that the parents of those who can benefit most from programming will lend themselves to it readily, although experience (in matters like birth control, for example) gives little indication that they will" (Weitzen, 1966: 143).

Behind the sense that parents ought to be subject to the persuasive influences and judgments of others, especially those of medical and scientific personnel, stood the assumption that there were clear social norms that determined which decisions constituted a "rational decision, based on facts and not fears" (Sarto, 1970: 21) and which did not. One of the functions of these early articles was to communicate to the public the parameters of those norms. This was done through vignettes that portrayed sample parents making representative decisions, which were then either endorsed or derided by the reporter or geneticist. Factors to be considered as "legitimate" concerns in these decisions included the age of onset of the disease, the severity of the disease, the odds of occurrence, as well as whether or not parents already had children and the parents' ages (see, e.g., Hammons, 1961; also Rusk, Swinyard, and Swift, 1969). This list of recommended factors indicates that there was not only a tension between describing lay parents as expert decision makers on a high technology decision, but also a tension between the preference for having parents rather than social agencies make these decisions and the desire to have those decisions fit within social norms.

Genetic counseling thus came to occupy center stage as the mode of genetic selection or heredity control in the United States, but it did so fraught with tensions and difficulties that were both important and unresolved (perhaps unresolvable). In addition to the immediate problems, there also lurked a problem with the applicability of genetic counseling to the future developments in genetic technologies. The slippery slope from preventing genetic disease to reshaping human beings might be a steep one. As philosophers would later point out, there is no clear line between that which is a genetic disease and that which is simply a condition that certain people find undesirable (Gardner, 1995; Murray, 1979). Responsible individuals therefore looked nervously toward the future, when genetic reengineering would make it possible to use genetic selection not only to cure disease but also to recraft human beings in multiple ways. The authors of the era reassured themselves, through the example of sex selection, that the future would not be out of control. They emphasized that sex selection practices had been "minimal so far," and they positioned doctors as gatekeepers, saying, "Most doctors flatly refuse to perform amniocentesis strictly for sex determination" (Bylinsky, 1974: 154). Such a stance, however, obviously violated client decisional autonomy, stepping beyond the bounds of the "nondirective" pact, or even informational suasion, into direct control and prohibition.

The difficulties involved in genetic counseling were given added significance by concerted efforts to expand the scope of genetic medicine. Many scientists began attempting to revise our understanding of disease so that most or even all diseases would be seen as "genetic" diseases. Numerous articles worked to define as genetic those diseases that had a "genetic component"—that is, all disease, since genetic interactions play a role in every substantive aspect of the construction and functioning of the human body. Hence, as the geneticist Joshua Lederberg suggested, "Any disease can then be said to have a genetic aspect" (1969: 80). Others described DNA as the "ultimate drug," and declared that all the rest of medicine was just a "holding action" until genetics could provide fundamental cures (McDougall, 1976: 529; Rosenfeld, 1974: 50). This movement toward generalizing the genetic model of disease was exemplified by the discovery that the viral theory of cancer causation was actually a genetic theory, since viruses seemed to cause cancer by altering the DNA of the host cell (Auerbach, 1975). This new model raised a key question for the decisional relations between genetic counseling clients and medical professionals: if all disease was genetic disease, then which model of client-physician relationships would hold—the nondirective model of genetic counseling or the expert-centered model of traditional medicine?

Public discourse about genetic counseling during this era described the central structure for implementing genetic selection among human beings. This structure did not enact the ideal visions made available by the coding metaphor—the rewriting of defective genes. Instead, the structure employed the technically feasible practices of amniocentesis and abortion. Consequently, it faced substantial difficulties in reconciling the egalitarian ethos of the era with a residually hierarchical social preference for "normality" and in reconciling parental choice with professional expertise. In addition to wrestling with these facets of genetic discrimination and perfectionism, the era would also struggle with a wide range of ethical concerns, including discrimination regarding race and gender.

7 ETHICAL CHALLENGES AND BIOHAZARDS

The Watson-Crick discovery of the double-helix structure of DNA was only one important highlight in a massive rush of scientific research on genetics that accelerated exponentially across the second half of the century. As the information from these discoveries and their spin-off technologies began to be applied in human, animal, and plant breeding, the discussion of ethical concerns increased as well. Forty-three percent of the articles in the 1960–1976 period included discussions of ethical issues, a statistically significant increase that also continued a long-term and relatively steady trend (see appendix 2, table 5). The increases were not merely numerical but also topical, for the media addressed a growing range of issues. The most sustained objects of ethical discussion were race and individual rights, but changes in the treatment of gender were important as well. These issues and the exploration of the implications of the broader program of substantially altering humanity were abruptly sublimated, however, by a fervent discussion of the health and safety implications of the new research on microorganisms. In the midseventies, concerns about "biohazards" would reconfigure the public discourse in a definitive fashion.

Ethical Challenges

RACE

Nineteen articles in the magazine sample contained arguments against the use of genetics to justify racial discrimination. Many of these arguments continued and extended the claim first ventured in the previous era—that there was no biologically sound definition of a race. Various advocates argued that races are "genetically open" systems, that there is a "genetic oneness of mankind" (Dobzhansky, 1966: 106), thus seeking to put to rest the persistent belief that persons from Africa, Asia, and North America lacked a common ancestry. Others emphasized that "no race can claim an exclusivity of intelligence, artistic or literary talent, or other inner traits" (Springer, 1958: 26).

This racially egalitarian advocacy became sustained enough to provoke an overt reaction that had been lacking in the previous era. William Shockley launched a campaign that was portrayed as claiming that race and IQ were genetically determined and that persons of a putatively homogeneous black race were born with lower potential IQ. Shockley was described as asserting that "black IQ increases on the average of about one point for each 1 per cent of caucasian ancestry," and it was reported that he had asked Roy Wilkins to recruit a hundred or so black intellectuals for blood tests to support this claim experimentally ("Shockley Asks Black Aid in Test," 1973). According to an editorial by Sandra Haggerty (1974) carried in the *Los Angeles Times,* the genetic racists also proposed financial incentives for black persons to consent to sterilization, providing $1,000 per point of IQ below 100.

Racist claims that were purportedly based on genetic grounds were developed and supported in two magazine articles (and five newspaper articles) in the sample. Most of the other dozen magazine articles and half dozen newspaper articles treating the issue gave Shockley and the others involved in the research or debate (including Arthur Jensen, Richard Herrnstein, Dwight Ingle, Edward C. Banfield) a mixed response, attacking the racist implications of the research, noting the deficiencies of IQ tests, and finding the research inconclusive, but leaving open room for additional study on the link between heredity and IQ. A few articles were completely hostile. These attacked claims made by Shockley and Jensen with a variety of counterarguments, especially that IQ tests were culturally biased and that it was impossible to separate socioeconomic from genetic contributions to IQ. They also defended compensatory education programs that were based on the assumption that children from deprived socioeconomic backgrounds could profit from additional assistance to compensate for environmental deprivation.

The conservative journal the *National Review* led the support for Shockley's assault on racial equality, and it was followed by miscellaneous writers of columns and letters in several newspapers. The press's defense of Shockley and his colleagues, however, was generally launched not as a defense of racist beliefs but in the name of "freedom of inquiry." Protest groups sometimes shouted down Shockley and others when they sought to present their arguments in campus forums. The press, with its obvious self-interest in free speech issues, highlighted this concern. The *Chicago Tribune,* for example, centered its coverage of the genetics and IQ debate on this side issue, singling out the comments of "one black C.A.R. [Committee against Racism] member," who purportedly shouted, "There is no freedom to teach racism" (Cassidy, 1974). The *National Review* similarly focused on accusing the opponents of Shockley and Jensen of trying to "censor" research in the area (Meyer, 1967). The relationship of genetics to the national value of freedom of inquiry, a relationship which in the previous era had helped transform an unfavored eugenics into the more popular family genetics, thus experienced a fresh and complicating binding in this period.

In total, Shockley and his allies clearly gave a strong presence to the genetically determinist belief that race dictated intelligence through genetics, but they cannot be said to have won their round. Sources outside the most conservative outlets assigned the debate at best a draw, and in general most outlets continued to support, at the conscious level, the view that genetic diversity was a positive value and that races were not unequal in their genetic endowment.

While the balance of overt declarations favored a belief in the genetic equality of races, strong residual subconscious racism remained. A picture in *Time* magazine offered the most egregious example. It featured a black male youth over the caption, "DOCTOR EXAMINING A SICKLE CELL SUSPECT: Incurable but avoidable" ("Detecting an Old Killer," 1971). No other article in the sample ever used the term *suspect* to describe someone who might have a genetic disease. The picture and its caption clearly drew on and resonated with the image of black males as criminal suspects. The "incurable but avoidable" subcaption conjured up the bases for the white flight that was a predominant response to integration in the era.

The penumbra of racism shadowed not only the iconography of the period but also the language used in the treatment of the issues. Sickle-cell disease, the genetic disease identified in public as most closely related to blacks, was the disease most likely to call up comments that "it can be prevented, provided that partners who both carry the trait avoid having children" ("Detecting an Old Killer," 1971). In order to enact that sentiment a screening program was launched to inform blacks of their genetic configurations with regard to the sickle-cell allele of the hemoglobin gene (an allele is simply a specific variety of a gene that codes for a specific substance). The program failed for numerous reasons. It tested young teenagers who had no immediate use for the information, and it did a poor job of transferring the technical information. People who had one typical version of the gene and one sickle-cell variant were carriers and were classified as having "sickle-cell trait," as opposed to those who carried two copies of genes that contributed to the sickling of blood cells, who were classified as having "sickle-cell disease." Not surprisingly, many people assumed that having the trait meant having the disease, although this is not true. Significantly, in other genetic diseases, those who carried one standard copy and one variant were likely to be called carriers rather than being labeled as having the trait.

The labeling of black carriers as persons who had the trait seems analogous to miscegenation laws that defined persons with even 1/32 of blood of African descent as being black. The threat of black deviancy was seen by whites as potent; hence, blacks who were carriers of a genetic disease were thought more threatening—already having "the trait"—than were whites who were simply "carriers" (a term with its own potential for stigma but which does not denote the individual as ill).

There were significant consequences to this linguistic discrimination and the policies associated with it. Blacks understood the purpose of the sickle-cell screening program as an effort to limit their reproduction, and they resisted it, labeling it genocide. Efforts to inform and limit the reproduction of those with the sickle-cell trait have continued to be ineffective for this reason and for others, including the fact that sickle-cell disease manifests itself in widely varying degrees of severity and that, according to some accounts, African American culture does not value rigorous quality control over reproduction in the ways of Western culture (Hill, 1994).

Sickle-cell testing served as an experimental project for mass genetic testing for carrier status. Because of this social experimentation, African Americans also were the first to face on a substantial scale the socioethical difficulties brought on by such testing. They experienced inadequate procedures for informed consent, were denied insurance (or assigned higher premiums), were denied flight status in the military, or were stigmatized in personal ways because of being labeled with the sickle-cell trait (Ausubel, Beckwith, and Janssen, 1974; Bylinsky, 1974; Lamberg, 1976). All these problems would become central ethical issues discussed in the press in the next era, when genetic screening moved to the mainstream, but the issues made brief appearances in this era because sickle-cell testing provided the experimental grounds for mass genetic testing.

A similar tension between overt efforts to attend positively to racial and ethnic difference on the one hand and subconscious racism on the other affected coverage of genetic issues facing Jewish groups. Ethnicity was "in" during the sixties, and genetic diseases lent themselves to classification by ethnic groups, given that some diseases have much higher frequencies in different ethnic groups. Several articles sought to give a positive cast to an ethnic approach to diseases: each group's unique identity, it was suggested, would be preserved and valorized (Yuncker, 1976). However, articles about Tay-Sachs disease also seemed readily to call up stereotypical linguistic patterns about persons of Jewish descent. In an article entitled "Jewish Diseases," *Newsweek* reported on six inherited diseases that have occurred chiefly among persons of Jewish descent. The article compared the symptoms to "an Old Testament plague," and concluded that the best solution to the problem was "to prevent affected children from being born." It ended with a statement ominously shadowed by the past: "The foundation hopes that the Jewish genetic diseases will eventually be wiped out for good" (Seligman, 1975). No doubt the authors and editors of this article did not intend to support a new holocaust, but the long shadows of our racial histories and stereotypes are difficult to control. Because of the life-and-death character of genetic diseases, these shadows were important ones.

The racial anxieties of the era manifested themselves even in those articles that engaged in speculation about radical modifications of the human species. Following science fiction of the era, these authors saw racial group relations as a model for a potential problem in cloning. They argued that racial conflict might be superseded by great battles for dominance between mass groups of clones (Rorvik, 1969b: 113). While this vividly depicted problem clearly does not lurk immediately around the bend, its imaginary presence suggests the broad significance of the racial model of genetics within understandings of heredity in the era.

SEX AND GENDER

Although the sixties and the seventies were an era of substantial change and agitation on women's issues, considerations of the implications of the new reproductive technologies for women were not as well developed as issues of race were. One tortured article in *Mademoiselle* tried to address the implications of the new technologies for women's psychology and self-esteem, but it could not extricate itself from presumptions that women should be baby makers first, last, and solely. The article argued that if women were to lose this basis of self-identity, they might launch a "wholesale invasion of men's last remaining island of emotional safety, their work" (which would, of course, destroy men's self-esteem) (Weitzen, 1966: 71). Many major concerns of women with regard to the implications of genetic testing received virtually no mention at all, and there was no discussion of the distinctive concerns of women who came from differing races, classes, and physical configurations. Moreover, background sexism abounded. When authors listed subjects for cloning, they tended to select women for their beauty (Raquel Welch), and they suggested the worth of cloning by noting that "great beauties could be cloned to cheer a dismal world" (Rorvik, 1969b: 112). While Einstein and Darwin made the list of desirable clones, women of science and female adventurers such as Marie Curie and Amelia Earhart did not. Barbra Streisand and Mahalia Jackson were the only female persons chosen for something beyond beauty, and the articles avowed that Streisand's physique would have to be modified before she would be truly clonable (no such mention was made of the shortcomings of Einstein's physique). One source found itself astonished at and threatened by the thought that women might clone themselves; after listing all the possibilities for "man" cloning "himself," such as "a police force cloned from the cells of J. Edgar Hoover, an invincible basketball team cloned from Lew Alcindor (Abdul-Jabbar), or perhaps the colonization of the moon by astronauts cloned from a genetically sound specimen chosen by NASA officials," *Time* noted that "a woman could even have a child cloned from one of her own cells." It mused further that "the child would inherit all its mother's characteristics including, of course, its sex" ("The Body: From Baby Hatcheries to 'Xeroxing' Human Beings," 1971: 39). *Time* did not play out the further thought that this might make men expendable, but others did (Rorvik, 1969b).

Though background sexism continued, and the perspectives raised by feminists were absent from the debate, discourse about genetics was distinctively favorable to women in some important respects. As in the previous period, articles continued to refute the tendency of men to blame birth defects on women. Some articles even described lay men (though not scientists) as more "emotional" than women or at least equally so. Though there was no parity in the portrayal of significant scientists as both men and women (outside of *Ms.*), and genetic counselors were uniformly and incorrectly referred to as "he," some women did make it into pictures of anonymous scientists at work. Moreover, several women were portrayed as heroes for bringing their family's medical condition out into the open where it could be dealt with. Important, though insufficient, was the portrayal of women as active partners in the process of making decisions about procreation where they had to be made. Given the proclivities of the era to assume that father knows best, genetic counseling might have assigned men the role of procreational decision makers. Instead, genetic counseling assumed that the couple was the locus of decision making. This accurately mirrored the actual practice of most usage of genetic testing by the white middle class, though it obscured the issues surrounding genetic testing for the increasing number of women who would later come to genetic testing through public welfare programs and who often did so without a male partner.

One article took the growing feminist movement even more seriously, attacking physicians as sexist because they used women's potential anxiety as a reason to deny them access to control over the technology (Etzioni, 1974: 41). Instead, this ethicist argued that women should control all their own reproductive-related decisions. Specifically he complained about the physician-imposed age cut-off of 40 years for the use of amniocentesis, suggesting that if a woman wanted to undergo amniocentesis at a younger age, that should be her choice. In the eighties and nineties, Barbara Katz Rothman ([1986]1993) would reverse the claim, arguing that such extensions of access to the technology constituted a sexist imposition on women. The reversal gives some pause for thought about the complexities of feminist positions on reproductive technologies, as did another important question raised in this period and still unsettled; both the *Los Angeles Times* and *Newsweek* asked, To what extent should these medical technologies be used to serve parental desires, rather than confined to use for disease? ("Genetic Engineering," 1972; "Pause in Test-Tube Baby Research Urged," 1972). The question becomes difficult for feminists when phrased not in terms of parental desires but in terms of women's desires to control their own reproduction.

THE INDIVIDUALIST FRAME

In contrast with earlier eras, concern over discrimination of a variety of sorts received explicit attention. Whether the issue was disability rights, race, or gender, however, in keeping with the proclivities of the day, each was understood within an individualist framework. There were clear advantages to this insistence on individual rights. Normalizing the assumption that individuals should control their own procreational decisions provided an important bulwark against a Nazi-style abuse of the new technologies. Insisting that regardless of his or her skin color, each individual was equal to every other provided a further bulwark against racist assumptions that some races were innately inferior to others. Emphasizing the fundamental human worth of persons with disabilities was also crucial to advancing a disability rights perspective. Moreover, focusing on the basic rights of individual human beings would be essential in the attempt to address a variety of new ethical concerns—including the effort to establish the principle that research on humans should be done only for the good of the individual involved and after complete animal research studies.[1]

In spite of all of these advantages, however, there are some grounds for concern that the focus on individual rights was interpreted in this era to mean that individual rights provided completely sufficient guarding principles for moral goodness to be maintained in the genetic era. The *New York Times Magazine* thus offered only humility and "basic respect for the life of the individual" as positive levers to "help determine whether our future will see the control of heredity as a happy dream or a tortured, endless nightmare" (Lasagna, 1962: 60). *Fortune* likewise was willing to rely on individual choice for ethical salvation, concluding: "It appears that answers to many of the new ethical issues will evolve only gradually, as a mosaic made up of individual decisions" (Bylinsky, 1974: 154). Reliance on "the democratic faith in personal liberty and personal responsibility" (Thorndike, 1973: 62) was also *Horizon*'s offering as a sufficient bulwark against the negative potentials of genetic technologies. In this era, there was almost no attention given to the ways in which individual decisions are always embedded in familial, cultural, social, institutional, and economic frameworks. The next era would offer a fuller perspective: an individualist ethic, it would be suggested, provided a good start for preventing genetic injustice and the violation of individual rights, but an individual rights approach left open the door to other major ethical problems. A concern with the influence of social structure, however, waited on the other side of a major discursive divide that radically realigned both symbolic and institutional structures by highlighting the fact that human beings are tied together by our shared vulnerability. The rhetorical event that precipitated that realignment was the debate over biohazards.

Biohazards

In the sixties, several scientists discovered and began to employ a potent new tool in genetic studies: restriction enzymes. These naturally occurring enzymes slice DNA strands at particular code sites; once scientists learned how to use them, they had a highly effective means of beginning to cut and paste genetic segments together from various organisms. The efficacy of the new tool, however, raised to consciousness an important new issue, that of public safety (see appendix 2, table 4; see also Turney, 1998: chap. 9; van Dijck, 1998: chap. 3). The safety issue was first addressed by a group of scientists led by Paul Berg, who had planned an experiment to insert a cancer-causing monkey virus into a bacterium that inhabited the human gut. After discussion among themselves, the scientists initiated a moratorium on certain types of gene transplantation experiments long enough to meet together to decide on the risks and on the procedures to use to minimize those risks. With modifications, their proposals were enacted by the National Institutes of Health as regulations limiting the experimental practices of its grantees, and further modified industry standards were adopted as well.

The "recombinant DNA" safety scare was a discursive event that condensed a number of crucial issues that had been bubbling throughout the period. By resolving them in more or less definitive ways, the scare closed the era, settling the important issues of whether the risks of continued research were worth the benefits and determining the social control structure within which the new research and its technologies would be normalized.

RISKS AND BENEFITS

When Berg and his colleagues proposed a pause in research while the social issue of public and environmental safety could be explored, they did something both unprecedented and profoundly sensible. Scientists were working with a powerful new tool, one that no one had stopped to consider conscientiously, and it was only reasonable that a short delay be taken to ensure that they were not naively playing with a doomsday machine. As it turned out, the scientists and their society were to decide that the risks were manageable and proportionately reasonable—just "another bite of Eve's apple" ("Minimizing the Risks of Genetic Engineering," 1976: 66). This eventual consensus lent credence to the posthoc claims of radical pro-science propagandists like James Watson that the moratorium had been a product of "intellectual sloppiness" fanned by "hysterics" and ultimately the product of "liberal guilt" (Watson, 1977: 12–13; see also Turney, 1998: 171). However, at the time the moratorium was initiated, the risks were not clear, and a few months' delay in research was hardly of serious consequence.

It did not take long for either the scientists or the society to decide that the risks were manageable and worth the potential benefits, though there were two cycles to the decision process. The public first received serious notice of the issue in 1974, when the moratorium was proposed. News magazines such as *Time* portrayed the pause as a reasonable response to fears that scientists "may loose upon the world new forms of life—semisynthetic organisms that could cause epidemics" ("The Andromeda Fear," 1974). The national press subsequently covered, with general balance and much sympathy to the scientists, the removal of the moratorium and its replacement with risk-reducing regulations. During this period, the press asked questions and used examples that were hardly sensationalistic and that did not attack the scientists or the enterprise of science. Most magazine titles and subtitles accurately represented the restrained tone of the articles: "Minimizing the Risks of Genetic Engineering" (1976), "Genetic Moratorium" (1976), "A Move to Protect Mankind" (McWethy, 1975). When *Time* magazine used the dramatic and freighted title "Andromeda Fear," it took the highly unusual step of providing a footnote that clarified and desensationalized the moratorium by distinguishing between the plot of Michael Crichton's science fiction novel *The Andromeda Strain* and the potential problems created by restriction enzymes ("The Andromeda Fear," 1974). At most, the article titles raised central and appropriate questions: "Is Science Creating Dangerous New Bacteria?" (Fried, 1975) and "Science's Newest 'Magic'—Blessing or Curse?" (McWethy, 1976). Subtitles clearly summarized the issues as raised in the articles, "The promise: man-made forms of life to cure disease, boost crop yields. The worry: that new organisms could be set loose and wreak havoc" (McWethy, 1976: 34). These articles gave extended lists of the potential benefits offered by the new technologies and accurately described the potential risks (albeit adapted to the level of technological sophistication of their audiences). The issue seemed well on its way to a resolution that favored the limited restrictions imposed by the National Institutes of Health and scientific self-regulation when a second wave of public attention occurred.

The second wave was generated by a group of northeastern political activists who convinced Senator Edward Kennedy and other politicians that scientific self-regulation was insufficient and untrustworthy. Congressional hearings then were held to explore the issues of public regulation and the level of risks involved in the research. Coupled with a debate by the city council of Cambridge, Massachusetts, about limiting such research at Harvard, this round of publicity intensified the discussion to the shouting point. Scientists on both sides of the issue made public statements, and many of these were far more intemperate than the sentiments expressed at the scientist-dominated Asilomar conference. Scientist opponents of unrestricted research joined the liberal anti-scientist activists in painting doomsday scenarios. Not to be outshouted, opponents of research restrictions labeled those who wanted restrictions as kooks and hysterics bent on destroying democracy and freedom of inquiry. As we will see below, in the end, the substantive conclusion would be the same as before the shouting began; limited regulations, unsupported by effective implementation procedures or review outside the scientific community, would become the accepted control mechanism. Much of the public media would conclude that a reasonable course had thereby been set: "the continuation—with reasonable restraints—of free scientific inquiry" ("Tinkering with Life," 1977: 45).

Satisfied with this outcome but shaken by the process of public debate, scientists would later rewrite the story of the recombinant DNA debate, focusing on the emotional intensity and blaming it on media sensationalism (Kornberg, 1981; Watson, 1977). This was not only a false scapegoating; it also missed the symbolic significance of this exchange of sensational discourse.

The biohazard stories wrapped up, contained, and sublimated the fears embedded in the debate about tinkering with human genes, which had so volubly opened the era, within the narrower issue of genetically altered organisms on the loose. This submerging of the broad fear of genetic reconstruction of human beings within the narrower issue of runaway bacteria was quite visible in several articles. The clearest example occurred in the coverage of biohazards offered by the *Chicago Tribune* in 1975. Ronald Kotulak, the *Tribune*'s science writer, began his article by explaining that the new "breakthru" technology had opened "the way to an almost unbelievable assortment of manmade mutants" (1975: 23). He indicated that the potential uses of these included the ability to "produce dramatic changes in the human species" (1975: 23). He alluded to Aldous Huxley's *Brave New World* and Hitler as examples of the potential totalitarian abuses of such changes and then moved on to detail the potential microbial hazards, including "the fear that gene-transplantation could unleash deadly new microbes similar to those depicted in 'The Andromeda Strain'" (1975: 28). He (correctly) described the fear of microbes as the source of the moratorium, but after doing this Kotulak made no more mention of the "frightening" potentials for manipulating humans themselves. Instead, the article settled down to focus solely on the microbial hazards and placed the context of regulation within the issues of safety rather than within questions of what it might mean to take control of and make decisions about reshaping the human species.

The *Saturday Review* accomplished this transition even more efficiently. Anthony Wolfe opened its report on the moratorium by noting the existence of new "biological techniques which give the scientists powers usually reserved for the Creator" (Wolfe, 1975), a concern that had been previously and amply expressed in most of the articles discussing the modification of humankind. Wolfe's reference is ambiguous, however, for he does not explicitly reference the human context. Instead, he uses the ambiguity in the general reference to begin from that previous context but then to redefine the context of public fear by referencing the benefits and risks solely in terms of the alteration of bacteria. By the end of the article, we are in a new public space: we have made the transition from an era where we asked whether humanity was wise enough to modify itself, to an era where we began to discuss safety restraints on the manipulations that might lead to such activities. An article in the high-circulation *Reader's Digest* moved in a similar fashion, containing language that evoked the fears of the earlier debate: "The new methods were already spreading to labs across the world and could unleash uncountable experiments in so many new directions that there would be no way to predict all the strange new organisms that might come into being" (Fried, 1975: 134). It then contained those broad-ranging fears by limiting them to the bacterial context in the rest of the article.

It is telling that after these articles about bacterial hazards, no more articles in the wide-ranging genre devoted to issues of "modifying man" (Rostand, 1959) appeared in the press for several years. Surely, there were many reasons for this fairly abrupt switch. Immediate concerns over serious safety hazards were more concrete and perhaps more reasonable than fears about grand-scale processes that could not yet be achieved. Additionally, some scientists had begun loudly to discredit such freewheeling speculation. However, part of the reason for the abrupt shift was symbolic: a suitable containment device had been found for the grander speculations and fears, concerns which had proven themselves undecidable and hence profoundly uncomfortable.

WHO DECIDES?

For the public consciousness, the containment process involved moving reflection away from the grand metaphysical and ethical issues of humankind's control over its own genes. For the scientists, the goal of containment was more pragmatic. Some scientists began to fear that the magnitude of their new powers would mean the end of their ability to operate unfettered by the public. Throughout the period in which the genetic code had been deciphered, many articles had asked the question now raised in the *Washington Post* and elsewhere, "whether some scientific decisions are too important to be left to scientists" ("Genetic Engineering Stirs Debate," 1975). No significant steps had ever been taken in that direction. Not only was there no significant legislation in regard to genetic engineering in the era; there was also relatively little in the way of concrete proposals for such legislation, outside some limited forays into publicly supported genetic testing. The reasons for this gap between a publicly expressed desire to contain science within consciously deliberated public agendas and any practical movement toward that agenda rest on the manifest difficulties of public regulation of technical enterprises (Farrell and Goodnight, 1981). These difficulties were explicitly expressed in the debate over biohazards.

Two broad loci were proposed for the regulation of biohazards: scientists and the public. Those who proposed that the public should provide regulatory control argued that scientists were too deeply self-interested in the outcomes and therefore too likely to promote their own interest in research over the public interest. *U.S. News and World Report* voiced that concern, asking, "[Are] the people deeply involved in this research the best judges of whether their work harbors more risks than society should be asked to bear?" (McWethy, 1976: 35).

Those who opposed public regulation countered that the public were ignorant and emotional. The head of a pharmaceutical company was quoted as saying, "What frightens me . . . is that a city council could prevent any kind of research. Science is not above public policy, but decisions should not be made by the uninformed" ("Minimizing the Risks in Genetic Engineering," 1976: 67). Notably, the public were portrayed as inherently and irremediably uninformed, even when testimony had been presented by scientists on both sides of the issue. This occurred because opponents of regulation successfully portrayed their side as purely scientific while tainting their opposition, even the scientists among their opposition, as merely political (see also van Dijck, 1998: 74–78). The success of this rhetorical move relied on the assumption that emotionality and ignorance were inherently linked in public actions. Several sources emphasized the emotionality of the public (and the way that emotion had tainted the scientists who supported public regulation), and they linked that fear to censorship. Thus, a molecular geneticist, Norton Zinder, linked the presumed emotionality of the public to potent historical exemplars: "We are moving into a precedent-making area—the regulation of an area of scientific research—and I must plead that this be done with extreme care and without haste. The record of past attempts of authoritative bodies, either church or state, to control intellectual thought and work have led to some of the sorriest chapters in human history" ("Tinkering with Life," 1977: 45). If science was self-interested, the interests and abilities of the public were also portrayed as suspect.

Those who decided that the public were a better risk for the vesting of control could cite the example of the Cambridge City Council's final decision to permit Harvard to pursue recombinant DNA research as a model of public deliberation. *Time* argued that scientists could "find reassurance in the experience of Cambridge. There, citizens patiently ignored political demagoguery, perceived the false notes in the voices of doom, mastered the complex issues and then cast their votes for the continuation—with reasonable restraints—of free scientific inquiry" ("Tinkering with Life," 1977: 45). As a counterweight, scientists pointed to the earlier, self-imposed moratorium as evidence that scientists were capable of self-regulation. *U.S. News and World Report* suggested that the imposition of the original moratorium "demonstrates that scientists are capable of considering the implications of their work in a context reaching far beyond the isolation of their laboratory walls" (McWethy, 1975). The popular, conservative magazine's science writer argued that this should be seen as a future model: "A scientist today must become an intimate part of the fabric of society, taking a special initiative to evaluate his work, to explain it and justify it to the public." He expressed hope that "the action taken by genetic researchers may well be a turning point for the men and women of science." Another article in *Time* concurred and quoted a Harvard Medical School biochemist, Richard Roblin, who argued that there was and should be a growing awareness of a balance between science and social consequences (Fried, 1975: 136).

There was no definitive outcome of this debate, as *U.S. News* noted, "The research is continuing, and so is the debate" (McWethy, 1976: 35). The problem for the public was analogous to that faced by individual couples in employing the technology. Just as individuals or couples had to rely on the information supplied by medical professionals in order to make their decisions, so was it necessary for the public to rely on scientific experts to understand what choices they faced. As Kenneth McDougall put the issue to the readers of *Intellect,* "Input from the public is essential, but to be of value it must be informed opinion," and he suggested that the scientific community should be the source of that information (1976: 530). Various experts in ethics proposed something like a council of wise persons as stand-ins for the public (Ausubel, Beckwith, and Jannsen, 1974; Etzioni, 1974), and these suggestions were implemented by various commissions. Once again, however, the practical consequence of undecidability, especially given the tight link between knowledge and power (Foucault, 1980), was to give scientists very broad autonomy. As the close of the century neared, very few substantive restrictions had been placed upon genetic research or the applications of genetic technology (Clayton, 1994; Milunksy and Annas, 1984). The few existing restrictions had been heavily influenced by scientists, because even supposedly "public" regulatory committees were numerically and discursively dominated by scientists. Arguably, this resulted in the committees' deciding that it is "ethical" to do whatever science can immediately do but not ethical to do what it cannot yet do. Perhaps because of the difficulties of following the technology, perhaps because of the balance of power and self-interest, the outcome of the biohazard debate was that the public ceded control of genetic research and biotechnology to science and industry, except at the most serious extremes of massive and relatively immediate, demonstrable threats to public health and safety. The association of genetics with "freedom of inquiry" that had inaugurated the change from eugenics to genetics in the first half of the century remained unmodified.

From the standpoint of democratic theory, there are problems with this model. There are also, however, problems with the practice of the model. Many scientists drew from Asilomar the moral that sustained and focused efforts at scientific self-regulation only stirred up the public and thereby threatened the funding and autonomy of science itself (Watson, 1977). Alternately, they simply found that they didn't have the stomach for the sensationalist attacks and maneuvers inevitable to politics, wherever it was conducted. As a consequence, scientific self-regulation came to be little more than a matter of individual choice. There would be relatively little collective deliberation by the scientific community. Scientists like Arthur Kornberg (1981) would reflect favorably on this locus of control, arguing that scientists, raised in the broader community, would reflect its values: "Whether scientists engage in improper activity will ultimately depend on the ethics and morality of the community," he concluded. However, in a radically liberal and individualistic milieu, such a restraint is no restraint at all. It might take only a few people, out of step with community morals, to break through the community's moral barriers.[2] Where there is no community control, individual scientists, and the biotechnology companies that fund them, are on their own to set the future. Market imperatives would therefore become crucial defining factors in the future.

In 1953 the discoveries of James Watson, Francis Crick, Maurice Wilkins, Rosalind Franklin, and many others had provided the crucial catalysts for molecular work in genetics that could lead to direct control of human genes. By the end of the decade, the public had become aware of the potential implications of this research, and many sources joined the *Los Angeles Times* in suggesting that "it's time now for society to ask itself whether we want to go through the door of this brave new world" ("Scientists Urged to Go Slow," 1972). The question as to whether humanity should deliberately undertake to remake itself was perhaps logically undecidable. It was at least undecided in the era; this major question about the remanufacturing of human potential was curtailed by the biohazard debate and subordinated to the issue of genetic disease and counseling. Given the dynamics of argumentative presumption in a laissez-faire democracy, this meant that the door was left open for individuals and corporations who wished to walk through, without much regulation or constraint by developed social norms or intellectual exploration.

By the midseventies, a distinctive social pattern for the development of genetic technologies and their application had therefore been established. It was a laissez-faire pattern. Parents and patients, scientists and corporations, were free to come together to do with genetic engineering as they wished. Parents and patients might have the last word in this decision process, but their access to the technology and their information about the choices were heavily influenced by the physicians and genetic counselors who were the direct manipulators of the technology. In the public realm, a similar structure existed. The public had putative control, but on the few panels that considered the issues for the public, scientists held sway. The only regular direct input of the public was control over the lump sum of the tax dollars they devoted to the various national research agencies.

These structures and the growing practices they supported were based on a new sense of human heredity, grounded in a new vision of the gene. The gene was not a particle but a fragment of code. Consequently, control of human heredity was not a class or a racial phenomenon to be manipulated by top-down political control of bifurcated categories of human bodies. Nor was it a fixed individual property—an unmalleable product delivered to each person by nature and fate. Instead, genes were messages, transmitted and controlled within families, always potentially rewritable, and ultimately comprehensible to the language-using animal. The new vision enabled less rigidly deterministic options, not all of which were utilized. The new era grappled more openly with but did not overcome racism, sexism, or discrimination against persons with handicaps. It toyed with grand visions of perfectionism and then set them aside.

This was an era of sweeping experimentation, for both scientists and the public. The intellectual revisions and practical outcomes were, however, quite modest. Both technical and social factors shaped the new meanings of human heredity and acted as restraints, staving off the nightmare of the era—that we would become too powerful, play God, and extinguish our humanity—for yet a little longer. Efforts at mass commercialization were, however, just around the corner.

PART FOUR

GENETIC MEDICINE

AND BEYOND

1980–1995

Scientific experimentation and research in genetics in the first three-quarters of the twentieth century were wildly successful. Bioscientists gained increasing prowess at manipulating genes within cells and producing organisms with desired hereditary lineages. Simultaneously, genetic research contributed to a massive expansion of the understanding of cellular processes. These successes enabled technically minded entrepreneurs to envision and undertake a transition to large-scale commerce in genetically based commodities. Consequently, during this era, the mass media began to discuss the biotechnological marketplace as avidly as it discussed bioscience. Regulatory issues dominated these debates, but these matters were quickly settled in favor of loose control by federal regulatory agencies based on quasi-scientific criteria.

Suitably for a business world built on a manufacturing mentality, the dominant public metaphor for envisioning the role of genes in human society during this era was the blueprint. With the focus on individual buildings rather than familial lineages, public discussion of genetic counseling began to be replaced by consumer guides for using genetics to the advantage of one's personal health. The simultaneous developments in business and medicine constituted something of a paradox: just as genetic medicine was gaining direct human applications, commercial imperatives drove the focus of biotechnology beyond medicine. The beginning of the era of actual applications of genetic medicine was thus simultaneously a time of extension beyond the vistas of genetic medicine. These cross-cutting changes were accompanied by sustained and expanding socioethical debates about the implications of the new genetics.

8 BLUEPRINTS FOR GENETIC COMMERCE

Just as the spirit of science is captured by metaphors of exploration, the spirit of business is captured by images of manufacturing.[1] To note this difference in spirit is not to claim that there is a yawning chasm between science and commerce, nor is it to deny that specific regions of scientific research are well funded because of their commercial applications, nor is it to ignore the fact that many leading bioscientists sought to parlay their discoveries about genetics into large capitalist enterprises. The observation means simply that scientific research and commercial manufacturing are not precisely the same activities and, therefore, that one might expect that the transition from an era when genetic manipulations were performed primarily in the research laboratory to an era when genetic manipulations are most frequently performed in commercial manufacturing establishments might require significant adjustments in public considerations of the meaning of genetics. And indeed, the shift from experimental laboratory to commercial factory was accompanied not only by the technical modifications and innovations necessary for scaling up and adapting exploration for profitable ends, but also by a reimagination of the fundamental concepts involved in genetics. As coverage of genetics migrated increasingly onto the nation's business pages, the dominant metaphoric representations of the biological basis of heredity shifted away from collections of autonomous genes carrying bits of rewriteable code toward depictions of a unified genome that constituted a blueprint for a living being. As this manufacturing metaphor realigned the gene to fit commercial processes, the business pages also efficiently and rather quietly managed the important issues regarding regulation of the new commercial processes. These new discursive elements not only enhanced the possibilities of the commercial profitability of genetics technologies, but also moved them further away from visions focusing on the manipulation of human heredity.

The Genome as Blueprint

The use of the code metaphor as a means for describing the action of genes did not disappear in the closing decades of the twentieth century. Instead, the code metaphor became increasingly literalized, that is, familiar and taken for granted. Magazine articles no longer placed the phrase *genetic code* in quotation marks or bothered to provide explanations of just how it was that genes could be understood as functioning like a language. There were frequent references to reading or deciphering the genetic code, but these near-dead metaphors were increasingly treated as simple descriptions rather than explanatory comparisons or active analogies. Meanwhile, the metaphor of the genetic blueprint took on increased importance. The blueprint metaphor had been used in the previous era, but its usage increased to a statistically significant extent in this period (see appendix 2, table 2). Not only was the blueprint metaphor the most frequently employed, but also this heightened usage often occurred in central places in magazine texts—headlines, lead-ins, and captions. One reason that this metaphor might have been increasingly centralized was that it helped to explain the new scientific emphasis on the human genome as a whole rather than on genes in isolation.

The coding metaphor had focused on bits and pieces of DNA, specific genes, and fragments of codes. In the early version of the metaphor, genes were independent messages to be read in separate sittings; therefore they could be rewritten in isolation from one another. As scientists began to turn their attention to genes in context rather than in isolation, they began to envision genetics in a more holistic sense. One component of this return to the organismic scale was the call to sequence the entire human genome, which gradually evolved at the end of the eighties and which became funded U.S. policy in 1991. The blueprint metaphor accompanied magazine introductions of this new project: "decoding the human genome" was said to mean "to write down the complete sequence of human DNA—the blueprint for a human being" (Gianturco, 1990). Not only did the new metaphor refocus attention on the ordering and contextualization of genes; it also established tighter links to the material characteristics of codes and their enactments.

In the late eighties and early nineties social critics began decrying the code metaphor of genetics for its decontextualization of the gene (Fox Keller, 1995; Kay, 1995). At the same time, in much scientific work and in the popular media, genes were being reenvisioned through the blueprint metaphor as more closely contextualized within both the cell and the larger environment. The blueprint metaphor invited more holistic thinking than did the more fluid and diffuse images of codes and language, which it incorporated and subordinated. Blueprints are complete, structured plans for whole entities rather than free-floating messages about making isolated bits of meaning. To imagine building a structure from a blueprint is to imagine a more material and contextual process than simply reading a message. Building requires bricks and mortar, time-sequenced activities, as well as the participation of multiple actors, whereas reading a message appears to require nothing more than the solitary reader, acting effortlessly and unconstrained.

The interactive nature of the genome and the cell was increasingly evident in the mass media's descriptions of genetics and heredity. In the most extreme case, DNA was now sometimes explicitly described as passive. Horace Judson, a highly successful science writer, noted, "The genes are passive, while the proteins actively build the cell" (1983: 13). *Time* magazine portrayed the gene-cell relationship in a similar fashion when it portrayed the cell as the active agent. According to this popular news magazine, "reading these genetic words and deciphering their meaning is apparently a snap for the clever machinery of a cell" (Jaroff, 1989: 64). The pacification of DNA was furthered by the concomitant increase in the use of computer metaphors (for a contrasting interpretation of this metaphor, see van Dijck, 1998: 123). DNA was now often described as a computer data bank, a metaphor that emphasized the active use of such data by other agents rather than the text itself as directing agent.

The activity of the cell and its regulatory mechanisms was further highlighted by articles pointing out that the same set of genes is present in every cell in the body, yet these genes act in radically different ways in different cells, and they produce radically different cells, depending upon cellular regulatory processes. Cells were thus increasingly described as regulating and controlling genes, as opposed to the earlier era, in which the descriptions went only in the other direction with genes firmly "in command." Indeed, such regulation of genes was now described as "one of the most important issues in biology" ("Mighty Mice," 1982).

This new set of descriptions did not simply and univocally demote DNA to insignificance. The degree to which genes were contextualized varied from article to article and even within articles. Many references to simple, singular genes "directing," "determining," or "commanding" the cell could still be found. Instead of offering a new dominant model, the new set of depictions supplemented the older, simplistic model of the autonomous gene with a sense of ambiguity about the relationships among genome, cell, and organism. That ambiguity was effectively captured in an account of cancer as an instance where the gene "declare[s] mutiny on the body" (McAullife, 1987b: 66). This image presented the gene as a servant or subordinate of the body in normal operation, but granted that the gene holds sufficient power for stepping out of its rightful place to cause havoc.

The complexities of the relationship between the genome and the rest of the biological entity were accompanied by the ever greater complexities revealed within the genome itself. Writers observed that interactions among genes aggregated in complex ways, producing a system of dizzying fluidity and variability. *Newsweek* summarized its coverage of recent developments in hereditary genetics with this claim: "Genes, it turns out, act so perversely they would have shocked the monk [Mendel] into a vow of perpetual silence." The popular news magazine quoted a geneticist who agreed, concluding, "It isn't nice and clean like it used to be" (Begley, 1992: 77). Moreover, it was not simply the code of the DNA that began to be understood as important; it was also the physical 3-D structure of genes and the proteins with which they interacted that began to be featured (Carpenter, 1989). All of this occasionally culminated in a developmental perspective, which noted that genes operate only in limited contexts and time frames, many only "during a very limited time of an embryo's development" (Cohen, 1993: 274). Thus, an increasingly dominant theme of the era of the genome was the complexity of the biological system at the molecular level.

There was some irony in this transition. For at precisely the same time that scientists and the popular media were increasing the complexity of their accounts of the genome and cellular interactions, academic critics were busily attacking older formulations and calling for the new developmental, systemic, interactive, and holistic perspectives that were evolving (Cranor, 1994; Fox Keller, 1995; Hubbard and Wald, 1993). Yet, this transformation to a more fluid, developmental perspective proved to be much less of a panacea than the critics of the era hoped. Individual units of discourse rarely bear their politics in such simplistic fashion. The systemic, interactive, developmental metaphors were wielded comfortably and quite often by the most conservative and profit-seeking journals in this era. A developmental perspective might be used, as the progressive science critics hoped, to heighten the importance of the interaction of environment and genes (such as the need for nutrition and intellectual stimulation in early childhood). But it might also be used to return to something like the old particle theory of genes, by indicating that there was no utility in later childhood compensatory education, since the basic neural pathways had already been laid down in early childhood and nothing further could be done. Similarly, a systemic perspective might initially seem to erase the vision of the gene as a magic lever to provide control over the species, but if molecular biology really functioned in a systemic rather than a gene-driven fashion, then recognizing the system would simply prove to be a necessary step toward control by would-be manipulators. The meaning of any component of the public discourse was thus likewise dependent upon the total formation. One fragment of the rhetorical formation concerning heredity that continued to be important was the range of available models for the interaction of genes and the macroenvironment.

As the blueprint metaphor was overlaid upon the coding metaphor, most articles in the era continued to emphasize the coactive role of genes and environment (see appendix 2, table 1). Thus, Cohen suggested that "the vast majority of common human illnesses are caused by multiple genes working in concert with multiple environmental factors" (1993: 272). Sixty-one percent of all articles in the previous era had taken this approach, featuring both environment and gene as coacting factors, whereas 66 percent took that stance in this era. This small increase was not statistically significant.

The relative stability in degree of determinism was accompanied by a tendency toward less specificity with regard to the exact models by which genes and environment interacted. Discourse early in the century had predominately placed the gene as specifying and preestablishing outer boundaries within which environmental factors operated. Discourse in midcentury had offered a range of models, most of which explicitly gave environmental factors substantial influence. At the close of the century, most articles tended to be ambiguous about the precise model underlying the relationship between the gene and environment. There was less thorough elaboration of these relationships. The era introduced, however, a new model based on normative comparisons.

In this era, there was a substantial consideration of genetic configurations that increased the susceptibility of individuals to particular diseases. These explorations included a radical reframing of the relationship of genes and environment. In place of a temporal frame (first you are given genes, then the environment operates upon what those genes provide) or a territorial metaphor (the genes establish the boundaries, and the environment establishes specific location within boundaries), these articles used a normative model. Instead of analyzing the interaction of genes and environment within the individual, they placed the risk of disease experienced by individuals against an imaginary norm for the species. An individual's genes were therefore described as increasing personal risk above a presumed norm, but behavior could increase or decrease risk. For example, *Forbes* noted that "a person's DNA makeup can exacerbate the effects of smoking or poor eating habits to compound the risk of contracting both cancer and heart disease" (Bronson, 1986: 142). In a similar fashion, *U.S. News and World Report* and *Reader's Digest* both reported that there were "patterns of genes that raise a person's susceptibility" to common diseases (McAullife, 1987a: 18).[2] Thus, personal features were measured against some average amount of risk, which could be raised or lowered for each individual depending on his or her particular DNA or behaviors and habits. Reporting about the cancer process also worked within this normatively based model, noting that DNA could begin as normal and then become damaged. This normatively based model included both genes and environment as contributing factors, without specifying interactions between the two factors.

Each of these steps toward the contextualization of genes whittled away in small and subtle ways at the deterministic power of the gene. The scope of the gene was further limited by a steady increase in differentiation. While articles were claiming that genes were implicated in all diseases, even in all life processes, they were also specifying that the degree of influence exerted by genes varied significantly depending on the trait or condition at issue. Thus, many articles noted that for some conditions, usually single-gene disorders, genes had a dictatorial power, determining the health outcome, whereas for other, usually complex polygenic conditions, genes exerted only influence. This tendency to differentiate among genes according to the degree of their causal influence had increased steadily across the century from 3 percent of the articles in the era of classical eugenics to 32 percent in this era (see appendix 2, table 8). That is more than a modest percentage, given that many articles addressed only one condition or dealt with narrow discoveries or with stock market conditions for investors (see appendix 2, table 3). Among relevant articles, differentiation constituted a fairly dominant frame. It was an increasingly small minority of reporters who wrote as though genes operated in a universal and uniform manner in all conditions.

These several trends in the public discussion weakened the imagined autonomy and power of genes, but judging where the public understanding rested with regard to levels of determinism was complicated by one additional shift. Articles in the era were becoming more ambiguous with regard to the type of characteristics to which they linked genetics. Physical features were still predominant (61 percent of all articles) over mental characteristics (33 percent of articles) and behavioral features (14 percent of articles), which were both continuing to decline (see appendix 2, table 9), but a fourth category of references grew dramatically during this period. Over a third of all articles made strong statements attributing ambiguous outcomes to genes. Some of these ambiguous references were modest, as in an article that simply referred to a desire to explore "how the gene manifests itself in growth" ("What's Big in Tomorrow's Barnyard," 1988: 3). In this instance one cannot tell what the force of *manifest* might be (a lot of influence? a little influence?) or to what characteristics *growth* refers. Though in this case the context of the article would encourage one to interpret growth in a physical sense, the limitation to physical growth is not explicit. Many other ambiguous references sounded grand and sweeping and could be legitimately read as describing the gene as having awesome powers: "the complete set of instructions for making a human being" (Jaroff, 1989: 62). On further analysis, even the scope of these apparently grandiose claims is not so clear. Does "making" a human being refer only to the process of conception? early development to birth? Does it include the soul? Does what is "made" include behaviors? the painting of the Mona Lisa? The same article quotes a scientist as saying that the Human Genome Project "is going to tell us everything" (Jaroff, 1989: 63), but it turns out that this scientist's view of "everything" is severely circumscribed by his own research interests. The scientist continues, identifying "everything" explicitly with evolution and disease. Other instances of grandiosity sounded all-encompassing, but proved to be mere tautologies, as in the frequently repeated claim that genes "control every facet of heredity" (Schulman, 1985). Of course genes control every facet of heredity (that is how they had been defined), but the statement left open the crucial question, Which elements are predominantly inherited and which are not? Consequently, though it is tempting to interpret these sweeping vagaries as a narrow and widespread ideology that presumed all-encompassing genetic determinism, it is probably more accurate to suggest that they represented the reporters' desire to continue to make the gene exciting for readers at a time when its exact scope and degree of influence were becoming increasingly specified and delimited.

A second journalistic procedure led to increased ambiguity in another manner. As genetic discourse became increasingly familiar, reporters (like scientists) began to adopt shorthand. Thus, an article might carefully describe the process of inheritance for the allele of the BRCA1 gene, which increases the risk that some women will develop breast or ovarian cancer, noting that not all women who have that allele will develop the disease. That is, the author might carefully delimit the power of the gene, but then later use the shorthand "the gene for breast cancer." In such cases, what level of determinism does the phrase *gene for* convey? On its surface, taken in isolation, the phrase sounds like a highly deterministic attribution. Does the reading public understand it in the context of the prior contextualization? That is, do they read it as a shorthand for the carefully hedged statements, and hence read it as relatively low in determinism? Or do they read it as overwriting the cautious hedges, reintroducing genetic determinism and effacing the limitations and hedges? And what about a later article that does not provide the hedges but only the shorthand? What does such a usage signify?

Critics of public discourse about genetics who were writing in the early nineties interpreted these shorthands and ambiguous statements as of a piece with the more deterministic ranges of the scale. They alleged that, overwhelmingly, contemporary public discourse about genetics was highly deterministic. Nelkin and Lindee, the critics who provided the most extensive study of popular discourse, concluded that "the images and narratives of the gene in popular culture reflect and convey a message we will call genetic essentialism. Genetic essentialism reduces the self to a molecular entity, equating human beings, in all their social, historical, and moral complexity, with their genes" (1995: 2). Another prominent critic, Abby Lippman, decried such public discourse as "geneticization." She agreed with Nelkin and Lindee, identifying the blueprint metaphor as the "predominant metaphor for genes and DNA fragments" and asserting that "in addition to the reductionism inherent in the blueprint metaphor is a problematic notion of determinism" (Lippman, 1992: 1470–71). Her complaints about the blueprint metaphor were conveyed by the mass media when she was quoted by Shanon Brownlee and Silberner (1990) in *U.S. News and World Report* as declaring, "Genes are not blueprints." The article conveyed her argument that factors such as diet and toxins contribute more significantly to diseases.

Instead of interpreting a blueprint as a preliminary outline, to which other elements would be added and which might be modified by the material limitations constituted by those other elements, these critics interpreted the blueprint metaphor as a reduction from those other elements.[3] In offering this interpretation, they raised valid concerns that deserved to be addressed. But when alternate critical interpretations are available, it is desirable to track down the reading public in order to get a better sense of how, in practice, they read the discourse.

In order to gain more insight into how such discourse might be understood by its intended readers, Melanie Williams and I completed an audience study with 137 college students. College students are clearly not representative of the entire population of the U.S. citizenry, and their interpretations might be diametrically opposed to other groups. But they are the public media's most important prospective target group about genetics. Not only will they constitute the well-educated readership that can afford to purchase glossy periodicals and newspapers (and the products of the advertisers who provide the bulk of financial support for such media), but they are also the group who will be the actual voluntary users of genetic technologies in the near future. They are, in other words, the intended audience of the genetic discourse produced by the mass media. Understanding how they interpret that discourse is therefore a priority.

When we asked such readers how they understood the blueprint metaphor for genetics, we found that they offered a fairly broad range of interpretations. Of the 137 audience members, 39 offered interpretations of genetics discourse that were explicitly deterministic; 58 responses were explicitly nondeterministic; 9 members offered interpretations with explicit components that included both deterministic and nondeterministic statements (the other responses were not germane). Thus, some of this target group read the blueprint metaphor in a fairly deterministic fashion, as suggested by the critics. However, a plurality of our readers did not interpret it in this closed fashion (for details of the study see Condit, 1999a; Condit and Williams, 1997). Instead, they interpreted the blueprint metaphor as a partial accounting, as a probabilistic rather than a deterministic forecast, and as malleable. Thus, one of our respondents noted that "this can shape what you can become, but there's always room to improve"; and another, that "a blueprint allows for variation—and leaves more gray areas to be filled in by the individual." A third suggested that a blueprint implies "that something can be done to alter the final state of health."

Most strikingly, many of these readers saw genes as malleable, either by technology or by the individual. Noted one, "A blueprint is a plan and not necessarily a reality—with the new era of genetic engineering nothing is impossible." This reader's sentiments were echoed by many others: "Genes can be altered, and there are now ways to predict defects." "Blueprint is more appropriate because biological technology comes closer every day to having the ability to alter specific codes. Blueprints can be altered; therefore, it takes the feeling of the inevitable away." Many of these respondents emphasized technological interventions to alter genes. Other respondents, however, emphasized the role of the individual, not technology. As one put it, "That's just the abstract—it doesn't have to hold the person back." In the words of another, "Just because one has a defect doesn't mean that person has 'lost' in life. That person can become whatever he/she wants to." Finally, some respondents recognized a role for both the individual and technology in remolding one's genetic foundations: "A blueprint, with scientific advances, perhaps lifestyle changes could be made or drugs/tests taken to prevent or lessen genetically related problems."

Thus many, perhaps the majority, of the intended audience members for journalistic metaphors about genetics interpreted the blueprint metaphor not as a closed deterministic system, but rather as partial, probabilistic, and distinctively malleable. The results of our qualitative and quantitative study have been supported by recent polls indicating that Americans hold a broad range of opinions about the relative influences of genes on human outcomes, but that the majority do not favor strong versions of determinism, even with regard to health issues (Singer, Corning, and Lamias, 1998). This raises the possibility that critics, like generals, are often caught fighting the battles of a past war. Perhaps excessive genetic fatalism was the primary problem of an earlier era, but an excessive sense of the malleability of genes might present its own, quite different, risks in an era where simplistic understanding of genes is being transformed into visions of readily available genetically based products. Perhaps the discourse of genes as blueprints and codes is equally likely to be problematic if genes are falsely perceived as too malleable and too open. Perhaps such a vision invites false prospects of infinite malleability and infinite interventions that are easy and quick, and it does not sufficiently account for the attendant costs and difficulties.

The failure of this vision of the malleable gene to address such restraints is signaled by one significant gap in the interpretations offered by these readers. They did not attend to either social support or social restrictions. These readers articulated a place both for technology and for the individual in constructing human beings, but they provided no explicit role for society and social structures in making the person what he or she becomes. Nor did they recognize the ways in which social structures would shape the accessibility of genetic technologies or impinge upon the abilities of individuals to reshape themselves.

These audience responses further reinforce the point demonstrated in part 2, that the relationship between levels of genetic determinism and levels of environmental interventionism is not one of simple and direct correlation. High levels of determinism are not directly correlated with heavy use of genetic technologies or with low emphasis on structural or cultural interventions. In the era of family genetics the level of determinism was relatively high (see appendix 2, table 1). Moreover, most articles of that era offered models that portrayed the gene as much more powerful than it was in most portrayals offered at the close of the century. Consequently the gene was viewed as so powerful and such a mystery that no human intervention was possible. As part 2 noted, because of that, people turned to environmental manipulations. In contrast, at the close of the century, the scope of applications available for using genetic knowledge widened, so even though the singular power of genes was being sculpted back in many ways, genetic intervention gained a level of focused attention it had not received in the earlier era. Beliefs in high levels of genetic causation ("genetic determinism") are not therefore equivalent to fatalistic responses or gene-focused solutions. Multiple ideological components and pragmatic factors play a role in influencing policy preferences.

The shift from the focus on fragmented bits of genetic code to an ordered blueprint for a whole and particular human being thus accompanied distinctive shifts in how the public discourse portrayed genes and heredity. In spite of the increased materiality of a "blueprint" as compared with a "code," this shift did not entail increased levels of determinism, but rather a more complex, differentiated (but often unspecified) relationship between genes and other factors such as nutrition, development, and individual activity. Simultaneously it offered a less specific role for culture and social structure. The increased sense of the malleability of genes was compatible with an expanded role for genetic intervention, and the increased possibilities for genetic intervention were made available by a rapidly growing commercial structure.

From Bioscience to Biotechnology

In the early eighties, the biosciences began to be transformed into biotechnology. The major site for genetic manipulation shifted from the cautious, out-of-the-way corridors of the academic scientist to the power alleys of industry. The business world greeted the scientific discoveries of genetic research with truckloads of tangible enthusiasm in the form of hundreds of millions of dollars in venture capital. The nation's business press celebrated the arrival of the "Age of Genetics" (Jaroff, 1992), noting that "it's hard to find a more alluring dream: to cure disease, feed the masses, safeguard the environment—and make a fortune in the bargain" (J. Hamilton, 1992: 74). Articles focusing on investment issues in biotechnology came to constitute a major category of public attention to genetics (see appendix 2, table 3).

Biotechnology was, of course, larger than simply genetics; however, genetics was then perceived as the pivotal component. As business writer John Carey put it, "The essence of biotechnology is exploiting genes" (1992). Over time, the definition of biotechnology would become increasingly broad. However, the tendency to define biotechnology in terms of genetics was encouraged not only by the symbolic impetus of the initial excitement about genetics but also because of the expanding applications of the technologies used in the effort to understand genes. Thus, the definition which proclaimed that, "in essence, biotechnology uses living organisms to make commercial products" (Shamel, 1985: 38), and which celebrated the potentials opened up by scientists' ability to "make bacteria function like miniature factories" (Wexler, 1985: 75), owed much to the symbology and the technologies associated with genetics, even if there were also similar debts to other facets of molecular research, ranging from chemistry to developmental biology.

A substantial part of the enthusiasm for genetic technologies arose from the potential for medical genetics. One of the most alluring promises was for the "rational" design of drugs. Popular medical writers deplored the hit-and-miss manner in which drugs had previously been designed; they asserted that biotechnological approaches, in contrast, would produce "'superdrugs' that will be discovered without the lengthy, trial-and-error approach used to find conventional treatments" (J. Hamilton, 1992: 67). The business reader was told that "with genetic codes in hand, scientists will be able to design drugs to attack the causes of disease rather than the symptoms" (Bylinksy, 1994). The medical dreams offered by population-scale genetic testing, gene therapy, and drugs produced by bacterial gene-splicing also provided visions of new, high-value-added markets with enormous, quickly garnered profits.

The business community's genetic dream began where the research scientist's dream had left off, with medicine. However, the business world rapidly overwrote the medical focus of the previous era. As genetic science entered the commercial realm, a thousand new uses for genetic technology proliferated in the prognostications of the business journals. The president of an international management consulting firm found that there were "more exciting" opportunities for biotechnology than human medicine. He predicted, "If pharmaceuticals were viewed as the short-range growth opportunity for biotechnology, the mid-range profit sector would lie in crop and livestock applications" (Shamel, 1985: 40). Lists of potential biotechnological products included insect repellent, antifreeze, animal hormones, energy supplies, detergent, animal feed and vaccines, and biocomputers, along with crops that were nitrogen fixing and healthier, not to mention better and cheaper plastics and chocolates. The biotech marketers promised "low-fat, high-fiber foods that taste good and that people really want to eat" (Jaroff, 1992), along with prodigiously productive dairy cows and microorganisms that could clean up everything from oil slicks to Agent Orange. They promised to feed the Third World with genetically manipulated seeds that could withstand drought and impoverished soils, would need no fertilizer or insecticides, but could deliver balanced nutrition. In short, they concluded, "genetic technology will give humankind an almost godlike power to improve its conditions" (Jaroff, 1992).

Not too surprisingly, some funny things happened on the way to the temple of self-worship. As with all the previous dreams about genetics, the biotech industry ran into what one business journal called the "things that go bump in the night" (T. P. Murphy, 1984). Biotech's boom was accompanied by constant cycles of bust. For each story about new potential miracles, there were stories of higher than expected costs, longer than expected production times, or products that failed to materialize. Many business analysts began to speak of the "far out field of genetic engineering" as "high on promise and short on profits" (Curran, 1987: 101) and to describe the biotech market as "an investment fairyland where hopes and dreams count for more than reality" (Bornstein, 1983: 112). *Forbes* worried that the difficulties with biotech meant "dream delayed, dream denied" (Teitelman, 1984).

This is not to say that biotechnology would not, in the long run, produce many of the products it had promised. It is simply to note that in biotechnology, as in other areas of genetics, the dreams were always of grandiose scale, neglecting the fact that grandiose changes generally tend to incur some form of grandiose costs. Consequently, as business companies set to work to spend the huge sums of venture capital inspired by their genetic dreams, they ran into various roadblocks. It turned out to be more difficult than imagined to scale up bacterial production of drugs and chemicals from the lab bench to the industrial vat. It turned out that physicians were not eager to "promote" large-scale genetic testing for incurable conditions with tests that were only 70 percent efficacious ("The Gene-Testing Boom,"

1994: 102).[4] And biotechnology required relatively highly trained and expensive workers, even on the production line. Thus, human insulin produced with bacteria turned out to be no less expensive than traditional mammal-based supplies of insulin for treating diabetes (at least in this era). Similarly, scientists had been interested in genetic tests for relatively rare diseases and were perfectly happy with compounds that were injectible, as long as they cured the condition or alleviated the symptoms. But injectible drugs and rare diseases did not carry enough volume to generate the overwhelming profits dreamed of by the biotech imagination. Thus, gradually, even the CEO's of biotech companies conceded that "the dream has been brought down to earth. . . . Reality is grudgingly sinking in" (Impoco, 1993: 59). Biotech would continue to boom and bust, and there would continue to be, simultaneously, enthusiasts and skeptics. Along the way, biotechnology would produce many new products in new ways. But successful commercialization required some substantial shifts away from the scientist-dominated visions. The concerns and skills of scientists had to be replaced or supplemented by other foci. As one business magazine put it, "People won't invest in professors anymore" ("Biotechnology Is Now Survival of the Fittest," 1982: 37). Biotechnology with its commercial structures and profit imperatives meant that "we can't just tell scientific stories anymore" (John Carey, 1994). If scientists, however, were no longer to be the sole or central players in the manipulation of genes, then the previous era's determination that work in genetics could be self-regulated by scientists was open for renegotiation.

The Regulation of Biotechnology

When the major sites for genetic manipulation were a few scientific laboratories and a few clinical trials in research hospitals, regulatory issues were relatively minimal and scientific self-regulation might have seemed sufficient. When the entire world became the potential location of genetic technology, when every patient might receive genetically engineered products, every consumer might eat genetically revised foods, and every yard might feature genetically engineered plants, the regulatory issues took on a far broader cast. This era featured a renegotiation of the regulatory issues, but this major renegotiation was accomplished with relatively sparse public attention. Commercial interests were able to transfer regulatory authority from scientists to "scientific criteria" executed by government agencies that were pledged to facilitate commercial transfer of technologies from the lab bench to the marketplace.

In the previous era, when scientific exploration had dominated, the major regulatory issue with regard to biotechnology had been the impact of the release of experimental forms of life into the larger environment. Scientists had asked the public to trust their judgment on this issue. Although some political figures had advocated skepticism about the ability of scientists to separate their own interests in advancing their careers and status from objectivity in judgments about safety issues, the scientific community had enough ethos in the public's mind to avoid stringent public regulation. This was not the case with the business community. The business community publicly espoused a "bottom line" philosophy that put investor profits as the highest value, and most commercial interests had long ago accommodated themselves to the idea that this value structure entailed a real and perceived lack of objectivity. Unlike the scientific community, business did not expect to go unregulated with regard to the safety or environmental impact of their products and production processes. Peter Huber, a science policy analyst, identified by *Forbes* magazine as working for a "free-market oriented" think-tank, put it this way: "The regulators' role is of course to prevent the fondest dreams of the optimist from being realized as the worst nightmares of the pessimists" (Huber, 1987: 369; Bailey, 1988: 138).

Unlike their scientific cousins, commercial interests did not fear regulation so much as they feared regulatory chaos. Huber told the National Academy of Sciences' Symposium on Technology and Agricultural Policy that "the gravest regulatory threat to the development of the new biotechnologies lies not in the stringency of regulation but in its disorder" (Huber 1987: 369), and *Business Week* deplored the problems created by the "absence of consistent federal regulations" (Rhein, 1985). *Newsweek* joined the chorus, lamenting the "regulatory void" created when, as biotechnology entered the commercial and regulatory scene, seven different agencies, including the EPA, the FDA, and the USDA, all claimed jurisdiction over the products of biotechnology, and none had a clear mandate (Rogers, 1983). Even when various regulatory agencies agreed to permit release of bioengineered products, public interest groups could seek court injunctions to delay or prevent the release of the product. Because of a lack of public trust and the regulatory disorder, various public interest groups in this era successfully blocked release of some potential products, and their efforts constituted virtually the only coverage of regulatory issues in the period (see also van Dijck, 1998: 73).

In the face of regulatory instability, industry sought a means for "quick and definite" regulation (Huber, 1987: 371). There were two prongs to their efforts. The first prong was to avoid the creation of new regulatory agencies. The industry succeeded in achieving this goal when the Reagan cabinet recommended at the close of 1985 that current agencies divide up authority for regulating biotechnology under old authorities rather than creating new legislation (Rhein, 1985). Subsequently, means for dividing that authority were developed and implemented, if imperfectly. Second, industry did not want the public or their elected representatives to play an active role in the regulatory processes. Like many of the scientists before them, they held hostile impressions of the public, portraying them as ignorant and liable to unfounded fears that would paralyze the industry. The public was, suggested the CEO of Monsanto, unable "to differentiate between facts and nonsense" (Mahoney, 1993: 401). According to business sources, not only was the public susceptible to the leadership of "crackpots" (Bailey, 1988: 139), but also they fell for "the science scare of the month headline and [are] unable to sort our [*sic*] real risk from just plain ignorance" (Mahoney, 1993: 401). The business magazines noted that some public interest groups had used "lengthy public hearings to delay approval of production facilities" and that profit in commerce was always reduced by delay (Levine, 1990: 177). Even the courts were attacked as unable to distinguish "solid scientific evidence from demagogic appeals" (B. Davis, 1984).

Because of its fear of public intervention, business made a pact with regulatory agencies to co-opt science and to employ supposedly scientific criteria, which were largely inaccessible to the public, as the sole means for judging its production processes and products. As *Business Week* put it, "The biotech industry might be better off enduring the probing of regulators than that of a nervous public" (Levine, 1990: 178). The industry thus promoted the "premise that regulations dealing with the clearance of new products are essentially science based. That is, if products pass a rigorously designed, scientific analysis, and are judged safe, efficacious, and of high quality, three criteria, they will be allowed into the marketplace" (Harbison, 1990: 725). The FDA and other agencies accepted this standard and immediately geared up to provide scientific experts as the bureaucratic regulators of biotechnology. Frank Young, the new head of that important regulatory body, noted: "FDA reviewers are scientists whose research expertise and interests are closely related to the types of products they review. . . . this science-based approach is essential to efficient technology transfer" (Young, 1988: 14). Some business leaders even called for the allocation of greater tax dollars to these agencies to ensure that review processes would not be delayed by lack of personnel (Mahoney, 1993: 403).

The selection of "science-based" criteria did not arise because business people believed in the methods and outcomes of science. Scientific outcomes are unpredictable, and delay is inherent. Within the ideals of science, if the research data are not clear, no decision is endorsed. Instead, business saw science-based criteria as using selected scientific procedures to provide regularized methodologies fitted to timetables rather than to correct answers. As one analyst put it, "If frontier developments in biotechnology are to be successfully commercialized our regulatory institutions must learn to answer scientific uncertainty with decisional certitude" (Huber, 1987: 371). Business thus backed "science-based" regulatory standards, but these were not equivalent to scientific standards of truth. Some of the forms of science were to be employed, but without the rigorous social controls constituted by the free give and take of the scientific community. The new combination involved bureaucrats in government agencies collaborating with industrial technologists to develop and approve products rather than to seek definitive and accurate answers about safety, quality, or desirability. It was this form of co-opted collaboration that the president of Monsanto endorsed when he declared, "I firmly support governmental/scientifically based review of new products" (Harbison, 1990: 724). Monsanto and other corporations invested heavily in producing what they claimed were science-based documents that should be recognized as "knowledge-based input" for regulation, even on an international scale (Harbison, 1990: 724).

This "knowledge-based" standard, with its limitations to three scientifically based criteria, responded to a European movement for the "fourth criterion" of social need (Bailey, 1990). The focus on a narrowly defined knowledge base was designed to prevent the public from bringing a range of criteria directly to bear in making a collective assessment of biotechnology (though expression of individual marketplace preferences would remain as a form of external restraint). Although some sources, including some business leaders, suggested the appropriateness of a public policy approach to biotech that would set public priorities for research and development (thus, for example, supporting research in pest-resistant crops rather than cures for baldness), such a system for allocating public research dollars in consonance with public priorities received virtually no consideration in the United States (Crooke, 1988: 734). Local groups were able temporarily to halt various releases of organisms, but otherwise there was little public regulation of biotechnology outside of these narrowly defined science-based criteria. At the close of the century it therefore remained true that there were "no established institutions or advocacy networks through which the public sector [could] make its voice heard on priorities in technological innovation" (Krimsky, 1982: 8). Instead, regulatory agencies were to shield industrial production from public preferences on the basis of putatively scientific criteria.[5] They were also to engage in active efforts to persuade the public that this arrangement provided adequate regulation. Frank Young, the head of the FDA, admitted that the FDA would use some of its taxpayer-based funds for an "educational effort to make sure the public understands and appreciates the safeguards in place for biotechnology products" (Young, 1988: 16).

The focus on "science-based" regulatory processes also had the effect of focusing regulatory attention almost exclusively on product-oriented concerns—the safety and efficacy of products. These were clearly important dimensions of public regulation, but an exclusive focus on these areas meant no regulation developed in other important arenas. While certification for genetic counseling developed in various medical professions during this period, there was no legislation regulating genetic counseling or even mandating its provision. Similarly, little regulation developed that would control the sale of gametes or clones for commercial purposes, an arena in which severe abuse, approaching slavery, seems possible. Additionally, only limited state regulations developed to protect the privacy of DNA records and insurance discrimination. There was one important regulatory initiative, the inclusion of genetic status within the Americans for Disabilities Act, but by and large the regulatory issues that were settled in this period were product centered.

The lack of attention to regulatory issues other than those related to product release might in part be traced to the skill that business groups manifested in focusing their public discourse on specific policy initiatives. When business authors and articles in business journals addressed social issues, it was usually with specific policy recommendations. In contrast, when other sources mentioned social issues, it was usually with vague ruminations about the ethical and moral "problems" or "dilemmas" and without any particular policy recommendations. The one exception to this tendency consisted of the debate over the patenting of genetically engineered or discovered materials. Under the instigation of Jeremy Rifkin and various religious leaders, significant protests were made against the Federal Patent Office's decisions to allow patenting of biotechnological products. These decisions included the patenting of an oversized mouse with a gene from rats that produced growth hormone and the patenting of stem cells. Following the U.S. Supreme Court's support for patenting in this new realm, the director of the National Institutes of Health sought patents for a variety of DNA sequences. These decisions of the Patent Office,

Supreme Court, and NIH director generated some opposition and turmoil among scientists,[6] but industry played a decisive role in the establishment of broad patent rights. The definitive rationale was stated by the widely quoted Peter Huber: "The creation of wealth requires a dependable law of private property" (Huber, 1991). Proposals were presented in Congress to prevent the patenting of life forms, but they never advanced far through the legislative process. In the end, the public was not roused to regulate. Polls showed the public was basically content with the development of biotechnology, though they continued to be edgy about its darker potentials.

The movement of genetics into a biotechnological paradigm was accomplished with substantial alterations in key public ideas about genetics. Stepping away from the laboratory benches, which had been heavily funded by agencies interested in human medical research,[7] allowed a focal shift away from human heredity toward a broad range of agricultural and industrial products that were less controversial and perhaps more readily profitable. This shift meant a decreasing concern with the perfection of human beings in favor of enhancing the utility of the world for human beings. It also loosened the linkage between genetics and heredity, for applications of genetically based information far exceeded the earlier question of modes of inheritance.

The commercialization of genetic discourse also included a shift in the metaphor-based understanding of the genome. More frequent and more central use was made of the blueprint metaphor, with its whole-genome focus and its more ambiguous, but less than fully deterministic, stance toward the relationship of gene and environment. These shifts constituted not the abandonment of older territories but rather the encompassing of new ones. Neither, however, was the change merely additive. The new powerful images of biotechnology and the social and financial structures constituted within this imagistic framework had significant impacts on the older territory of human genetics as well. These changes were most pronounced in the reconfiguration of genetic counseling within personal health-based genetics.

9 TOWARD A PERSONAL HEALTH GENETICS

The commercialization of genetics that began at the end of the twentieth century did not eliminate genetic counseling as the major interface between genetic experts and the public, but it did begin to change the character of this counseling as well as to narrow its significance and its scope. There was less discussion of it in the mass media and a substantially different attitude toward it. Instead of being portrayed in a straightforward, favorable manner, genetic counseling would now be portrayed as a consumer product to be obtained through savvy use of the medical marketplace. Simultaneously, stories about genetic counseling directed at reproductive control were supplemented or replaced by coverage of genetic surveillance as a personal health practice. This new presentation of genetic medicine was framed as part of the larger consumer health movement, and the shift from reproductive counseling to personal health further loosened the previously tight association between genetics and human heredity.

The Limiting of Genetic Counseling

Only four articles in this era addressed genetic counseling as a central activity. The topic thus dropped from its place as one of the major features of the discourse of the third era to secondary status in the final decades of the century. This decline did not result from any decrease in the use of genetic selection technologies. Manipulation of the genes of one's offspring through the technologies of genetic selection had begun in the previous era, but it continued to proliferate rapidly, assisted in this last era by additional technological innovations. Amniocentesis with abortion was supplemented by chorionic villus sampling (CVS served the role of amniocentesis but at an earlier stage of pregnancy) and preimplantation selection (which allowed fertilization of eggs outside the body, selection among embryos, and then implantation of selected embryos).[1] These tools, along with a growing number of genetic markers and tests and a reduction of the age at which medical authorities deemed testing for Down syndrome to be appropriate, substantially increased the number of pregnant women who were receiving genetic screens and tests.

In spite of the resulting increased need for genetic counseling, the practice was deemphasized in public discourse and also reduced in clinical settings because of the commercialization process. Good genetic counseling that matched the image of nondirectiveness and full information provision was a time-consuming, repetitive task that required relatively highly trained but narrowly specialized personnel. As numerous studies have demonstrated (Bartels, Leroy, and Caplan, 1992; Kolker and Burke, 1994), to inform a woman or a couple effectively about the implications of any prenatal genetic test required a substantial time investment. Forty-five minutes were the minimum investment, but with some clients and some conditions it could take multiple hours. When prenatal screening was experimental, this kind of expense could be absorbed in the costly research process. When prenatal screening was applied to rare cases, this kind of expense could be absorbed in a broader practice. However, if prenatal screening was to become a routine component of medical commerce, this substantial expense was a major stumbling block. Rather than hiring the genetic counselors needed to do this work, the medical establishment adopted two strategies.[2] First, in California, as a matter of policy the state resorted to the use of pamphlets to mandate the information women would receive about prenatal screening. The backup to the pamphlets was information provided by the excessively busy attending physician and staff. Second, across the nation, attending physicians became the "genetic counselors" for most women receiving prenatal testing. As *Parents Magazine* noted about genetic counselors, "Now every gynecologist and obstetrician is expected to be one" (Andrews, 1982: 93). Gynecologists and obstetricians did not receive any extended training in genetic counseling or any certification to perform this function (in contrast with NSGC-certified genetic counselors, who generally received two years of focused training). It was merely their status as medical professionals that supposedly qualified them to undertake this task. This lack of training was accompanied, at the structural level, by a lack of respect for or interest in the process. Physicians were accustomed to directive interactions with their patients. They were trained to assume that accumulated medical knowledge dictated medical choices and that individual values played little role in such decision processes.[3] Their general tendency was to tell patients what tests they needed. In contrast, partly because of its involvement in reproductive rights, the practice of genetic counseling prescribed a nondirective (or "client-centered") approach. In general, physicians tended not to see the need for or appropriateness of this nondirective approach (Bernhardt, 1977; Capron et al., 1979: 106). Moreover, physicians were accustomed to limiting their interactions with their clients to 10-minute chunks. There is no evidence that in typical prenatal care, women who received genetically significant tests were apportioned any more than these 10-minute chunks, and where physicians prescribing other genetic tests were formally surveyed, they reported providing shorter counseling sessions than did trained genetic counselors (James et al., 1995).[4]

The physicians' traditional lack of focus on the decisional autonomy of the patient and their chunking of patients into 10-minute productivity units thus coalesced to constitute commercial forces that were squeezing sufficient versions of genetic counseling out of the prenatal screening market. This routinization of screening meant that its users had no formalized opportunity to reflect upon and make critical choices about its use (Katz Rothman, 1993). Instead, physicians implemented screening within their practices according to their own preferences and their commercial interests.

These commercial forces were legitimated by a growing critique of genetic counseling that attacked the ethos of the nondirective approach in two ways. In the public literature, nondirectiveness was attacked for failing to provide guidance, leaving the client with only numbers and information. Whereas earlier articles had positively emphasized the practice of leaving decisions to the couples themselves, the few articles that addressed the subject in this period bemoaned the fact that couples had "few places to turn for support," finding genetic counseling unhelpful because, "while experts have no problem bandying statistics about, the average couple finds them abstract and unhelpful. Obstetricians and counselors take a nonjudgmental stand, informing couples of the options but making no recommendation" (Sandroff, 1989: 256; see also Bosk, 1992). Meanwhile, in the critical literature, leftist scholars operating from sophisticated accounts of the inherently persuasive character of language, and perhaps from a desire to impose their own agendas on clients, denied that pure nondirective counseling was possible, implying that nondirectiveness should be abandoned as a guide (Brunger and Lippman, 1995).[5]

This delegitimation of genetic counseling by physicians and critics was further bolstered by attitudes of patients themselves. Because patients were unlikely to be aware of the information they lacked, they were often satisfied to accept testing without being informed of the consequences of the tests (Callanan, Cheuvront, and Sorenson, 1995;

Lyon et al., 1995). Furthermore, it was authentically difficult, even for trained counselors, to identify simultaneously the informational and emotional needs of clients and to provide the information they would need without some directiveness stemming from their own perspectives, cultural frameworks, or institutional routines (Rapp, 1988; Wolff and Jung, 1995). Some articles tried to defend genetic counseling, saying, "I'm worried that the counseling will be lost in the shuffle" (Gorman, 1995: 62; Wexler, 1985), but these proponents of autonomous, informed decision making were working against commercial forces and entrenched medical routines of great strength.

As prenatal genetic counseling began to be compressed, therefore, few seemed to remark that the goal of enhancing the reproductive choices of parents was already being lost. Instead of women or couples choosing what form of genetic testing they would have for their children, individual physicians and institutions such as the American College of Obstetricians and Gynecologists, independent hospitals, and insurance companies were deciding which pregnant women should have which genetic tests.

Personal Health Genetics

These small steps away from reproductive autonomy were not widely debated but instead were obscured by the displacement of the media focus from prenatal genetics to the genetic health of adults. In the late eighties and early nineties, genetic testing of adults became one of the dominant ways in which the public imagination envisioned that genetic technologies would be deployed. Under the model of personal health genetics, genetic testing was portrayed as a tool for discovering information about one's own predispositions so that one could make life-style changes, including alterations of diet and exercise patterns, as well as decreasing exposure to carcinogens such as cigarette smoke and the sun (e.g., Cimons, 1985).

The arrival of the boomlet featuring genetic testing for adults in the eighties and nineties coincided with the fitness boom for upper-middle-class adults, and coverage of genetics was slotted effectively within this personal health framework. Articles oriented to personal health argued that individuals should know their own genetic predispositions so that they might adapt their diet, exercise, and exposures to stress and other environmental elements to provide maximum self-protection. Such articles often began by relating stories about individuals who narrowly escaped death because they learned of their family health history or took a genetic test and thus discovered information about a serious health condition "before it's too late" (McAulliffe, 1987b: 64). *Good Housekeeping*'s coverage of the issue, for example, began with the story of a woman who was rescued from heart disease: "If it weren't for knowing her family's medical history," counseled the magazine, "Cindy Lefew might be dead" (Krause, 1994: 161; see also Neimark, 1993). More fanciful scenarios imagined that individuals would soon have "access to a computer readout of their own genome, with an interpretation of their genetic strengths and weaknesses . . . [which would] enable them to adopt an appropriate life-style, choosing the proper diet, environment and—if necessary—drugs to minimize the effects of genetic disorders" (Jaroff, 1989: 67). Several articles argued that this precise individual targeting would improve the public health because individualized information about risks would be more effective at motivating people to enact healthy behaviors than generalized public health messages were. One scientist, clearly not a close student of human behavior or of the research in health communication, argued that "if it were me and I knew I was predisposed to alcoholism, I wouldn't take my first drink" (McAuliffe, 1987b: 67).

This personal health approach to genetics was undergirded by the moderate level of genetic determinism of the era, which postulated a moderate, if varied, level of stability between certain genetic configurations and particular health conditions. Such statistical correlations allowed the creation of "genetic information," which then allowed individuals to modify their physical environment. This vision was occasionally contested. Critics of the approach of personal health genetics pointed out that people would not necessarily act on such warnings. *Glamour* told the story of a woman who simply rationalized that whether she smoked or not, she was going to die at the same age as her mother had died of lung cancer (Cimons, 1985: 269). So, the woman concluded, she might as well continue to smoke. The efficacy of targeted personalized genetic data was thus put into question. These challenges, and other possible challenges to personal health genetics, were relatively sparse.[6]

Genetic testing was thus increasingly envisioned not solely as selecting one's children, not solely about heredity and heritage, but also as a matter of ordering one's personal life to be maximally healthy. Such a focus had commercial advantages, since adult health testing could be more easily and economically subsumed under regular medical practices than could genetic testing of fetuses that was oriented toward reproductive selection and abortion. This vision of adult genetic testing for personal health was more widely imagined than implemented. There were yet few commercially marketable tests available, although family histories could still be informative. Relatively rare genetically influenced diseases such as some types of breast and colon cancer were slowly beginning to move toward implementation in accordance with the personal health model. However, genetic markers for more common conditions such as heart disease and widespread cancers were nonexistent, so mass marketing of personal profiles garnered relatively little practical implementation. In practice, genetic selection remained the most common direct use of medical genetic technologies.

Genetic selection of offspring was thus on the increase in the period, even though public representation of it was declining. Genetic testing for personal health applications was rare, but its public representation was dramatically heightened. In the absence of sufficient institutional structures for thorough genetic counseling to cope with the information demands of these technologies, the public media began to provide consumer guidance to assist average readers in dealing with the new practices.

Consumer Guides

Articles taking the form of consumer guides included both genetic counseling and personal health issues. An article can be described as a consumer guide if it treats a practice as a purchasable commodity and fosters the attitude of savvy consumerism toward it (Condit, 1996b). Thus, rather than describing the practice of genetic testing uncritically as a new and useful option (as had been the case in the previous era) or exploring the social forces involved in the practice (as critics in both eras advocated), media reports in this era introduced readers to the strategies necessary to best employ the new technologies according to their own personal preferences.

The relatively few articles that discussed prenatal screening and testing tried to provide readers with information that would help them locate and use genetic counseling to their own best benefit. These articles both familiarized the audience with some of the routines of genetic counseling and provided some advice for those who were considering using such tests. The articles told readers to prepare in advance, to ask their own questions, and to make decisions before a crisis arose. They also warned readers about the economic costs of the procedures. In reaction to the commercialization of counseling and testing, much of the advice prepared consumers to be assertive in locating the kind of assistance they wanted and in resisting the pressures of medical personnel. Readers were told (perhaps contradictorily) to be insistent about their own wishes, to have all available tests, and to make their own choices. They were also warned to select only physicians who had been board certified by the American Board of Medical Genetics (e.g., Andrews, 1982; Schaeffer, 1987). These articles tended to be more critical of genetic counseling than the earlier articles had been, but they did not oppose the practice; instead, they sought to create critical consumers of the practice by warning readers of both its pitfalls and its potentials.

Articles focusing on building one's "family health tree" provided the same sort of consumer guidance with regard to personal health. These articles generally began with a frightening story about someone who either avoided or failed to avoid death or serious debilitation on the basis of his or her familiarity or lack of familiarity with family susceptibilities. They moved from such motivating scenarios to instructions on how to make a family health tree and how to interpret it.

In contrast with earlier articles about family health weaknesses, the consumer guidance articles did not presume that those families with genetic disease were inferior or abnormal. Articles about personal health presumed that everyone has genetic weaknesses rather than that some few persons are "abnormal" and the special objects of genetic scrutiny. Personal health genetics claimed to offer information for everyone, so that each person could know his or her own personal health risks and tailor his or her life-style and medical scrutiny to those particular risks. Everyone, equally, became the object of genetic scrutiny. Thus, as general health instead of birth defects became the domain of genetic medicine, the older dichotomy of normal versus abnormal human beings was weakened. In place of the old superior/unfit dichotomies, it was now increasingly presumed that all human beings have weaknesses, and they therefore can all be protected and improved by what genetic medicine has to offer. This new egalitarianism functioned not to delegitimate genetic manipulation but rather to ensure the largest possible market for the new products offered by the biotech industry. In contrast with the hierarchical discourse of eugenics, the new egalitarian vision of personal health genetics held that human beings did not need to be selected so much as they needed to be corrected individually.

Gene Therapy

The use of genetic testing for establishing personal health regimens provided one means for correcting individual health weaknesses. A more fundamental form of human correction offered an even more compelling dream for the era. True genetic engineering appeared to offer an obvious way out of the philosophical quagmire presented by genetic selection, and it appeared to be more permanent and complete than life-style alterations based on testing. Consequently, many writers expressed hope that "in the next several decades, drug treatment will be supplemented or replaced by genetic engineering. Doctors will insert good genes into a patient's DNA, where they will take over the function of defective ones and actually cure the disorder" (Jaroff, 1992). Gene engineering, usually now reframed with the more benign and medicalized label "gene therapy," was thus imagined as a simple magic bullet. The science writers quoted the scientists as prognosticating that genetic research would bring us to the point where "any physician can take a vial off a shelf and inject an appropriate gene into a patient" (W. French Anderson, quoted in Jaroff, 1993). The biggest enthusiasts saw this next application of the technology as immediately available; a chief executive officer of a genetics firm noted, "The 1990s will bring a new era of genes as drugs" (Hal Broderson quoted in John Carey, 1993). Even the more cautious estimates, however, would prove to have underestimated the costs and difficulties involved in gene therapy. Trials would be slow; roadblocks would be enormous. The suitability of gene therapy would vary by disease. As the new century neared it seemed increasingly unlikely that gene therapy would be a panacea for all human health ills delivered like a pill from a bottle. The hopes for gene therapy, in the eighties and early nineties, however, were very high.

Though it would prove to be errant, this substantial propagation of a vision of genetic engineering as a magic bullet invited reconsideration of the issues of genetic enhancement, which had been so vividly explored in the previous era. A small set of articles followed this lead. A few of these featured the full science-fiction flair of the previous era, as in the instance of the writer who imagined human beings who could photosynthesize ("Bizarre Crimes and Punishment," 1993). In contrast with the previous era, however, most of the articles dealing with genetic enhancement of non-health-related human features were largely negative. This almost uniform negativity may have been related to the commercialization process, which was making it increasingly evident that genetic technologies would not be turned exclusively toward the higher goals of lofty human improvements, but more toward whatever the market would bear. In this new context, popular writers ridiculed the application of these expensive new high technologies to such apparently trivial problems as baldness. Similarly, worried magazine authors decried the values reflected in a poll suggesting that a quarter of Americans thought it acceptable to use genetic engineering to improve physical appearance (Brownlee, 1994). There were, of course, some exceptions to this negative attitude toward nonmedical enhancements. Richard Seed and Randolph Seed, who would later become notorious because of Richard Seed's for-profit promotion of human cloning, were quoted as arguing that "just generally trying to improve the human race is a good thing" (Johnson, 1992: 82).[7] In addition, there were some neutral treatments of enhancement where the line between genetic enhancement and genetic health was particularly murky, as was the case with substantial extension of the human life-span. Clearly, the goal of improving human health might well culminate in an indefinitely extended life-span. This likely consequence of the widespread application of the technologies of medical genetics was mentioned briefly, but recurrently, and usually in a supportive fashion. In general, however, a recognition of the tenuousness of the line between genetic "therapy" and genetic "enhancement" and the eagerness of some to cross that line troubled the public conscience in this era (Joseph Carey, 1985; Crooke, 1988: 733).

The process of commercialization and further scientific developments had significant ramifications for the way in which genetics was placed on the public agenda in the last quarter of the twentieth century. The previous era had emphasized genetics as a family matter, focused on genetic selection through genetic counseling. In this era, reproductive potential became an increasingly minor component of the genetic vision. Instead, genes were increasingly viewed as important because they could provide functional information to individuals, allowing lifestyle modifications to improve health, or because they could provide mechanisms for therapy. Such perspectives did not completely sever the linkage between genes and heredity; the information in the genes was still believed to be handed down from one's parents. However, the meaning of the gene was drastically different when it was merely an informational marker or functional vector for improving one's health or life-span, rather than a personal identifier linking one to a heritage of family, race, or nation.

This major shift in attitude required no shift in levels of genetic determinism. Genes were still envisioned as having a moderately strong influence on individuals, especially with regard to health, but this influence was not believed to determine the final fate of individuals. One's genetic blueprint could be studied, the public writers of the period believed, and actively reshaped.

This favorable attitude toward genetic manipulation was also not linked to increases in discriminatory attitudes. If anything, the era was more egalitarian than ever. Genetic defects were located in genes rather than in people, and unlike the era of classical eugenics, the link between the microscopic part and the organismic whole was less totalistic. Additionally, everyone was viewed as equally susceptible to genetically based problems. Though different families carried different diseases, all families, and hence all individuals, needed to attend to their own specific genetic weaknesses. Abnormalities were now the norm. Arguably, this was a perfectionistic attitude, but primarily with regard to the relatively narrow parameters of physical health.

The progressive dimensions of the era thus combined with forces of commercialization and technological advance to support a set of public attitudes that were largely favorable toward genetic manipulation. As genetic manipulation came increasingly to appear as an acceptable, imminent, and even mundane reality, however, both its advocates and its opponents turned greater attention to the establishment of ethical boundaries for these practices.

10 ETHICAL CHALLENGES FOR GENETIC HEALTH

The movement of genetic information and technologies out of the scientific research laboratory and into general commerce was accompanied by a substantial increase in the quantity of ethical debates about the applications of genetic information. Almost half (45 percent) of the articles in the new era raised ethical concerns of some sort (this was a statistically significant increase from the previous era: see appendix 2, table 5). These articles also included a greater number of different types of issues, and there was also a statistically significant increase in the number of articles arguing against discriminatory stances (see appendix 2, table 4), and a general decrease in expressed discriminatory attitudes toward persons with various atypical genetic configurations (see appendix 2, table 10). This increased attention to ethical issues did not amount to a debate of these issues in any sense that might be approved by traditional standards of argument. In the postmodern mass mediated era, ethics were not debated. Instead, different positions received a hearing; concerns were voiced.

Whatever the insufficiencies of its form or quality, discussion of ethical concerns covered a wide range of issues. One major group of these concerns addressed dimensions of equality and discrimination, including disease and disability as well as race and gender. A plethora of more operational issues also were addressed, by conservatives as well as liberals. These included concerns about health insurance, privacy, stigmatization, and environmentalism. Little consensus developed, and almost no concrete protective measures were adopted. The one exception was a broad social consensus that the undertaking as a whole was morally acceptable, if not imperative.

Ethical Imperatives for Genetic Manipulation

The fundamental question of the ethical appropriateness of genetic manipulation per se was infrequently raised in this period, probably because most people treated it as settled. It was, however, asserted directly by the president of research and development of Smith Kline and French Laboratories, who asked, "Is it ethical to alter genes?" (Crooke, 1988: 732). He concluded that it was not only ethical but also "essential," a conclusion reflected in the presumptions of the overwhelming majority of mainstream publications. Stanley Crooke argued that the need to alter genes arose from the Darwinian drive for fitness. But he also indicated that its ethical appropriateness depended upon "whether you are satisfied with the current state of the world" (1988: 733). He suggested that as long as children were born with genetically transmitted diseases, as long as people suffered malaria and malnutrition, as long as they died from AIDS, we would need to pursue genetic technologies as a means of erasing those evils. The head of the FDA, Frank Young, concurred with this view, arguing that "the alternative to controlling nature is to allow nature to control us" (Young, 1988: 16) and that such passivity was simply not the human way. There was no direct argument against this conclusion in any of the discourse in the sample. There seemed to be an overwhelming consensus in the period that there was "plenty of room for abuse" (McAuliffe, 1987b: 65) but that the advantages made it essential to guard against such abuses rather than to abjure the technologies altogether. Discourse about genetic ethics was therefore directed primarily at locating potential areas of abuse and encouraging approaches that minimized potential ethical problems.

Discrimination, Disease, and Disability

The potential of genetic technologies to generate or exacerbate discrimination was probably the most frequently addressed ethical issue. The desire to modify human genetic configurations implied that some configurations are better than others, but this implication was contrary to a widely shared ethical principle that human beings should be treated as equal. This conflict was not resolved by deemphasizing egalitarian principles with regard to the specific status of persons with disability. In fact, the transformation away from blatantly discriminatory vocabularies for discussing persons with handicaps or diseases, which had begun in the previous period, became relatively complete in this period. Terms like *defectives* and *idiots* and *the unfit* were no longer used in the public arena to refer to people who had a low IQ or physical disabilities or mental diseases. Instead of denying either egalitarianism or genetic preference, the concepts of genetic preference and equal human dignity were accommodated to each other through careful reformulations of key terms and concepts.

The accommodation of genetic preference and equal humanity was accomplished in large part by moving negative descriptors to the molecular level. Instead of defective people, we now had "healthy genes" and "defective ones" (McAuliffe, 1987b: 64). Where an article in the previous era had described a woman with Huntington's chorea as a "culprit" who had passed her genes to many children, articles now referred to "culprit genes" rather than placing blame on individuals (McAuliffe, 1987a: 23; Wexler, 1985: 75). Instead of abnormal persons, we heard about "abnormal genes" (Joseph Carey, 1985). When mental and physical diseases became understood as the product of faulty proteins, and when it began to be imagined that these proteins could be corrected, the locus of illness and disability was shifted from the essential identity of the person to a microscopic phenomenon hidden, but manipulable and changeable, within the subcomponents of the person's physical body. Genetic technology was therefore understood as manipulating small parts of individuals rather than as altering their essential identity. One scientist directly expressed this contrast between essential qualities and genetically manipulable components when he denied the possibility of true chimeras, saying, "The essence of a particular animal is something you don't change. . . . You merely enhance certain aspects of it" (Wallis, 1987). Thus, disability and disease were to be conceived of not as essential traits of individuals but rather as accidental, malleable traits.

This movement of potentially discriminatory descriptors to the microscopic level was accompanied by changes in descriptors associated with the patients of genetic medicine. Persons with genetic disorders were now most often described as afflicted or affected by a disease, rather than as diseased or unfit. When terms like *defective* were used, they were often qualified with quotation marks to signal their marginality ("The End of Insurance," 1990). Notably, where article titles in the previous era had asked, "Will the Baby Be Normal?" article titles in the contemporary era asked, "Will My Baby Be OK?" (Mereson, 1988; Schaeffer, 1987). The force of the normal/abnormal distinction thus was losing some of its strength. Everyone was normal; we just wanted to ensure that each person's genes would be "OK." The focus on health and disease presumed that problems of infants were temporary and secondary rather than permanent and central aspects of identity. Thus, instead of saying that a child had a 25 percent chance of being deformed or defective or retarded or abnormal, articles in this period were more likely to say "the child has a 25 percent risk of being ill" (Wexler, 1985: 76).

While there were some exceptions to this movement away from directly stigmatizing vocabularies, this shift in terms was also accompanied by favorable reflection on the role of persons with handicaps in the society. As in the previous era, several voices argued that the lives of persons with handicaps were worth living and supporting. *Parents Magazine* declared, "Many people with handicaps cope with them quite well, leading productive and rewarding lives" (Andrews, 1982: 184). In contrast with the earlier era, where visits to children with Down syndrome were prescribed as a means of encouraging parents to abort pregnancies if the fetus had been diagnosed with Down, the alternative result was now expected from such visits. Noted *Vogue* magazine's Ronni Sandroff, "Once they visited children with Down's syndrome, they decided to carry through the pregnancy" (Sandroff, 1989: 246). People with various genetic conditions spoke out to say that their lives were worth living in spite of the medical challenges they faced. One man with Huntington's disease articulated his choice against using genetic testing and abortion by saying, "Would I have wanted them to abort me? Hell, no, I'm glad I'm here. Screw everything else" (Cohen, 1993: 275). Other persons articulated a desire to avoid being forced into normalizing themselves through genetic technologies. They worried that they "may be coerced into undergoing genetic treatment for inherited disabilities" (Brownlee, 1990; Leo, 1989). The focus on disabilities was significant in its own right, but it also opened onto a broader concern with the construction and reification of human hierarchies.

Many participants in the public discourse worried about the implications of genetic technologies for exacerbating hierarchy. *U.S. News and World Report* expressed the fear that genetic technologies would provide "one more excuse to divide the world up into superior and inferior, us and them" (Leo, 1989), while the Nobel Prize winner David Baltimore suggested that the new information provided by genetics "will challenge basic concepts of equality on which our society is based. . . . society will have to come to grips with diversity—and we are not prepared for that" (Baltimore, 1983: 52). Alexander Capron, a law professor at Georgetown University, even asked whether genetics implied "mandatory genetic alterations to give everyone the same (or equally functional) genes" (Capron, 1985: 13). Several sources replied to these worries by emphasizing, as they had in the previous era, that everyone is genetically flawed, if in different ways. *Vogue* urged its readers to remember that "we all carry a number of genes for disorders, and that, given a conducive environment, all of us are in some sense unfit" (Sandroff, 1989: 256). The chic women's magazine also quoted a health care ethicist: "We need to value our blemishes and to insist on the worth of people with different abilities and disabilities" (Sandroff, 1989: 256). These voices thus encouraged comfort with diversity, noting that "one person's 'defect' is another's difference" (Mosedale, 1989). They also encouraged resistance to the "narrowing of normality" (Brownlee, 1994). They noted that "today's norm may become tomorrow's deficit" (Capron, 1985: 12), and they expressed concern about the possibility that genetic technologies "may lead to a very narrow notion of what a physically normal person is" (Leo, 1989). A general concern with "genetic discrimination" was thus relatively widespread in magazines of many political hues (see appendix 2, table 4).

Discrimination and Social Structure

In contrast with previous eras, this anti-hierarchical, anti-normative notion of genetic differences did not take the form merely of naive individualism.[1] There was substantial discussion about the role of social structure in the maintenance of equality and individual freedom to choose within a society where genetic manipulation of humans was a routine possibility. The defendants of disability rights suggested that regular use of genetic tests could "alter our perception of what society owes to children, particularly those burdened by handicaps" (Capron, 1985: 12). They worried that if handicaps began to be perceived not as God's will or as the product of random events, but rather as a matter of parental choice, this might reduce social generosity to such children and their families. This led Chicago attorney Lori Andrews to ask, "Will society cut back on all of its support to handicapped children?" (Andrews, 1982: 97).

Various authors warned that pressure to use genetic tests could come from social attitudes, institutional structures, and economic pressures. Jeremy Rifkin highlighted problematic social attitudes, claiming, "The implication is that women have an obligation not to allow a fetus with certain abnormalities to develop into a baby" (Sandroff, 1989: 248). Others worried that the testing establishment might "push parents into abortions they don't want" (John Carey, 1990: 82). The press also conveyed the concern that such pressure could arrive through a lack of economic support for true choice. The readers of *GQ* were told that "the government could simply start cutting back on special-education programs for people with Down's, and, before you know it, parents wouldn't be able to *afford* not to go along" (Cohen, 1993: 274).

The concern about social support for children with disabilities extended readily to parallel issues facing the general population. The first worry was that competitive social pressures would drive everyone to use genetic technologies to produce children who were not only free of seriously debilitating diseases and major handicaps but who were "superbabies." As John Leo explained to the readers of *U.S. News and World Report,* "If a child risks an array of biases because of a minor physical weakness, the pressure will be on to produce the ideal child" (Leo, 1989). He pointed out the possibility of eugenics arriving as a "mom and pop operation"; in order to give their children good opportunities, parents would feel compelled to employ genetic technologies because other parents were doing the same thing. In such a scenario, few people would really want to use genetic technologies for child selection or modification, but everyone would feel pressured by the logic of the competitive marketplace to employ the technologies to avoid disadvantaging their children. *Vogue* argued that it was already the case that "the desire to have the healthiest baby possible is translating into pressure to take advantage of everything medical science has to offer" (Sandroff, 1989: 246).

This worry about competitive pressures raised an additional concern over universal access to the technologies. The direct use of genetic technologies on human beings was not cheap, and there was little reason to believe that it would become so through time, given the intensive involvement of medical personnel. Therefore, critics worried that the new technologies would create a biologically inferior class, because "the offspring of the rich could get richer not just by receiving trust funds and computers by their fifth birthday but by getting higher-intelligence, better-looks and longer-life genes by the third morning after conception" (Cohen, 1993: 274).

Such a class-based biology threatened the presumption of equal opportunity in the political system in a profound way. The society already provided unequal schooling and nutrition to children of different races and classes, but its "Horatio Alger" myth argued that meritorious individuals could overcome these disadvantages, so that individuals of merit still had opportunity to succeed. To the caring and discerning eye, that may not have constituted a very satisfactory meaning for "equal opportunity," but most of the members of the society had hidden the inequities from themselves well enough to live with the terms of the agreement. If, however, the individual was molded for inferiority (whether by conscious intent or by a lack of access to technologies), then any pretense that individuals could overcome socially imposed inequalities by individual merit was invalidated. The game would be rigged in every respect. In order to avoid facing that prospect, the theologian Richard McCormick articulated as a basic principle of genetic ethics that the benefits should be "generally available (not coercively) to all, regardless of geographic location, economic ability or racial lines" (McCormick, 1985: 342).

The focal concern about genetics had clearly undergone a profound shift. Rather than worry about eliminating the "unfit," in the new era a major issue was the access of the less privileged to genetic technologies. This change of emphasis was well captured by a shift in iconography. In the first half of the century, genetic issues had been visually summarized for the public with the photograph of a black sharecropper's family with polydactyly. At the close of the century, the person pictured with polydactyly was a queen, Anne Boleyn. Related shifts in many images signaled substantial changes in the character of the problems that were addressed in the two eras.[2] At the beginning of the century, the poor had been the primary targets of genetic scrutiny and control; ethical arguments attacked the classism of such targeting. At the close of the century, access to genetic technologies threatened to become a rich person's advantage; the focus of ethical concerns thus shifted from the targeting of the poor to their exclusion from opportunity.

Issues about egalitarian access to the technology also translated to attention to the construction of egalitarian priorities for research. Some sources suggested that distributive justice required that public priorities needed to guide research agendas, especially since most of the research was publicly funded. For example, it was suggested that provisions should be made for the development of "orphan drugs"—drugs for which there was no lucrative market but which were desperately needed by small groups of people or by low-income groups. Similarly, it was urged that policies needed to be developed to protect small farmers from a market dominated by high technology products produced most efficiently on a large scale. And resources should be directed toward the development of items such as renewable energy sources through biotech.

Quite predictably, many industry leaders and scientists bridled at the specter of government control of research. They did not, however, deny the desirability of egalitarian outcomes. Rather, they suggested that the technologies were inherently prone to produce equal and widespread benefits. They argued that biotech was "the most equitable, accessible, and transferable technology to reach agriculture since the Industrial Revolution," and that it would "produce the greatest good for the most people tomorrow" (Harbison, 1990: 724, 726). They claimed that unlike other technologies, biotech offered a product—improved seeds—that could be easily integrated into low-tech agriculture in the Third World and thus provide "a harvest whose scope and abundance we can scarcely imagine" (Young, 1988: 16). Genetic technologies, they suggested, would "better feed the world's needy" (Crooke, 1988: 733). These arguments assumed that the social structure was adequate to deliver these wonder goods in an equitable manner, and this assumption was related to a reliance on individualism that countered the numerous voices of the era that were focusing on social structure. As in the previous era, this group of individualists argued, "We should devolve the ethical decisions to the level of the individual patient and the individual physician insofar as possible," because "humankind will be best served by decisions that support the individual" (Crooke, 1988: 734). Notably, however, there was one area in which conservatives eschewed individualist frameworks. That was with regard to the group issues raised by questions of race and gender.

Discrimination in Race and Gender

Discussions of race and gender in this era were framed largely through a popular representation of evolutionary biology that took a form often dubbed evolutionary psychology. The self-proclaimed media expert on the subject was Robert Wright, and his views were widely published in news magazines such as *Time* (Wright, 1995a), in journals devoted to sociopolitical discussion such as the *New Republic* (Wright, 1994), as well as in more specialized periodicals such as *Psychology Today* (Wright, 1995b). These explanations of evolutionary biology asserted that there are fundamental human behavior patterns, the context of which has been rigidly determined by narrow environmental conditions uniformly experienced by human beings as they evolved, especially in the hunter-gatherer stage. In place of the genetically given quality of plasticity or dependency on learning that the geneticist Dobzhansky and others had previously described as humanity's chief evolutionary characteristic, evolutionary psychologists claimed to have identified types of behavior that are biologically pregiven to certain categories of people. Sex was openly identified as such a category, and race was covertly identified in this fashion. Only publications with specifically targeted audiences, such as *Christian Century,* directly challenged this general perspective (Jones, 1995).

RACE

As in previous eras, the relationship between race and heredity suffered from a dual—and dueling—public consciousness. On the one hand, most sources that explicitly addressed racial topics insisted on hereditary equality among the races: "Every ethnic group is prone to inherited imperfections," they emphasized (Schaeffer, 1987: 97). On the other hand, through indirect means, the purported flaws of subordinate groups garnered more attention than any flaws assumed to belong to the dominant race. At the close of the twentieth century, the flaw of interest was violent crime, something that media scholars demonstrated was tightly associated by the mass media with African Americans (Entman, 1990).

In the previous era, William Shockley had argued that African Americans were inferior in IQ. In this era, James Q. Wilson and Richard Herrnstein took a different tack, arguing that violent crime was a hereditary phenomenon, resulting from inborn traits such as more massive bodies, lower IQs, and less social restraint (Leo, 1984). Many in the press, along with the proponents of this link themselves, tended to discount the additional step linking crime to race. However, it was not white-collar crime but "impulsive" crime such as "rapes and muggings" that were emphasized (Leo, 1984), and mugging was certainly associated in the popular imagination with racial division (Hall et al., 1978). A few sources did make this link, if circumspectly (Herrnstein and Murray, 1994).

The linkages among race, crime, and heredity were not as direct and overt as in the previous era. Not only was race reached indirectly through crime or intelligence, but also the aims and language used to establish the relationship between crime and heredity were carefully qualified. In an interview with *U.S. News and World Report,* for example, Wilson emphasized the partiality and probabilistic nature of the link he was asserting, saying, "Criminals tend to be—*and it's only a tendency*—individuals who have a rather athletic build that runs to fatness. . . . with the proper care, nurturance and reasonably good luck, these need never result in crime at all. Nobody is predestined to be a criminal" (J. Q. Wilson, 1985; my emphasis). Further, Wilson argued that children with aggressive tendencies should be identified early to help them learn to control those tendencies; they did not assert that mothers of such children should be paid to sterilize themselves (as had been suggested in the previous era). Though the linkages were more attenuated and the remedies less draconian, these arguments continued to provide fuel for racially discriminatory attitudes and policies. This perspective suggested that crime, particularly black crime, was not the fault of social structures, unequal opportunity, or historically sedimented disadvantage, but rather was inherent to certain types of people.

Issues of race and ethnicity were further complicated in this era, however, by the developing attention to the cultural side of racial and ethnic difference. In one important stream of stories, respect for cultural diversity clashed with concern about eugenic uses of technology. Dor Yeshorim was a program developed by Jewish groups in New York to ensure that children with Tay-Sachs and other serious diseases would not be born within their ethnic community. The procedures, chosen by the community's religious leaders, included heavy pressure on individuals to participate in genetic testing along with influence over individual marital choices by religious authorities. Many persons outside the community saw such strong community normative pressure as a coercive invasion of individual rights. They warned that "the coercive practice of eugenics has a habit of creeping in from unexpected quarters" (Brownlee, 1994). This different culture was thus threatening to the dominant community, albeit in different ways from those who were putatively inborn criminals.

Relatively little attention was given to this cultural practice, other than as a curiosity. The case nonetheless signified a serious problem about the usages of genetic technology which had long-term and fundamental implications and which was then being faced more routinely within genetic counseling venues. The technologies of genetic selection had been sold and promised under the banners of individual rights and the rational improvement of human health and well-being. In a world that respected cultural diversity, however, including cultures that allowed more limited rights for their members and different conceptions of rational decision making about health and well-being, genetic technologies would be powerful tools used in ways not approved under their original rationales. Racial and ethnic issues thus would continue to be difficult points to be negotiated in the era of biotechnology. Race and culture, however, did not constitute a singularly prominent focus for debates about the relationship between heredity and identity groups.

GENDER

At the close of the twentieth century, debates over gender issues were more prominent than ever because of the influence of a major backlash against the successes of the women's movement. The dynamics of movement and countermovement played themselves out clearly in the issue of genetics. On the progressive side, women were portrayed by the press as having a more prominent and independent role in articles about genetics than ever before. Women appeared as scientists in the pictorial images of the era, and women were sometimes assigned primary decision roles for the usage of genetic technologies rather than being always submerged in "the couple." Even the relatively conservative news magazine *U.S. News and World Report* phrased the discussion in a more female-centered way than had been typical in the past, predicting (somewhat errantly) that the new alternative to amniocentesis, chorionic villus sampling, "offers a woman the option of choosing a safer abortion." The magazine also quoted a physician's claim that "evidence suggests that it will be as safe for the fetus and even safer for the woman should she choose intervention" (Joseph Carey, 1984). This article put women, not couples, as the central choosing agent, and it also gave the safety of women in a medical procedure equal billing with the safety of a fetus, something that has been too often a rarity in press coverage of pregnancy issues (French, Frasier, and Frasier, 1996; Kline, 1996). Additionally, some articles actively advocated for women's rights and against fetal protection policies, though these articles appeared primarily in women's magazines such as *Vogue* and *Glamour,* rather than in the mainstream news establishment press (Kirp, 1990; Marshall, 1987; "When Will Some Companies Stop . . . ?" 1988). Not surprisingly, *Ms.* magazine went even further, campaigning against the "medical commoditization of women's bodies" and for the protection of "those most vulnerable," particularly poor and Third World women (Johnson, 1992).

Even as the progressive movement established its case for women's interests, however, the evolutionary psychologists constructed a countercase, arguing for the "difference" of the category woman and explicitly linking these purportedly scientifically established differences to claims that legal measures such as affirmative action and sexual harassment statutes ran against evolutionary demands. In the *New Republic,* Robert Wright engaged in what can only be called feminist bashing when he labeled the arguments of feminists "nutty" (Wright, 1994a: 37) and asserted that men could not be sexually harassed and implied that all men were inherently profligate philanderers. Other sources described the evolutionary psychology of women by saying that they couldn't "handle many projects at once" or by noting that women and old people were more altruistic, sympathetic, and nurturing than men (Wright, 1995b; Zuckerman, 1985).

These articles did not praise men's bad behavior, but they did excuse it by claiming that all men inevitably behaved in these fashions and by tracing such behavior to a biological imperative. In contrast, women's behavior, whether praised or belittled, constituted a justification for inequity and an argument against expecting women to have equal power in the society: women were simply not genetically programmed for it. Wright and others occasionally offered progressive palliatives for these unfair evolutionary tales. They suggested, for example, that greater economic equity would help prevent some men from monopolizing so many women (though the concern seemed to be as much about the poor men who ended up without access to women as about women's needs) (Wright, 1994b). They empathized with women's struggle with the two-shift day (children and work) and noted that some solution needed to be found, but they excused men from any responsibility for assuming a fair share of the responsibilities, indicating simply that evolutionary history meant that it could never happen. Men, they implied, just weren't genetically constituted to worry overmuch about children, and no law or social structure could overwrite that programming (see the critique by Jones, 1995).

No mass media outlet provided any critique of this male-favoring version of evolutionary psychology, in spite of the possibility that none of these positions was necessary or true. Popular accounts of evolutionary biology in previous eras had emphasized that animal populations were diverse rather than uniform, and that populations tended to stabilize not with uniform genetic configurations but rather with genetic diversity. It would have been fully possible (and reasonable, it seems) to apply these same principles of evolutionary biology to humans: to assume that there were a variety of men who used a variety of evolutionary strategies to ensure the survival of their genes, included among them relatively intensive care of their own young. Similarly, it might reasonably be assumed that women came with a spectrum of different evolutionary strategies available as well, ranging from producing a relatively high number of children to a relatively low number of children and including both those women who were more adventurous in the struggle for territory or food resources that would assure the survival of the kin group and those women who were more timid and stayed around the campfire to protect the reproductive resources in their bellies. Given that evolutionary biology had catalogued diversity among other animal species, it seemed peculiar, in scientific terms, to posit that human females were all identical and that human men were all simply philandering no-accounts.

Whatever is the real case with evolutionary psychology (and the picture remains sketchy even today), the evidence was hardly well-developed at the time. Scientific studies had not been undertaken that explored the ranges and varieties of human behaviors within groups. Because they were locked in an early struggle to establish the impact of their fundamental notions, evolutionary sociobiologists were exploring similarities across cultures rather than differences within and among groups. The existing studies were relatively primitive and simplistic. But, of course, the mass media writers who composed these articles were actually interested not in exploring the state of research in human evolutionary psychology, but rather in providing a backstop against the egalitarian progress made by women. Had it been otherwise, the same evolutionary factors could have been used to argue in favor of affirmative action instead of against it (e.g., if the only thing men had going for them was excessive ambition, wouldn't society be better off if affirmative action put more women in managerial positions, so that the good of the institution would be fostered rather than the personal status needs and advancement of the hormone-dazed male individual?).

The predominance of such ulterior motives in public discourse about evolutionary psychology was further evidenced by press notices of research which was, at that very moment, exploring the potential for genetic diversity within human sexuality by exploring homosexuality. Although interesting and publicly debated in its own right, that research was also confined to men (Cohen, 1993: 274), probably because conceptualizing diversity among women ran counter to ingrained assumptions of women's supposedly inborn passivity and domesticity. Clearly, the goal was to paint a particular picture of women—as evolutionarily doomed to tend the children and to be empathetic to everyone else's (i.e., men's) needs.[3]

Inequities with regard to women and genetics showed up elsewhere as well. Most troubling in the long term was the continued portrayal of women as the special objects of medical scrutiny. Health researchers have long recognized that women are more likely to come under the scrutiny of the medical system (originally because of pregnancy and now because of their longer life-spans, but also because of male resistance to medical scrutiny). This pattern received endorsement by the mass media with regard to genetics. When sample cases of the use or refusal of genetic tests were presented, the models were more likely to be women if the test was to be accepted and more likely to be men if the test was to be refused. This pattern had its most dramatic instance in the case of surgery for cancer. One article featured a story about a man who had been diagnosed with prostate cancer. The article noted that genetic tests would provide ever earlier diagnoses of such cancer; however, it endorsed the man's decision not to have surgery to remove the cancer, even though that choice meant risking substantially shortening his life. Concerns about incontinence and impotence were presented to justify this risk. The article concluded that "such tests can warn of dangers that are so far off, they may not be worth worrying about" (Gorman, 1995: 60). In contrast, when the cancer at issue was breast or ovarian cancer, the presumption was reversed. A woman *who did not even have* breast or ovarian cancer, but had a genetic configuration that increased her likelihood of developing breast or ovarian cancer, was expected to consider favorably an ovariectomy and/or a double radical mastectomy (Krause, 1994).[4] It seemed that men should refuse medical scrutiny, whereas women were heavily encouraged to undergo not only scrutiny but also massive surgical alteration on the mere heightened chance that they might develop cancer. While, depending on one's opinion of the advisability of surgery, one might see either the male or female as being unreasonably treated in this case, the distinctive difference in expectations should be of concern to egalitarians either way. As the century closed, therefore, there remained much in the mass media over which a feminist might wring her hands.

Pragmatic Issues

Discussions or debates about ethical issues relating to discrimination based on sex, race, and disability were prominent in this era, but they were not the only important concerns. More pragmatic and immediate issues related to direct implementation of genetic testing or other technologies, such as employment and insurance discrimination as well as privacy rights, stigmatization, and environmental issues, were also widely discussed.

Issues relating to employment and insurance discrimination based on genotype had received some attention in the previous era, but they received even greater debate at this time, perhaps in part because these were major foci of the Ethical, Legal, and Social Implications research branch of the Human Genome Project. Warnings about insurance discrimination also received a boost because concerns over the revocation of insurance helped support the ongoing effort to gain a comprehensive national health insurance program. As the *New Republic* put it, "The more we know about our genes, the less sense health insurance makes" ("The End of Insurance," 1990). However, conservative forces also began to reply to these arguments, positing a "company's right to know" about an employee's genetic configuration and offering the case of an airline pilot with a hereditary tendency toward heart attack as an example of why the company's right to know might be of general interest (McAuliffe, 1987b: 69). Conservative publications also argued that if people were to know their own health profiles but insurers were not allowed to know, that would lead to "massive insurance fraud" (Leo, 1989), since such people would be able to stock up on insurance at unrealistically low rates.

Other issues relating to individual rights also continued to be discussed. There was a general concern about maintaining the privacy of genetic information and a specific concern about the inability to maintain the confidentiality of such information once it was stored in computer data banks. The concerns about stigmatization in employment raised in the previous era were also extended to stigmatization in schools. The right of children to make their own decisions about genetic testing, as opposed to parental rights to control the medical choices for their children, also were debated (Monmaney, 1995). These issues further generated discussion of the stigma associated with genetic labeling and its potential consequences.

Other psychological concerns were also explored, and most of these explorations weighed in against genetic testing. Noted one women's magazine, "Advances in prenatal testing now allow parents-to-be to reject genetically 'unfit' babies," but, they asked, "at what emotional cost?" (Sandroff, 1989: 246). The article claimed that parents who used the procedures felt "victimized, not empowered, by the new technology" (246). *U.S. News and World Report* quoted the conclusion of a molecular biologist after both research and personal experience: "Our discovery is great for science—but on a personal level, I'm no longer sure it's an advantage to know these things" (McAullife, 1987b: 69; see also Schaeffer, 1987). These articles thus weighed a new set of variables against the use of genetic testing, charging the utilitarian benefits of genetic testing with new sets of costs on the personal level.

One ethical concern transcended individual and social concerns. Extending beyond the previous era, when concerns with the release of genetically engineered organisms had been widespread, exploration of environmental issues continued and expanded. Opponents to biotechnology still worried about a bacterial "kudzu" spreading uncontrollably and disrupting natural ecologies. In addition, organic farmers and others expressed concern about increased usage of herbicides while agriculture companies developed herbicide-resistant crops. The crops would enable the use of greater quantities of herbicides to destroy the weeds around the crops, with worrisome environmental effects from the increased volumes of chemicals used. Others argued more generally that biotech continued a trend of "high-input, industrial agriculture," which they saw as both socially and environmentally debilitating (Keehn, 1992: 33). Setbacks in the bioengineering of hogs gave critics grounds for challenging the claims of companies that they could "manage" or control the products of their engineering (Crooke, 1988: 734). These hogs were leaner and provided more meat for less feed, but they also suffered infertility and ill health (Johnson, 1992: 82).

The concern with the environment thus led to more specific concerns with animal welfare. As animals were patented and their forms manipulated, a critique developed on behalf of the animals. Empathic individuals worried that certain animals that were being engineered would live their lives in "continual agony" for the sake of greater productivity (Gorman, 1988; Wallis, 1987). Not only were engineered pigs less healthy, but so were genetically engineered mice ("What's Big in Tomorrow's Barnyard?" 1988). The earlier claims about humans—that genetics was a zero-sum game in which improving some aspects led to injuries in other aspects—was presented as an increasingly evident pattern in animals. In addition, a few critics linked these environmental and animal rights arguments together to attack patenting of new animal forms on the grounds that such manipulations violated "the biological integrity of species" and thus were not only morally unacceptable but also threatened nature's basic control mechanisms, with unknown environmental consequences (Rifkin, 1984). In reply, proponents of patenting and manipulation argued that the glass was half full. Though they didn't refute the problems with bioengineered animals, they denied that catastrophic environmental results were likely, pointing to the lack of evident problems to date and concluding that "early fears proved unfounded" (Rogers, 1983).

Many of these ethical objections to genetic testing clearly had a politically liberal cast. However, conservatives and liberals agreed on at least one point; an article from the *Economist* reprinted in *World Press Review* concluded: "Patenting humans would amount to enslaving them, and the idea is widely frowned upon" ("With Billions in the Balance, Many Moral Questions Remain," 1988: 30). Conservatives also had their own concerns about the technologies, including the fear that genetic testing would lead to "many more abortions" (Leo, 1989) and that overreliance on genetically produced drugs or even gene therapy could produce a more "drug-dependent, drug-using society" (Baltimore, 1983: 53), one in which people expected to be constantly altered, chemically or genetically, to maintain happy feelings.

In general, however, conservative magazines, led by the business journals, tended to defend the genetic technologies from ethically based attacks. Some of the defense consisted simply of ridiculing the opposition. For example, *Newsweek* mocked "the alarmist fantasies of the superrace critics" (Clark, 1984: 68), and others called opponents Luddites or crackpots. This ridicule appears to have been somewhat effective, for authors who wished to raise concerns about the technologies positioned themselves carefully against a "crackpot" position that would be concerned about human enhancement technologies or other more serious issues (e.g., McCormick, 1985: 338–39). The effects of such distancing were to mute debate about these broad issues for the long term and to leave the nation more or less unprepared for the soon-to-arrive eventualities such as cloning. Although they used ridicule with frequency, however, the biotech supporters more often defended genetic technologies by citing the positive side of the utilitarian cost-benefit sheet and by manifesting a faith in technology and society to make all come right in the end.

The advantages of biotech touted by its supporters included not only feeding the world's needy and cleaning up the environment but also protecting American hegemony. Several business leaders emphasized that Japan was likely to catch up soon, or that international competition drove us to need to be best at biotech (Gorman, 1988; Miller and Miller, 1983: 51). Defendants of the belief that biotech would ultimately provide a good ethical balance relied on a positive faith in the American and technological systems of which they were a part, and supporters used this system as their ultimate argumentative ground for forging ahead with the technologies, largely without regulation, in spite of the worries. They argued that one simply had "to have a fundamental belief that the interplay of forces within the society will, at a minimum, not be disastrous" (Baltimore, 1983: 53) and that "experience has shown us that advances in knowledge are worth struggling for and are more likely to produce good than ill" (Young, 1988: 16). As one source made clear, failure or catastrophe was simply unthinkable: the investment in biotech had been so huge, the dreams associated with it so large, that it was "almost impossible to believe that the investment being put in is not going to pay off" (J. O'C. Hamilton, 1994: 86). In contrast with the pessimistic pro-genetic technologists of the eugenic era, these biotech enthusiasts relied not only on the good of the system and its purported self-righting capacities but also on the good of individuals. As Horace Freeland Judson explained with regard to germline gene therapy, "This is not something that any responsible medical investigator would do with humans" (Judson, 1983: 17).[5]

As the century closed, the American public debate was increasingly occupied with the particular forms that ethical and social concerns might take when genetic technologies became commercial forces and began to be structured into human life practices. One could not say that we were sleepwalking into the brave new era. However, equally one could not say that we were moving into the new forms of life with our eyes wide open, after thorough public consideration of the issues and careful policy debate. Genetic control remained a muddy area without the "clear lines" researchers such as W. French Anderson hoped to maintain (Glick, 1989), and it remained "difficult to know when to stop" ("Going Nature One Better," 1986). As John Fletcher, then chief of the bioethical program at the National Institutes of Health maintained, it remained fair to say that "our society is starved for creative debate on these questions" (Wallis, 1987).

Summary of the Era

As the first century of human genetic technologies came to its close, humanity was adopting, by fits and starts, by great discoveries and minor innovations, ever more technological intervention in the common gene pools of life on earth and in the genetics of individuals and human families. The metaphors that guided this process included increasingly literalized coding metaphors, which addressed the individual components of the process, and a highly visible blueprint metaphor, which addressed the holistic way that the various components of the genome came together to generate a particular person. The blueprint metaphor was more materialistic than earlier coding metaphors, but it was not rigidly mechanical, as the earlier atom or particle metaphors had been. Instead, the materiality constituted by blueprints was one that was probabilistic, but not dauntingly so, one that was open to human intervention—real but malleable. Within this world view, genes influenced human behavior, but not exclusively so, as the vague and largely unarticulated components of "environment" were admitted to be important, and both gene and environment were portrayed as contributing to every individual's deviations from population norms. Within this framework, genes seemed to be increasingly promising candidates for manipulation. And, when genes began to be manipulated, many ethical questions began to have a reality and an immediacy that had previously been ignored or displaced. This was all the more the case when powerful economic forces sought advantage from the new technologies. Now that the consequences of the technology were increasingly real, it could no longer be the object of innocent dreams of grandeur.

Discourse about genetics in the closing quarter of the century was thus increasingly focused on preventing discrimination. It was more cautious in its perfectionism, and its determinism was imagined as harnessed by human will and technique. Commercialization of genetic technologies and increasing scientific research added one additional conceptual consequence of great significance. Genetics became increasingly independent of heredity. It is not that the heredity-genetics link was severed but that genetics began to cover terrain that was not exclusively or primarily heredity. One portentous example of this decoupling was provided by the treatment of cancer. As several authorities noted about cancer, "We are coming to understand that it is a genetic problem." This statement, however, did not merely refer to the 10–15 percent of cancers that were inherited. Rather, it argued that all cancers, whether they were the product of inherited familial syndromes or not, were tied to genes: "It's not that cancer is an inherited disease—few cancers are heritable—but that the agents that cause cancer often do it by modifying genes" (Baltimore, 1983: 52; cf. J. O'C. Hamilton, 1990). This stretching of genetics away from heredity was also a consequence of the increasing focus on personal health as a product of genetic information, in place of the older focus on the hereditary health of one's offspring. Moreover, the closer gene therapy came to being a real possibility, the more genetic diseases became treatable in a fashion similar to all other diseases. These developments provided the potential for a reduction in the use of prenatal and preconceptual selection. Ironically, gene therapy thus provided at least the image of a genetic alternative to living with the endowments (whatever their scope) of biological heredity.

Obviously, the extensive and growing commercial applications of genetics also loosened the conceptual ties between genetics and human heredity. Genetic products and the benefits of genetic research and information proved to be far more about nonhereditary factors than had been imagined throughout most of the century. Genes were indeed inherited, but they were also active components of cells and organisms, and the latter facet of genetics began to hold greater significance. Instead of genetic research giving us ways to manipulate human heredity, as had been widely anticipated, its primary benefits turned out, at least by the end of the century, to be techniques to improve our research into cellular processes and to produce a variety of drugs and commodities that were neither "hereditary" nor even "genetic" in a direct sense.

As the twentieth century closed, the meanings of the gene—which had originally been born of hereditarian thinking—began to extend far beyond that early thinking. The more humans became conscious manipulators of genes at the microscopic scale, the less significant the natural hereditary process became. The meaning of the gene was now as much about its functioning within individuals as it was about replication—the transfer of forms from one generation to the next. But more than that, the meaning of the gene was now dispersed to a thousand sites, public, commercial, personal, and scientific, that had not figured in either the dreams or the nightmares of its earlier imaginers.

This shift away from heredity was only a strand, not a dominant feature, of the rewriting of the meaning of the gene at the close of the century. But even the characterization of heredity had changed dramatically, the focus having shifted across time from race and nation to family, to the individual, sometimes to an individual embedded in social structures. Even issues beyond the individual and the social, such as the environment and our genetic relationships to other animals, were also being given some attention, but they were not by any means yet integral to thinking about genetics. Perhaps that would change with time. Further chapters about the meanings of the gene, the code of life, and the gene's life in human coding are already being written.

CONCLUSIONS AND SPECULATIONS

We are often tempted to look back to our past in order to gain lessons for our future. Genealogies, like other histories, however, sometimes tell us that the past offers no simple lessons for our future. History is as much about particularity, unique confluences, and multiple forces that can produce multiple effects as it is about grand theory and uniformity and predictability. Genes, as the products of evolution, have in them as much the particularity of human history as they do the generality and uniformity of physical science. It would not be surprising if the same proved true of the confluence of our biological and social histories—that the story of our self-conscious reflection on and understanding of our genes lacks predictable paths and simplistic recurrent effects.

Yet even if we cannot make simple predictions or find unifying patterns that are never contradicted by other patterns, perhaps we can still learn something of value by looking back at our first 100 years of conscious effort to control our genes. Perhaps if there are no universal laws about the social meanings of genes, there are sets of recurring motifs and tendencies about which we can be aware as we take our next steps. In these closing pages, therefore, an overview of key features of the twentieth century—its determinism, discrimination, perfectionism, and metaphors—will set the stage for some speculations about the relationship between scientific and public discourse and future public discourse about heredity and genes.

Heredity and Genetics in the Twentieth Century

Contemporary critics have urged us to worry that thinking about genes would lead us to be discriminatory, perfectionistic, biological essentialists. A close and careful look at the ways in which public discourse has framed the meaning of the gene in the twentieth century suggests that our choices are much broader and more difficult than that. Determinism, egalitarianism, and perfectionism have had changing relationships to discourse about genetics and heredity, and their impact has been more a function of their specific formulation and interrelations than primarily a matter of degree.

VARIATIONS IN GENETIC DETERMINISM

Across the century, changes in the degree of determinism, models of the relationships between genetics and environmental factors, and levels of interventionism have not been tightly correlated with each other. The degree of determinism in discourse about genetics throughout the century has remained surprisingly constant (see appendix 2, table 1). There has been no statistically significant change from one era to another if we define genetic determinism broadly as the assignment of exclusive influence over human outcomes to genes. In most periods, most sources have concluded that genes and the environment interact to produce human characteristics. Yet the range of models available for specifying the relationships between genes and human outcomes have expanded substantially across time. In the classical eugenic era, eugenicists envisioned a unified "germ-plasm" that carried traits from one generation to the next. The germ-plasm was an almost mystical entity, similar perhaps to the Nazi concept of "blood" (Frank, 1995), whose properties determined all facets of a person—mental, physical, and behavioral. The germ-plasm of some families was good, and of others, bad, and this biology was functionally interchangeable with the family's ability to raise children properly to contribute to the upward progress of the race. The eugenicists' view was strongly contested, so public sources did not uniformly endorse government control of breeding, but when nature and nurture were discussed explicitly in the press, the majority of sources combined eugenic and euthenic accounts.

When scientific research and geopolitical events substituted the concept of genes for the notion of germ-plasm, there was only marginal change in the degree of determinism implied by the new model. Genes were now imagined to be beads on a string. They were indivisible particles deposited in persons at conception, and these primordial particles worked their will without much interference from the environment. It was believed that genes determined the "outer boundaries" of a person's characteristics and that environment determined where, within those boundaries, one might fall. The outer boundaries set by the genes were imagined as quite constrictive. Most often the nation's press suggested that bad environment could harm a person's genetic heritage, rather than indicating that environment was necessary to determine which abilities would be realized among the many potential abilities that a person might develop.

Though the degree of determinism and the gene-environment model did not constitute significant changes from the previous era, these models were overtly applied to a much narrower range of human characteristics. Most advocates agreed that physical characteristics could be traced to genes. Far fewer than before described the linkage between genes and mental or behavioral traits in such a tight fashion. Diseases of the body began to dominate the images of genetics.

In the next period, the era of genetic experimentation, the research flood that bore with it the metaphor of the genetic code allowed a dramatic shift in the model of the relationship between genetics and human outcomes. Although the numerical frequency with which the gene and environment were mentioned as causal agents did not change, the vision of the nature of genes was altered significantly. Genes were no longer primordial particles, deposited at conception and immovable throughout the life of the individual. Genes were now merely fragments of code, susceptible both to corruption by environment and rewriting by conscious human design. The explicit models of gene-environment interactions did not fully exploit the potentials in this image, but they did loosen up the old "gene as boundary" model. Genes might still be envisioned as setting the boundaries within which environment operated, but for some writers these boundaries were now seen as nonrestrictive. Genetic potential was excessive; multiple possibilities could be created through social structure and parental nurture. Thus, environment played a more important role in determining which of many possible pathways made available by genes were actually opened up. Genes were more like a starting point than like a fence determining absolute limits. Physical characteristics were still the most important and direct outcomes of genes, and the disease model gained increased force. However, the coding metaphor encouraged increased linkages to mental disease and mental ability as well. Also of importance was a change in the terms applied to describe those forces that provided alternatives to genes. In the earlier eras, terms like *nurture* or *family upbringing* had been predominant. In this era, multiple factors including family setting, social structure, culture, gene-cell interactions, and even gene-gene interactions were submerged in the broad term *environment.* The alternatives to genetic influences thus became more diffuse and ambiguous at the same time as they broadened to include social variables beyond the frame of the family.

By the close of the century the linkage between genetics and human outcomes had reached a relatively pliant point. Scientists revealed enormous complexities in the genetic interactions among genes and between genes and environment. Genomic systems, like all complex systems, were understood to harbor far greater contingency and unpredictability than had been imagined under earlier models. These scientific facts were reflected in much of the cultural discourse. The new model of the relationship between genes and environment did not posit genes as a starting point from which nurture or environment then took over. The older temporal and territorial metaphors were to some substantial extent supplanted by a normative model. The new metaphors, substantiated perhaps by the scientists' new vision of a single "human genome," presumed that there was some normative value for each human characteristic; both an individual's genetic legacy and one's social and familial environment either added to or subtracted from one's abilities and features as compared with that normative point. This did not become a truly developmental model that recognized the interactive pathways between genes in development and environmental stimulation and nurturance, but some developmental perspectives were aired.

The progressive potentials of this new genetic model stirred a determined resistance. Political scientists and psychologists, along with some evolutionary biologists, reached back to the old simple models of genes and heredity to insist that behaviors like crime and capacities like IQ were deposited in individuals at birth. They argued that people were preprogrammed in simple types—such as women, criminals, men, and racial groups. Evolution, like God casting humans from Eden, had decreed that some types of people would suffer (Avise, 1998). Wise social policy, they suggested, would respect these biological decrees. Like the eugenicists before them, however, many of these advocates were motivated more strongly by their belief in social castes and their due than by well-developed science, for the scientific research on sociobiology was primitive at the time, and its hypotheses would change substantially as the young science developed (E. O. Wilson, 1998).

These sets of changes in genetic discourse and the battles about them across these eras indicate that the problems associated with biological determinism do not result simply from whether or not we believe that genes exert an influence on our human characteristics. In all these time periods, most advocates assigned roughly the same proportion of influence to genes and environment. Most articles assigned somewhat more influence to genes than to environment, but through time they increasingly varied this role by type of characteristic. Genes were more likely to be described as having greater influence in mental retardation than in personality characteristics, more likely to dominate in physical disease than mental, more likely to dominate in simple physical traits like eye color than in physical diseases like cancers. One major dimension of change through time, therefore, has been the types of characteristics assigned to genes. In the eugenic era, complicated complexes such as "prostitution," "crime," and "unfitness" were routinely assigned single-gene linkages. By the close of the twentieth century, the attributes associated with particular genetic configurations had become simpler and narrower and most often focused on physical disease.

Even more important to the progressive social potential of a vision of genetics than degree or type, however, is the particular model of the interactions assigned to genes and other factors. If genes provide restrictive boundaries and environment can only detract from genetically programmed potentials, then the model is heavily biologically deterministic and the progressive potentials are low. In contrast, if genes and environment interact to produce a learning animal with substantial plasticity, then genes still determine what kind of animal we might be, but that animal is, by its biological nature, a cultural one. Similarly, genes may be portrayed alternately as tightly sorting people into castes—women, the child bearers; white men, the strong hunters; black men, the deviants—or as providing broad potentials for all individuals on the basis of the unique evolutionary characteristics of human beings, specifically their large brains and their symbol-using propensities. The latter, more open view is not anti-genetic. It does not urge that we study environment instead of genetics. It does urge, however, that we study genes as they are contextualized in environments, and that we remember that humanity is the species that remakes its environments, and a species therefore that lives in an enormous range of environments. Scientific descriptions of the relationship between human behaviors, genes, and environments are therefore enormously difficult to get right.[1]

The portrait of these relationships offered by most critics of genetics has so far held that the paradigm of genetics has always and perhaps necessarily related genes to other organismic and environmental components in a reductionist manner. The critics have urged more "holistic" and developmental perspectives. Such complaints themselves seem based in excessively reductionistic foundations. Early visions of the "germ-plasm" were, as chapter 1 indicated, holistic in their character. But that overly tight identity between the genetic part and the organismic whole was obviously not politically liberating. In the second era, the gene was envisioned as a discrete and deterministic part, but that most reductionistic portrait did not result in genetically driven social policies. In the third era, the shift from gene to genetic code introduced a far more plastic view of genetics than the critics have noticed, but this was also the time when social structures again began to be linked to genetic programs. In the most recent era, though visions of the gene as a discrete and autonomous entity still persist, a fourth alternative was introduced through the concept of the genome. This vision of the integration of part and whole seems to offer a far more sophisticated perspective than critics have recognized, one that opens up possibilities for developmental and interactive components. Yet this new integrated vision hardly guarantees favorable social policies in itself. Each feature of the discourse gains meaning only in relationship to other components. Hence deterministic frameworks work in conjunction with values about equality and normalcy.

VARIATIONS IN DISCRIMINATION AND EGALITARIANISM

The story I have told here of the change in models of determinism through time is largely a progressive story; the public discourse in each successive era has offered a somewhat broader range of somewhat more open models than in each previous era.[2] The story about the discriminatory potentials of genetic discourse has its progressive moments too, but it is more clearly a story of struggle. I have little doubt after reading this discourse and after subjecting it to social scientific analysis that in the most recent periods there has been more substantial attention to a wider range of ethical issues raised by genetic knowledge and understanding than in the previous eras. But the most important part of the story is not, I think, about the growing prominence of egalitarian visions. It is very important that we have shifted our vocabulary from *defective* in the eugenic era, to *abnormal* in the second period, to *healthy* in the third and fourth periods. It is even more important that the public discourse evidences clear signs that we have decreased our readiness to classify someone as inferior merely because she or he is different from us, and that we have increased our collective ability to appreciate the essential dignity and worth even of those we label as having handicaps or illnesses. The medicalization of abnormality has been accompanied by a view that treats such differences as substantive but not determinative of a person's essence, which is crucial to egalitarian perspectives. However, as important as these trends are, the persistent difficulties in our search for equality deserve even greater attention.

Even as egalitarian discourse has become more explicit and elaborate across the century, discriminatory icons and comments have recurred at the subconscious level. Black children being tested for sickle-cell disease were subconsciously, iconographically equated with criminals. When it was no longer publicly acceptable to proclaim that blacks were genetically inferior, the "criminal" came to stand as a surrogate for African Americans, and crime was linked to genetics. And some journalists as well as scientists, in direct and explicit discourse, have persistently and resolutely resisted the egalitarian consensus, arguing that blacks were biologically destined for crime, and women, for housework. Of equal importance, throughout this history, from the eugenic era to the present, too many genetic scientists and genetic advocates have been guilty of a failure to attend to the egalitarian potentials of a genetic vision. Perhaps this is because, deep down, too many geneticists believe that it is only their genes, and not at all their social privilege, which makes them smart and successful. Perhaps, however, it is simply that at some inchoate level geneticists have believed the conservative claims that the only story about genetics which makes genetics interesting and important is one that proclaims that genes validate existing social classes and categories. Because they were focused on protecting and advancing their discipline, therefore, geneticists may not have aggressively pursued the egalitarian visions opened up by a few scientists such as Theodosius Dobzhansky. Perhaps, however, the failure is equally the fault of humanists, who have lately tended to cling to idealist (i.e., nonmaterialist) and outmoded visions of what equality could mean and have therefore failed to provide a viable egalitarian framework into which genetics might reasonably be incorporated.

Whatever the cause, scientific research on genetics today is heavily imbalanced. The model of genetic action which sees our genes as opening up the broad range of human possibilities rather than narrowing them is still a subordinate image in both the scientific community and the public discourse. Perhaps because the disease model has dominated their thinking, scientists are more likely today to ask what mean pernicious differences separate groups of persons rather than what range of healthy differences are available to human beings. But the disease model has already been recognized by scientists as too simple to capture the way in which nonaberrant genetic systems function. Disease occurs when only one key feature in a genetic pathway goes awry.

Normality, in contrast, generally requires multiple successful interactions between genes encoding many regulatory factors for many different gene products. Slight and contingent shifts in all these factors necessarily produce different ranges of function that are responsive not only to environment but also to individual developmental histories. Normal function, by its very nature, is therefore highly variable.

Normal variability has not interested scientists as much as simple categorization of defective products has (although recent studies in ecology are redressing this imbalance). It is indicative, I think, that human geneticists distinguish rigorously in their discourse between variant alleles and disease mutations, but outside of social frames of preference there is no distinctive difference between a gene with a mutation and an alternative "naturally occurring" allele. The genetic configuration for what we call sickle-cell disease is both a "disease mutation" (especially in the context of particular impoverished environments) and a naturally occurring allele with its own environmental advantages (in the context of environments with high malarial risk). Embedded in this seemingly benign verbal distinction, there is not only a strong bias for preferring some kinds of genetic configurations over others but also, more important, a preference for examining genes in dichotomous terms (e.g., sick/health) rather than in terms of ranges and forms of functions within ranges of environmental contexts.

Whatever the causes, this lack of scientific investigation of normal variation is probably one important reason that the increasingly egalitarian public discourse and the dramatically increasing discussion of ethical problems in the implementation of genetic technologies have produced very few substantive social effects. Without the support of the scientific community on a scientific topic, egalitarian social perspectives are not likely to become predominant. At the end of the century, therefore, the debate over genetics and discrimination remained largely stalemated in a configuration that assumed that mention of genetics was tantamount to anti-egalitarianism and that egalitarianism could be manifested only by the erasure of knowledge of genes.

PERFECTIONISM AND GENETICS

The critics of genetics have also urged us to beware the tendency to adopt perfectionist expectations about our children, and hence about our species. In a strict sense, this study shows that we have never been perfectionists and are increasingly less prone to perfectionism. The term *perfect* almost never occurs in this discourse set. When it does appear it is most often used negatively by a critic of perfectionism. To focus narrowly on the use of the term *perfect* might appear to be a straw argument, but even casting the net more broadly, stated preferences about children have become increasingly less stringent through time. In the eugenic era, only "superior" children were wanted. There was little talk of average or normal children. One's only options seemed to be to produce a superior child or one who was unfit. After the eugenic era, expectations were significantly altered when normality became the desired goal. In the final two eras, health rather than broad normality was the expressed goal for children, with terms like *normal* increasingly reserved for the molecular level of the genes rather than used for whole individuals. Moreover, in the latest era the concept of health was democratized even further when it began to be argued that everyone had abnormal genes and hence we were all prone at different times in our lives to be unhealthy. When everyone is potentially ill, then it becomes more difficult to stigmatize individuals for such illnesses.

This ever narrower path of the public discourse gives some grounds for hope that the common sense of parents will provide a bulwark against the siren call of genetic perfection. However, the fact that we have experienced a fairly positive trend during this century should not be taken as a guarantee that such a positive trend is inevitable. Although few people will assert a desire to have "perfect" children, it would be possible to interpret the standards of the term *healthy* in such a rigid sense that the concerns that gave rise to a debate about perfectionism could be activated by that term as well. Seduced by commercial advertising or perceived competitive threats from other families, it is surely possible that some parents would be driven to excessive genetic testing and to discarding fetuses or embryos who have any significant potential for any major disease at any time in their lives. That would be a gruesome outcome for genetic testing, given that all of us die, most of us from some disease or other.

While most parents have many good motives in the enterprise of childbearing, the goal of commercial testing companies is to make as much money as possible from testing. Given that the companies and the physicians control the information about testing and that many would-be parents may not have a good understanding of that information or access to reliable genetic counseling, abuse of genetic testing becomes a distinctive possibility. (I say this fully aware that defining the boundaries between abuse and use is itself enormously difficult.) Even though Americans have so far managed to dodge the perils of any severe degree of perfectionism resulting from genetic information, because of these potentials, the critics' concerns should continue to animate our self-scrutiny. As genetics continues the transition from a scientific face to a commercial face, and as our abilities for genetic selection and alteration move beyond single-gene-disease disorders to a whole panoply of human physiologically based characteristics, this scrutiny will become even more important.

METAPHORS

In addition to concerns about determinism, discrimination, and perfectionism, critics have spent substantial effort critiquing the metaphors used to describe genetics. Hundreds of metaphors have been applied to genetics, from the *New York Times*'s ill-fated description of DNA as a wire hawser to more popular renditions of DNA as a computer disk, a recording tape, a pool, or a recipe. I have here explored only the four most dominant metaphors. Each of these metaphors has proven to be far more than a simple stylistic frosting slathered on top of a scientific cake of facts. The metaphors have functioned as powerful and intricate devices for connecting up social interests, social fears, and scientific discoveries. Thus as science moved from germ-plasm to gene to DNA to genome, the new findings were related to social understandings and practices by the metaphors of stock breeding, the atomistic all-controlling gene, the code, and the blueprint.

The stock-breeding metaphor fit wonderfully (if perniciously) with the interests of the professional class, who most frequently wielded it. It salved their fears over the Darwinian revolution, making them into the breeders and other humans into the bred. It collapsed the frightening diversity of the new American multicultural scene into a simple hierarchy—superior self and unfit other. It borrowed from the common experiences of a still heavily rural population, familiar with the practices of plant and animal breeding. But it also conjured up the images of powerful, rational control by the scientist carefully and effectively breeding his peas and flies. These images fit well with the notion of the germ-plasm as an indeterminate mass of hereditary material. There was no micromanipulation of particular genes yet in these images. Instead, one moved the stock around in their various pens, segregating one set of animals from another in order to control their breeding. The consequences of seeing and treating humans as animals subject to such control were also what reflection on the metaphor might have predicted.

The vision of the controlling, commanding gene—a string of atomistic self-contained particles—was somewhat different from the stock metaphor. This metaphor lacked unity, depth, and detail. Such a diffuse metaphor was needed for an age that reacted against genetics yet recognized its power over humanity as absolute. This gene was terra incognito. We knew it well enough to recognize its power but not well enough to offer any full description. We were content in this era to leave its features to the scientists, appropriately cloistered in their labs. In our own lives we ducked and dodged its lottery as best we could, but we recognized its cold, lifeless, characterless grasp over our fate.

How different was the gene of the DNA era, insubstantial bits of code, strung together on the crystal staircase formed by the incredible double helix. How excited we were to decode the book of life. How effectively this metaphor opened up our fantasies and speculations. We could, it seemed, rewrite nature at will. We could rewrite ourselves, not to be a superrace, but to be a diverse transspecies filling a variety of niches around the galaxy. But we could also, we feared, produce runaway microbes that would destroy our ecosystem. How glad we were to settle our thoughts into more orderly paths by transposing the image of the blueprint over the notion of the code.

The blueprint metaphor coordinated the disparate parts of the code into a whole and particular text. Instead of seeing the individual as the product of messages passed down from generation to generation, the blueprint envisioned a unique individual with his or her own distinctive pattern, which was a product of the particular interactions of the particular bits of text that chance had brought together in the genome. Moreover, the blueprint metaphor projected a vision of design. Blueprints were designed by someone, and through gene therapy we could redesign unhealthy blueprints that would rewrite and reorchestrate the genomes of individuals with disease. Genetics did not have to be solely about heredity anymore.

Science, Society, and Rhetorical Formations

In this century, American public discourse has imagined the gene in a variety of ways. It has focused on genetics as a product of race, family, or individuality. It has seen genes as providing narrowly restrictive boundaries or as constituting infinitely rewriteable slips of text. It has argued that the character of genetics mandated unequal social structures and that the nature of the genes required us to provide maximal developmental opportunities to everyone. Our thoughts about the gene have almost always been associated with hopes for control, but the range of control and the purposes for control have been enormously variable. Just as we have come to see genes as playing roles within complex series of interactions, so it seems that we might view discourse about genes in the same fashion. The meaning of the gene is not static, nor is it the predictable product of a simple set of uniformly acting variables. Rather, the meaning of the gene is a product of multiple interactions in a developmental cascade that is responsive to crucial external variables along the way.

One of the key relationships that shapes the outputs of these interactive variables is the respective roles played by science and public values in producing a social formation. Some scholars have been at pains to argue that science is merely the predictable product of regnant social values. Thus, Baconian science has been identified as the product of a patriarchal world view (Harding, 1986). On this account, science is merely a subset of social values in operation, so that the rejection of eugenics was a product solely of the growth of democratic discourses in the public realm, and increasing complexity within genetic science had nothing to do with it (Paul, 1998: chap. 7). These views emphasizing the role of social values in science and its social relations have developed in response to equally simplistic perspectives that have portrayed science as utterly separate from public values. Such views of science as fully autonomous would deny that genetic research has been influenced at all by world views outside the narrow confines of the scientific paradigm of data-driven hypothesis formation and (dis)confirmation. There is, of course, a third option, that the interactions of science and society are complex, variable, and contingent.

In the early years of the century, scientific research in genetics enhanced the public acceptability of eugenics by supplying an image of a control lever for regulating human "breeding." While it is indeed true that the scientists who espoused this view were probably influenced by political values that favored eugenics, the public response to eugenics was not a simple adoption of a technology that the public values had already called for. The public rhetorical formation was too contested and variable for such a simple circular production-and-adoption cycle to explain the science and society relationship adequately. The rhetorical formation of this era was highly polarized in many respects, and advocates on both ends of the spectrum selected from what were then competing scientific accounts, including among others Lamarckianism, non-Mendelian genetics, and reductionist Mendelianism. Similarly, the end of the era was caused not by a singular factor—solely the rise of democratic values or solely the new and more complex findings of genetics—but by the coincidence and interaction of these factors. Had one or the other of them not been present, a more contested transition would surely have occurred and a different rhetorical formation would have followed.

Instead, in the second era there was a highly nonpolarized rhetorical formation, but one that exerted little social impact. There was wide social and scientific agreement that genes were important and biologically fundamental. That agreement had little social consequence, in substantial part because the science of the day and the social values of the day did not cohere in any functional applications or in any instructive value propositions. Thus, the particular relationship between the specific features of the science of the moment and the social values of the moment were contingent and coproduced a particular formation. Nothing makes this equation clearer than the substantial shift that occurred when the science changed.

Scientific change on its own was not sufficient to introduce a new era. The new discoveries of molecular genetics were at first not communicable to the public. Science was mute in the public sphere until it formulated the coding metaphor. This metaphor worked both for scientists and for translating science to the public. Once both a new set of scientific possibilities was opened up and a vocabulary that could broadly communicate those possibilities was developed, then the century's third rhetorical formation of genetics rapidly developed. In this third era, a disjunction between public values and scientific possibilities was a dominant feature. The public value set explicitly and repeatedly wished for gene therapy. The scientific formation could not provide that. Instead, genetic selection through amniocentesis and abortion became the central application of genetic science. In this respect, the state of the science drove the character of the rhetorical formation in a crucial area. But the state of the science was not the sole factor in shaping the rhetorical formation. The tension between public values and scientific possibilities fostered a rhetorical formation that featured radical undecidability as its central feature. This failure to come to a social conclusion left open the forward movement of the science, and to that extent the scientific discourse preserved its own ability to operate in the social sphere. In this way, science itself had a great deal of power. But science and public values were neither simply opposed nor simply isomorphic. Simple opposition would have shut down the science; simple identity would have led to far less indecision and would not have fostered the growing ethical debates.

The fourth rhetorical formation about genetics in the century featured it as a relatively settled component of the public discourse, though there were some increased contestation and polarization around the formulations offered by genetic sciences. Ethical questions were increasingly visible, but these were largely subsumed as management problems within the new formation rather than as alternatives to it. Clearly, the prominent notion of genetic influence in this era was heavily driven by the well-funded scientific sphere. But the science itself was at least partly responsive to public discourse, as is evident in the evolutionary changes in the work of public scientists such as E. O. Wilson across the last quarter of the century. Moreover, commercial forces began to generate yet another set of powerful influences that would further alter the mix in the future.

Across the century, therefore, the relationships between science and public values and their influences on the public sphere have varied on the basis of the particular character of the public values available and the visions and products offered by the scientific realm. At all times, science is not independent from the public value structure, but there are multiple options in the public value sphere that various scientists can draw upon. Similarly, there are within the scientific sphere different resources for varying public advocates. The complexities of these relationships make simplistic theoretical pronouncements about the relationship of science and public values perilous. They also make predicting what will come next or specifying what should come next extremely difficult.

Future Discourses

Whenever I talk about the genetic discourses of the past, people ask me what comes next. To address this question requires that at this point I shift hats from the social scientist and critic, describing and interpreting what has been, and become more fully an advocate. From such a standpoint, it is relatively easy to imagine both dream and nightmare visions for the future.

The nightmare future includes the routinization of genetic selection as a part of pregnancy with stringent standards for what kinds of embryos or fetuses shall join the human race. It includes massive advertising for genetic testing of healthy adults to stimulate disproportionate fears, thereby wasting health dollars as well as generating unproductive anxiety in the populace. It includes the virtual elimination of genetic counseling by genetic counselors certified by the National Society of Genetic Counselors, except for unusual applications in research protocols. It includes, alternatively, the endorsement of directive counseling. The nightmare future is one in which there is genetically based discrimination with regard to race and class, not only through differentials in financial access, but also through institutional protocols that serve the needs of some cultures and classes well but that employ genetics as a means of social control over other cultures and classes (Condit, forthcoming). Given the increasing interest in behavioral genetics, the dark vision of a genetic future may also include an early and sustained sorting of persons into tracks on the basis of their genetically predicted predispositions and abilities. The worst versions of the nightmare even feature cadres of rich men cloning themselves to gratify their narcissistic urges, and clone shops promising parents and speculative business conglomerates the likelihood of spectacular athletes, singers, and beauty stars born from the copyrighted cells of the latest popular celebrities.

The dream future is more pleasant, if less spectacular, to imagine. It includes significant reductions in much illness and suffering due to new drugs and procedures opened up by our understanding of genes and genetic interactions. It includes pest-resistant seeds at reasonable prices rather than pesticide-resistant crops at outlandish costs to farmers, consumers, and environments. It includes innovative new products and substitutes for old products to produce a less toxic environment.

Will our future be nearer to dream or nightmare? The best guess is that it will include some of both elements. The way in which groups and individuals take up the issue will tip the balance in many cases. The Center for Disease Control is formulating a public health initiative on genetic testing, emphasizing adult testing and responsiveness to phenotype, rather than the selection among genotypes at the pre-embryo or prenatal stages (Khoury, 1997). If that effort can be maintained, it may mean that we have avoided publicly supported massive prenatal genetic selection, though limitations on the CDC are obviously not enough to prevent private commercial ventures from pushing such an agenda.

In order to avoid specious mass commercial testing, we need first to restrain commercial entities from misleading advertising about their products, that is, advertising that generates disproportional or unreasonable fears and anxieties. We also need to avoid pressure on physicians to offer these tests. Physicians' resistance to the tests has been an important brake on massive genetic testing. Proponents of mass testing often attribute such resistance to ignorance. While primary-care physicians may indeed lack expertise in advanced genetics, they are not necessarily deciding against genetic tests out of ignorance. Genetic tests are proving to be more expensive and less informative than many had hoped. Physicians tend to learn rapidly about new and unfamiliar tools that are of major utility in diagnosis and treatment. Genetic tests currently appear not to offer cost-effective diagnoses or avenues for treatments on a mass scale.

In spite of their generally reasonable preferences, physicians could be herded into offering genetic tests by fears of liability. If patients who later exhibit diseases, or whose children exhibit diseases, successfully sue physicians for failing to offer technologically available tests, then physicians may move to mass testing for self-protection. We need to develop cultural resistance to the expectation that a physician should offer all available tests to all persons in the absence of strong and particular indications for such tests.

To avoid our nightmare futures we need also to develop more stringent regulations and economic supports for some medical practices. We need regulation of commercial applications for cloning. I would prefer that we make cloning of complete human beings illegal, though we should protect the rights of persons who are born as clones through technological manipulations in other nations or through illegal activities by their "parents" (whether that activity itself is punished or not).[3] The need to protect the civil rights of those who have been reproduced through cloning should be obvious. The need to ban cloning arises from its narcissistic nature: it is an activity undertaken by parents for their own benefit and not for any benefit to children. There is much potential harm in being treated as a narcissistic object of a parent. It is true that parents may treat nonclones as narcissistic objects, but we cannot regulate all such behaviors. We can easily regulate cloning (it is, after all, a crime whose manifestation is both enduring and traceable), and so we should.

The major potential benefit advocated for cloning of full human beings is that it appears desirable for infertile couples and lesbians. It allows those couples to have genetically related children. However, this asexual form of reproduction actually allows a genetic link to only one member of the couple, and that is no advantage greater than other, sexual, forms of reproduction (through donor gametes). As a person who was unable to have children and for whom cloning could provide a personal genetic link to a child, I understand the attractiveness of the option. But my personal desires seem far less significant than the broad social implications. This is one case where society needs to consider the good of children and to protect the general principle of the worth of individual human lives over parental desires. We should therefore restrict those parental desires.

Unfortunately, I have doubts that we will be successful at prohibiting cloning of full human beings. If cloning were to be pursued by only a few wealthy narcissistic men and a few wealthy lesbian couples, the damage to society would be minimal. If, however, cloning were to become a mass market in "star" children, the situation would be otherwise. Because genetic potential only opens up possibilities and probabilities, clones of highly marketed celebrities would be bound to disappoint their buyers. When children become expensive, disappointing commodities, the consequences are far too likely to be dire. Even children who bear out their marketed potential are being treated as slaves and investments rather than as people. Moreover, our current laws provide no restrictions on ownership (parenthood) of large numbers of children pursued as lucrative investments by individuals, corporations, or individuals fronting for corporations. We should therefore make it illegal to advertise or sell human cells for cloning, and we must devise restrictions on purchasing children as investments. Regulation of such enterprises will be difficult but not impossible. The goal is not to eliminate every last instance of such cloning (all laws are broken sometimes), but to prevent the development of a mass market and the attendant market-driven cultural values. I do not expect that we will be able to do this before cloning of humans becomes technogically feasible, though it would be easier to do it now, before industries have developed to fund resistance to such legislation.

These needs touch on the more fantastic side of future developments, but there are also mundane steps that need to be taken to bring us closer to genetic dreams than to its nightmares. The most pressing need is to require the funding of substantial client-centered genetic counseling by counselors certified by the National Society of Genetic Counseling when any genetic test is offered or results are provided. That is, it should be illegal to offer, perform, or interpret the results of DNA-based tests without providing, at the same inclusive price, genetic counseling. The major mechanism for this will be to require all insurers to provide genetic counseling by these certified professionals when they provide coverage for genetic testing, but the rules should obviously apply to all sources. Individuals who want to use testing cannot and should not be coerced into accepting counseling, but it must be made legally, pragmatically, and financially available. This will make genetic testing more expensive, but it is the only way to assure even an approximation of informed, autonomous decision making. Without such relatively informed and autonomous decision making, we have simply developed a new and pernicious form of social and reproductive control. Without access to this key safeguard, genetic testing is clearly nightmare rather than dream. If insurance companies and individual medical professionals were to replace government as the agents of reproductive and personal control, the character of the nightmare would be changed, but it would hardly make genetic testing a beneficent activity. Genetic testing without some reasonable degree of informed, autonomous, individual control is clearly a form of eugenics we should avoid.

Preventing the production of cloned human beings, ensuring adequate genetic counseling, and forestalling commercial pressures for excessive genetic testing are all policies that I hope will be on our agenda. But in order to achieve policies like these it will be necessary that we think about, and hence talk about, genetics in an appropriate manner. What then are the ways in which our future discourses should negotiate the problems of genetic determinism, discrimination, and perfectionism?

DISCOURSE ON GENETIC INFLUENCES (DETERMINISM)

Social critics of genetics and genetic discourse tend to write as though the only acceptable public discourse is a genetics-free discourse. They clearly prefer an intensive focus on social structures and cultural factors rather than on biological components, especially genetic ones (e.g., Hubbard and Wald, 1993; Katz Rothman, 1998; Lippman, 1992, 1993). As a few observers have pointed out, however, social determinism does not offer a much more heartening image than biological determinism (Kaye, 1997: 100; Peters, 1997: 28). If we are merely automatons shaped in every detail by external forces, what difference would it make if these forces were the products of social structure or of genes? If we can manipulate genes more readily than we can manipulate social structures, then why not manipulate genes?

Social structures may at first glance seem to offer a less mechanistic, less material and binding framework than biology, but the vision of overdetermination by race, class, gender, age, and so on, that is offered by many social theorists is rarely warm and reassuring. It is a vision of oppression and unwarranted privilege, of individuals "always already written" by their society and their social position within it. Neither the social nor the biological version of determinism is progressive (or sufficiently descriptive) in my view. This is not to argue that false images of absolute individual freedom from influence and constraint are the progressive alternative.[4]

Instead, I think, the progressive view holds that words like *determinism* are epistemologically outmoded and that singular factors such as social structure or biological substrate cannot adequately describe our natures and our choices. Rather, the characteristics of human life are the product of historically sedimented but dynamically ongoing interactions among our biology, our culture, and our physical environment. These interactions produce a "self"—a person who makes choices, some consciously, some unconsciously. This self is not static, for we each change through time, but the self has some degree of stability (of a nature best characterized by chaos theory) and some uniqueness. These selves blend our biological and cultural heritages in complex interbraidings that emerge through time. Because of our advanced culture, for example, nearsighted people survive and their genetic configurations are reproduced. Our biological characteristics are thus influenced by our cultural characteristics, but each cultural characteristic is in turn a product of the biological possibility for developing advanced culture and a product of previous cultural configurations. Biology and culture thus interact in recursive loops.

That same dynamic recursivity exists within our biology. Genes are not static dictatorial particles. Genes don't even exist as discrete physical entities. DNA exists in a physical sense, and segments of DNA interact with each other and with other cell components, which interact with other cells in a given organ, which interact with other organs, which interact with other organisms and other external environments. Genetic recombination in the hereditary process does not follow Mendel's elegant laws because of the messy complications of linkage, imprinting, and developmental contingencies. Both heredity and functional genetic processes in the life of an organism are historical, contingent processes. Consequently, they are not simply lawlike. They can be probablistically measured on a population basis, but they cannot be accurately predicted on an individual case-by-case basis. The law of gravity dictates that all apples fall to the ground. Genes are more unruly than falling apples.

Human beings are, in part, products of these unruly human genetic configurations. These genetic configurations establish the limits of our material abilities; we do not have the necks of giraffes, and we cannot calculate pi to 10,000 digits in our head like a Cray computer. Our genetic configurations also enable our possibilities; we can build cranes to reach to the trees and spaceships to reach to the stars. We can design Cray computers, write symphonies, and dance salsa. But our genetic configurations do not determine the precise outlines of how we enact our possibilities. That is, on the output end, genetic maps have low resolution. Our genes allow us to make music and dance, but some cultures do jazz and others do salsa. The pyramids of Egypt, the Mayans, and Las Vegas all visibly share a motif that arises from the interaction of human genetic potentials and the physical potentials of the inorganic world. However, this same motif has been executed for different purposes and to different effects. The differences are substantive and not the products of variations in racially distributed genes. They were also never predicted, and the uses to which pyramid motifs will be put in the future are also not predictable to any fine degree of precision. Culture and biology interact in human beings to produce a developmental cascade which opens up some cultural avenues and not others.

The process by which genes and culture interact has been called coevolution by Edward O. Wilson (1998). But Wilson's account of coevolution is still too genetically deterministic because his sociobiology still does not yet understand two basic facts about human genetics. The first is the multipotentiality of human genetic configurations, and the second is the variability of those configurations.

Wilson's most recent elaboration of the role of evolution in human life is far more progressive than previous sociobiological accounts, in part because he has made a serious effort to integrate culture into the account (rather than merely making culture an epiphenomenon of genes) and in part because he incorporates, to a point, the dynamic version of theory enabled by chaos theory. Wilson's account stops short of a truly integrated approach because of his inaccurately truncated account of epigenetic rules. He uses the examples of snakes and color vocabularies to argue that there are basic epigenetic preferences installed in human bodies by way of evolutionary pathways acting on and through genes. Because some such regularities exist, Wilson argues that research in genetics should be the central focus of social science, and this research should be focused on describing those epigenetic rules that determine (all? most? the only interesting?) human cultural forms. His account of epigenetic rules is useful and interesting, but it neglects a crucial factor in these rules: human epigenetic rules (i.e., human genetically based scripts) are capable of conflicting with each other. In fact, not only are they capable of such conflict, but they are virtually destined to conflict with each other.

Consider the inherent conflicts between our altruism and our competitiveness. In a major breakthrough for an egalitarian paradigm of genetics, for example, Wilson essentially concedes that there is such a thing as group-based evolution. That is, human beings have the genetic potential for altruism because human groups in which altruism operates provide greater survival value to the genes of the group members than groups do where no altruism exists. Fine, humans have a genetically based potential for altruism. Unfortunately, however, humans also have a genetically enabled tendency for self-protection and advancement (status climbing in dominance hierarchies being one of our favorite activities). Wilson's paradigm would have us believe that there would always be epigenetic rules for deciding when altruism will come into play and when status seeking will come into play. He may be partly right; there may be some instances of such selection defaults in strong cases (e.g., mothers may be more likely to choose altruism vis à vis their young children than self-aggrandizement at the expense of these children). In general, however, the potentials for altruism and self-advancement are broad potentials that come into play at both the individual and the social levels in situation-specific and culturally learned fashions. In any particular situation, the choices that will be made between these two potentials for any particular individual or society are not predictable on the basis of epigenetic laws and rarely predictable on the basis of epigenetic rules. Generally, the choices will be the product of the particular history (development path of culture + physical resources + biological characteristics) of the individual or the society that is making the particular choice. In other words, human genetic predispositions are nonlinear in their character. Human genetic features have the characteristic of multipotentiality. This characteristic is what enables variability in human culture, and it is a necessary concomitant of human language. Without this multipotentiality human beings would be merely instinctual creatures rather than the most skillful learning animal.

The multipotentiality of the genetic inheritance of human beings does not deny that one will be able to provide some descriptions of uniformities in human cultures that can be linked to genetic configurations, or even that one will be able to make some predictions about what kinds of conditions will favor particular cultural manifestations (of selected genetic potentials) over others. It does mean that one will not be able to explain human characteristics fully through a genetic account or even a scientific account (as Wilson understands scientific theory [1998: 255]). Wilson's dream that all human life can be explained by his version of science is simply wrong. In chaotic systems, history matters. History means particularity, and particularity is incompatible with Wilson's vision of scientific theory as a process of generalization. We might, therefore, be perfectly willing to entertain Wilson's suggestions for further research into the linkages among the physical, biological, and social sciences. There is no doubt interesting material there. But we should resist firmly the false notion that such research will tell us all that there is worth knowing about human beings.

The substantial effects of the multipotentiality of human genetic propensities are related to and exacerbated by human genetic variability. It is difficult to predict what human cultures will do, because the individuals in them are not genetically uniform. Although environmental studies (especially genetic studies) may be slowly weaning them from this notion, biologists have had a devilishly difficult time incorporating into their theoretical structures the fact that there is not one optimum gene or collection of genes for most human needs. Works by sociobiologists and evolutionary geneticists routinely make the assumption that, at any moment in time, the particular genetic configurations that exist in a population must be optimal and that only one genetic configuration can be optimal. This is an ironically static and monotonic assumption to exist within a theory that depends on change processes and variability. In order to argue for evolution and the importance of genes, one does not need to argue that evolution is finished or that evolution functions only with a static environment. Those are artifactual assumptions of early discourses of Darwinian theory and its self-defense against competing theories (and perhaps assumptions of the political biases of many sociobiologists). Instead, what is true for other animals is true for humans as well: there are a range of environments, cycling through the seasons and also through yearly variations, and different traits are favored depending upon which microenvironment is temporarily dominant. This means that survival and growth of human (and animal) populations as a whole are favored if they have genetic diversity. Different cycles of conditions can allow different configurations to prosper.

The existence of genetic variability and genetic multipotentiality means that Wilson's version of reductive science as an all-sufficient basis for human self-understanding is inadequate. In Wilson's account, the only interesting fact is the gross similarity of color language among various human groups. But why are these similarities more interesting and worthy of study than the many differences in parsing color among these groups? Why do we "understand" human genetics only when we wipe out the differences that our genes enable? Wilson's view commands us to view color-blind individuals, as well as the one known color-blind culture, as though they are not really a part of the range of human beings. But upon serious reflection that appears to be an arbitrary and unscientific approach to understanding humanity.

Because human beings have always created their own environments to greater and lesser degrees, we feature substantial genetic variation. This does not deny that we are influenced by our genes, but it does complicate the picture considerably. It is past time that these complications begin to be taken seriously in the research agendas of genetics. This will require substantially greater self-consciousness by geneticists and those who fund genetic research to ensure a greater balance in our research efforts by providing greater support of research into human variability that describes ranges of human potentials and interactions with environments rather than merely supporting research into disease, which prefers a disease model and which tends indirectly to support less egalitarian views of genetic potentials. This does not mean that we should support faulty research designed to produce predetermined outcomes that favor our politics. Counter to the implications of some science critics, the fact that science is ideological does not make it *mere* ideology. The real world does push back on scientists, and it is difficult for a scientist to prove a truly false proposition (e.g., that the earth circles the moon) if the methods of science are rigorously applied. Anyone who has worked in a lab where genetic research is done knows how truly recalcitrant the nonhuman world can be. Nature, however, cannot answer questions we do not ask. If no scientists pursue questions about the ranges of human variability within categories, as well as across categories, we will not have the data to attend to this variability. Society therefore needs to fund research in areas that would support egalitarian paradigms, to discover what degree of support for those paradigms might be found (just as it has liberally supported paradigms that test hypotheses in which more conservative formulations are explored). Currently, through historical accident, scientific convenience, and funding biases toward disease, geneticists, biologists, psychologists, and sociobiologists do not conduct research on the questions raised by egalitarian perspectives. Instead of studying variability among women, we study men versus women (Condit, 1996a). Instead of studying the variability of fat-preservation and utilization mechanisms, we study what genetic inputs predispose one to gross obesity. Instead of studying varieties of learning style, we study what racial group has a lower mean of one kind of learning style. Genetics can fit into an egalitarian paradigm, but to do so, it will have to incorporate questions about variation and multipotentiality into its theoretical frameworks and hypotheses.

How scientists address issues of genetic determinism will be important for using genetic research in a fashion that encourages progressive paradigms rather than discriminatory ones. However, scientists are not the only important group who will shape the meanings of the gene. *Ecotimes,* the on-line list-serve calendar for people interested in various ecologically related topics at the University of Georgia, always ends with this quotation, attributed to the cartoon "Dilbert": "Science is a good thing. News reporters are good things too. But it's never a good idea to put them in the same room." This cynicism about the interaction of journalism and science is widely shared (Taylor and Condit, 1988). I am less critical of journalistic treatments of genetics than are either scientists (who accuse the press of fear mongering) or social critics (who accuse the press of not fear mongering enough). However, there are things journalism can and should do to forward the dreams and discourage the nightmares of genetic determinism. The most important is to present the relationship between genes and behavioral traits and mental abilities accurately.

The reporting of the press about deterministic linkages between genes and human characteristics currently features some strengths and some weaknesses. Genes clearly have some influence on human characteristics, and it is reasonable for the press to report findings that describe those linkages. As Peter Conrad (1997) has observed, however, public writers need to avoid framing the issue in terms of a simplistic contest between genes and environment. Too many headline stories about genetic discoveries bear the flavor of a triumphant celebration of the genetic view over the environmental view (Conrad and Weinberg, 1996).[5] The important questions, however, for at least 50 years, have not been whether genes or environment dominate. All reasonable and knowledgeable people concede that both genes and environment have their roles. The interesting questions are where and how genes and other factors interact.

The national magazine press has been becoming increasingly sophisticated about these relationships. Close to a third of all articles in the past 20 years make clear differentiations between the strengths and singularity of genetic influences in various conditions (see appendix 2, table 8). Newspapers, especially more local newspapers, are less likely to make such distinctions, but they are also more likely to report on a single disease or condition, so that such differentiations are less immediately called for. However, the press as a whole needs to continue to develop a vocabulary that capture the variability of genetic influence and the indeterminate influence of genes. A variety of sciences, social sciences, and humanists have been employing the vocabularies of chaos theory, systems theory, and postmodernism to develop ways of talking, which are also ways of understanding, that capture the sense in which complex factors interact. It is difficult to translate the esoterica of these vocabularies into popular prose, but that is precisely the responsibility of the press. These new world views recognize that the term *determine* is often an inaccurate descriptor, belonging to an older, mechanical age. In complex, dynamic systems outcomes are "contingent." That means more than to say that multiple factors have influence. It means also that each factor can have a different influence depending on which other factors are in the mix (in the same way that a child might be quiet and orderly when its parents are around and loud and boisterous around particular friends). Moreover, as chaos theory emphasizes, small differences in initial input can make large differences in output. The contingency of complex systems also may include a developmental or historical component. That is, the influence of a particular factor or set of factors can depend on what has happened before, namely, on the order in which various factors are brought into the mix. (It may be harder to settle the child down to quietness if the child has first been with the boisterous friends and then the parent is brought in than if the parent is around first and the friends are then brought on the scene.) Because genes are involved in complex and contingent nonlinear systems, terms like *determine, command,* and *direct* are inaccurate in describing most genetic processes.

A good metaphor for understanding how it is that genes can be important, but not have linear impacts, is available through a comparison of genes to the members of a football team. Clearly, the outcome of a particular team's season is heavily dependent on the abilities of individual football players. There are no football games without individual players, and the quality of the games depends very heavily on the players. However, except in cases of gross teamwide deficiency, the outcome of a particular game, let alone the particular patterns of a game or a season, is not predictable. Players are key inputs, they have a major influence in the system, but they do not result in simple or repeatable and predictable outcomes. Genes, like individual players, are crucial inputs into human systems. But they do not produce regular, linear life histories. Our knowledge of the statistics of a player, or even of all the players, does not necessarily enable us to predict how plays, or games, or seasons will take shape. Likewise, our knowledge of the coding sequences of genes does not enable us to predict, judge, or reliably influence life courses.

Many press sources have shifted to more accurate vocabularies that capture the variability in genetic systems, such as *influence* or *predispose* or *susceptibility.* However, when providing shorthand labels, the press, like scientists, tends to use labels that might be misleading. There is no such thing as a "breast cancer gene." Instead, whenever any of a series of mutations in the BRCA1 gene occurs or is inherited, the likelihood of breast cancer developing increases. The shorthand of the phrase *breast cancer gene* is therefore highly misleading. Horace Freeland Judson has argued that the press needs to teach the public the scientific term *allele.* One could then speak more accurately of the predisposing allele for breast cancer. The press seems to believe that this additional vocabulary would tax the public too far. If so, another possibility is to use combinations of more ordinary words that convey the same information. For example, one could speak of the genetic cofactor for breast cancer, or even the BRCA1 cofactor for breast cancer. Creative journalists will no doubt be able to come up with even better common-vocabulary solutions to the problem of simplification. It is worth emphasizing, however, that many of these writers, in their full discussions, refer to susceptibility and predisposition rather than causation. It is often the need to correct the shorthand, both in headlines and core texts, that currently presents the greatest problem, and the same problem exists in the talk of scientists. In my observations of their public and scientific discussions, geneticists are at least as likely to use misleading shorthand phrases as journalists are.

This is not a matter of political correctness. It is a matter of getting the story straight. Getting the story straight also means more than avoiding false or misleading wording. It requires choosing what elements are needed for a full and accurate story. Key elements need more attention in reports of those studies about the linkage of genes to human behaviors. The press has reported studies purporting to link genetic factors to alcoholism, homosexuality, and mental disorders such as manic-depression and schizophrenia, among others. Critics have accurately pointed out that many of these studies were reported in the press prematurely, that the findings have often been retracted, and that the retractions come with far less attention than the linkages (Conrad, 1997). It is difficult to redress the structural forces that encourage newspapers to try to break news stories first and to be reluctant to emphasize "negative" news (such as disconfirmations). But whether or not the scientific studies are presented prematurely, they need to be presented more accurately. Two features of such studies need to be carefully explored in every popular press report about them to avoid inaccuracy.

First, press reports need to avoid misrepresenting the concept of heritability. Measures of heritability do not compare the relative inputs of genetics and environment. Measures of heritability tell us that genes have some greater or lesser degree of influence on different traits, within a specified set of environments and within a specified group. Thus, even if alcoholism is shown to have a moderate heritability in a particular population in a particular environment, there would be no alcoholism within a society where there is no alcohol. Thus, all that a measure of heritability tells us is whether, within a given environment, genes make some difference. Such measures do not tell us that genes make more or less difference than the environment, partly because the contributions of genes and the environment are not simply additive factors that together always equal 100 percent of the variance. Interaction effects between environments and genes are buried within the studies and incorporated into the major variables on both sides. Measures of heritability and variance thus give us a very restricted kind of information, not the global measure of "genes versus environment" for which it is often mistaken. Those who want direct comparisons of gene and environment need studies of "norms of reaction" (see E. O. Wilson 1998: 139–43), which are generally unethical in human populations.

The only studies which can be said to have made efforts to compare directly the components of genetics and environment are twin studies. However, these studies rarely have included significant variances in environment. Since environments are enormously complex, and we know almost nothing about what the environmental boundaries of various conditions might be, we really have only the most tenuous grasp of the comparative influence of any genetic cofactor. This is probably one major reason that many studies of genetics and behavior have not been successfully replicated and that the reported variances range wildly. What such twin studies tell us, therefore, is that when the environment varies only mildly—to some imprecisely measured extent—there is a moderate correlation of some behavioral conditions between twins. That does not tell us much about what happens with those traits when we vary environments dramatically. Raise one child in a monastery, another around lead paint chips, another in the home of a successful chemist who reads to the child at an early age and sends the child to Stanford, and environment might well dramatically affect the outcomes generated in the presence of predisposing alleles for behavioral components such as alcoholism or homosexuality or mental components such as SAT scores. In other words, current studies do not tell us that genes account for 30 percent or 60 percent of the variance and that unspecified environmental factors account for the rest. Instead, they tell us only that when environment remains roughly constant or ranges over a moderate span, genes exert a moderate influence on individual characteristics. Such findings are not mere tautologies, for they tell us that genes do exert some influence, but they do not tell us anything significant about the relative influence of genes and environment or about their interesting interactions.[6]

These arguments for more moderate portrayals of the role of genetics in human characteristics might be read as implying that genetic studies are socially insignificant. Why, then, should we bother to fund, report, and read this research? When all is said and done, I think it is a matter of healthy curiosity. For many of us, it is simply fascinating to try to unravel the many forces that participate in our being. It is unfortunate that scientists and journalists have come to believe that these understandings are of value only if they reshape our social policy or education options or drug treatments. This belief that knowledge is valuable only to the extent that it has practical payoffs contaminates the critical literature as well. Such assumptions, however, lead us to exaggerate, misstate, and misrepresent findings about genetics, because it is only if genetic factors are simple, uniform, and overwhelming that they will lead to obvious and definitive medical and social treatments. But genetics is interesting even if it does not lead to an Archimedean lever by which we can uplift the human world. Let us simply perform and report these studies in wonder at our world, with the goal of knowing ourselves. If we are a society rich enough to spend billions of dollars on baseball, hockey, and wrestling, we are wealthy enough to enrich ourselves with wonder at our molecular being. No need to claim that baseball cures cancer. No need to claim that understanding genetics will eliminate all the world's ills. Perhaps such an attitude will give us the space to report the role of genes in a vocabulary that recognizes genes as influential, but not deterministic, in their functioning.

A MORE EGALITARIAN GENETIC DISCOURSE

As it is with humanity in general, so it is with social groups and individuals. The future possibilities for any individual at any given moment are a product of the ways in which his or her biology, environment, and culture have interacted in the past. This dynamic perspective is important, because it is the fulcrum that permits the formulation of a revised conception of political equality, one compatible with our understanding of ourselves as a biocultural species.

Political equality has often been conceptualized as uniformity, similarity, or identity. To say that all people are created equal has been associated with the notion that all people are the same—that they have the same potentials and abilities. But this older conception no longer holds sway. The notion of identity has been attacked from the left by both culturalists and feminists for its inadequate recognition of the differences among peoples. It has long been attacked from the right because it effaces biological differences among individuals or groups. Today we are increasingly comfortable with the notion that persons have different potentials, abilities, and preferences, and therefore it is less threatening to consider the possibility that some component of these abilities is inborn or biologically based. Our notion of equality today is becoming one drenched in differences, not simply in difference. My personal history has repeatedly driven home to me the acceptability of a difference-based equality.

At the age of 20, after torturing numerous music instructors, I finally conceded that no matter what training or nutrition I might have received, I would never be and could never have been a great musician. This admission came with difficulty to me, for I loved music, and I never lacked for self-confidence in any endeavor. I could not, however, overlook the fact that my next-door neighbor, who lived in a home less supportive of music than my own, was a natural musician who played beautifully with ease. No matter how diligently I practiced, the music I produced was lame. On many a summer's evening, after I had struggled through my lesson, I could hear through the shared wall of our townhouse, my neighbor replaying my pieces beautifully, by ear.

All my life I have been able to hear the difference between bad music and good, but I could not produce good music. This does not mean that my neighbor and I were unequal as human beings. I was a better organizer and math student than he, and I was much better at two-square. My neighbor and I were not at all the same in any single area, but we were equal in our different potentials.

The grounding of equality in difference does not replace older bases of equality but adds to them. The older principles held that people deserve equal political rights because of a universal, shared basic dignity that we recognize inhering in human beings qua human beings. But in addition to that minimal bit of shared identity, today we also recognize each other as equal because we recognize that we each have enormous and rich potentials that are rich precisely because they are different from each other. The range and combinations of human abilities are dizzying (as any tour of your television menu will reveal). These differences require us to treat each other as equal in part because they are incommensurable with one another. Is a major league baseball player superior to a second-tier violinist? Is the best snow boarder in the world better than the best ice dancer? Does a good mathematician contribute more to society than a good nurse? At each moment of our lives, our future potentials are contingent. We cannot know of what value each of us will ultimately be to another, which talent will be needed, which production will be valued. Each of us may be superior to another person in one talent, but there is no way of adding up our worth or potential worth in all talents as compared with one another. Even a cumulation of mediocre abilities in the right combination can prove to be pivotal or at least immensely valuable (as Winston Churchill's life may perhaps suggest). Moreover, the presence of that diversity is itself an enormously valuable social resource, so that our differences one from another also deserve the protection of equal political valuation because of the value in their social aggregation.

The value of human diversity explains why equal protection and equal rights demand that we nurture the growth of every person. My neighbor, in spite of his prodigious talent, did not go on to be a musician. Neither music nor ambition was particularly valued in his lower-middle-class home. Thousands of children in America have genetic configurations that would allow for musical genius, but they do not have access to music when they are young, or they are exposed to lead paint, or they are not fed enough nutritious food, or their parents discourage excellence, or they follow some other talent and some potentials are not developed. Because genes enable so many different options for us, they are not the sole factor in individual human contributions to our social pool. The nation's media need to remind us constantly that what each of us is at any moment in time is a product not only of our vast genetic potentials, not only of our personal will, but also of the nurture provided in our family and the social structure in which our family, our schools, and our neighborhoods are embedded.

We are beings who are constantly changing through time. We do not ever have a finished identity, determined by any single input. We are all forever works in progress till the moment of our death. It matters that we get the right food and love at the right time. We are not simply thrown together; we develop. Our genes are unique because they are there from the beginning, but they are not static or all-powerful. They change in response to environment, and their meaning varies depending on the developmental sequences in which they participate. Moreover, because we are the symbol-using animal, the learning animal, the deterministic resolution of the genes is distinctly limited. Beethoven clearly had the genetic potential to write music that I could not write, but the First Violin Concerto is not a product of genetic programming. It is an invention, a dazzling creation, that is obviously the product of multiple social and biological factors, but also obviously a contingent marvel that cannot be reduced to those factors.

When we appreciate the full richness of the biological symphony, we will talk differently about the important parts played by genes in the music of life. The research that grounds this book has suggested that we have made important strides in that direction across the past century. However, such social processes are not simple products of inevitable social or biological forces. They are the result of a contest between (or at least the interaction among) multiple conflicting drives and agendas. To continue to develop a progressive cultural meaning for the gene will be enormously demanding as the applications from genetic science pile up. To spend our time denying the inputs of genes seems less fruitful than working to incorporate those findings within a new egalitarian understanding of humanity's potential. This requires, however, that we consider not only egalitarianism and nondiscrimination but also the specter of perfectionism.

PERFECTIONISM

The siren call to perfect ourselves is, as I suggested at the opening of this book, both a dangerous one and a source of human strength and creativity. The positive side of the use of genetic technologies to improve ourselves is obvious. Genetic technologies will reduce the incidence of serious illness or provide treatments for these illnesses.[7] When we go beyond this basic affirmation of improving ourselves through reducing illness, however, the picture rapidly gets murky. The use of genetic technologies for human perfection, at both the individual and social levels, will require our attention to a long series of social paradoxes.

A logical paradox arises from a strict logical opposition built into a set of statements. In contrast, a social paradox arises from competing social demands or interests, each of which would ideally be fully maximized, but each of which cannot be fully maximized because it is in some ways in opposition to another essential social value. Some of the most fundamental social paradoxes of genetics arise from the tension between the importance of placing a high value both on the individual and on the social.

With genetics, the connection between the individual and the social is an inherent one. A person's genes are his or her own, but they are also linked to the genes of everyone else. The choices one makes about one's own genes therefore always affect other members of the human species. Because of this as yet indissoluble tie between the social and the individual, genetic policy is dangerous whether it runs to the extreme of individualism or of social interest.

The dangers of excessive focus on the social were vividly illustrated in eugenics. But the dangers of excessive individualism have recently been given increased attention. If people are free to follow their individual whims, then children may be produced in a faddish manner that may not be good either for those children as individuals or for the larger society in which they live. While eliminating the baldness gene or the obesity gene from future generations may not prove disastrous, mass preference for a large size for males and a petite stature for females might have significant consequences, as might mass preference for any singular attribute at the expense of others. Moreover, given that individuals always live in a social structure, leaving decisions to individual choice means in substantial degree leaving the reproduction of the species to the interests of commercial entities whose primary interest is financial outcome. If this results in the production of relatively rigid biologically based castes, it is hard to believe that we would have anything near an ideal social system. Though one might have to concede to Adam Smith that such an outcome might arguably be maximally efficient and economically productive, we might not find those goals of the invisible hand to be sufficiently encompassing in this fundamental case.

A second layer of paradox, the impossibility of truly rational choice in the case of genetic self-manipulation, also has both social and individual dimensions. At the level of the individual the problem is a practical one. In order to make a fully autonomous and rational choice, one must have expert knowledge of genetics. Practically, it is impossible for all lay persons to have sufficient depth and breadth of knowledge in genetics to make fully autonomous and rational choices. Even now, genetic counselors complain of the difficulty of covering all the important information in an effective way in a mere hour or two of counseling. In the end, most would admit, they do not and cannot cover all the important information. They select what they and their client see as most important among many important issues. If commercial entities achieve their dream of multiplex testing, where prenatal scans include dozens or hundreds of potential genes, there will obviously be no way that the lay person can be said to have made an informed decision to choose those tests. Rational genetic choice for the individual will therefore become a practical impossibility.

Some experts are tempted to believe that their expert knowledge therefore puts them in the position of appropriate choice maker. After all, they reason, they know the scientific or medical information. The nature of choice in human affairs, however, is such that rational autonomous choice always requires a weighing of values and preferences, not merely a series of scientific equations. If medical or genetic experts act as choice makers, they must also employ values in their decisions. These values will arise either from their own personal perspectives, or from cost-benefit calculations by the institution in which they work, or from their perceptions of dominant social values. In other words, at the least, they will be imposing other people's values on the person receiving the genetic test. At the worst, they will simply be practicing eugenics—controlling the reproduction and genetics of an individual to fulfill the interests and values of the society. While some few persons see that as laudable, most of us do not. Moreover, such a value set clearly sacrifices the value of the individual for the goals of someone's perceptions of the society. It does not escape the paradox.

Several ameliorative measures have been proposed to balance the tensions in this paradox. The most widely endorsed is a broadly expressed interest in improving genetic education. Such initiatives must, of course, be weighed against other social values and other possible educational endeavors, but in any case the best that such education can be expected to do for the general public is to improve their ability to understand good genetic counseling. One cannot teach the details of every possible genetic test to every person. Genetic counselors must provide these details. They can do so more effectively when people have a basic understanding of genetic interactions. However, even then, genetic counselors will always continue to face the difficulties raised by the role of information bearer. As much as genetic counselors strive to be client-centered, there is an essential tension between the body of knowledge about genetic tests and their social consequences and the uninformed desires of the client (which must be respected). It is to be expected that new expert information might alter those desires, or there is relatively little point to the counseling. However, there is a need to provide information that might alter the choices of clients in a way that respects the client's autonomy. Counselors can only manage these tensions in better and worse fashions. They will never be able to overcome them.[8] Thus, in a genetic era, we will live with the paradoxes of choice at the individual level, even with better education in genetics.

A further version of the paradox of choice manifests itself with regard to the poles of individual autonomy and cultural diversity. Genetic science has been developed primarily within Eurocentric science, where the value of individual autonomy has been highly prioritized. Genetic applications have been portrayed as desirable to the public that funded them in terms of individual benefits. The justifications for genetic choice have been based on its ability to improve the well-being of individuals. Moreover, all of this has occurred within the framework of positive valuation of reproductive autonomy. In the forties, genetic science and technology sold itself to the American public in large measure as an opposite of eugenics. Genetic technologies would enhance individual choices, not serve the interests of a collective such as state, race, or nation.

As genetic technologies expand, however, they are applied in an increasingly multicultural context. Women from cultures where females are subordinate are now subject to genetic technologies controlled by their husbands or families. Young people from cultures in which religious leaders control major aspects of life are now subject to genetic technologies controlled by their fathers or religious figures. Genetic technologies thus partake of the broad tension between respect for cultural diversity and the value of the rights of the individual.

My own belief is that the rights of the individual outweigh the preferences of the culture unless other individuals suffer in the process or unless immediate and substantial harm comes to the community. However, I recognize that I am influenced by my own cultural preferences in supporting such a principle. I would argue, however, that use of genetic technologies outside this framework violates the pact with which they were developed. Transfer of this technology outside its own original cultural context transforms its character in ways that its originators cannot feel are benevolent, and so they are obliged to try to restrict such transformations of the motif. They are, however, also limited by respect for the boundaries of other cultures. We can try, therefore, to ensure individual autonomy within the boundaries of our own nation, but we are severely limited in the tools for doing so both here and, even more to the point, abroad. We live with this rather damning paradox as well.

There are even more profound dimensions of the paradox of choice at the social level. The project of genetics—to remake ourselves—appears at first sight to be the culmination of the Enlightenment's values and goals. At the same time, genetics demonstrates the limits of that project. The goal of using scientific knowledge to make ourselves into a better species (or to make our children better children than those nature or chance might provide) would seem to be the ultimate achievement of rationality and science. Not only can we remake nature around us to suit our needs and values, but we can also remake ourselves to suit our needs and values. That shift in focus from the world outside us to our own selves, however, harbors a crucial problem for value selection. When we choose to make other entities over to suit our needs, our needs are a relatively stable criterion, and we may debate what those needs truly are, but the debate is internal to our being. When, however, we choose to make our selves into something else, we cannot be saying that our new selves will meet the needs of our old selves, for we have denied the sufficiency of our old selves. We have made the needs of the old selves secondary to the needs of new selves. But how can our old selves, as inferior beings, possibly know what our new selves need? Any fundamental change in our species' being, therefore, cannot be a matter of rational choice. It can only be a matter of emotive or valuative desire.

This paradox is true about any effort to remake our species and our selves, but it becomes radically debilitating only when we reach for relatively fundamental changes. Consequently, the more effective genetic choice is imagined to be, the less rational are the grounds for making these choices. The conditions of the paradox do not mean that we cannot make choices, and the existence of this paradox certainly does not mean that we will not make choices. It does mean that the more fundamental our choices, the more fundamentally they depend on rationally unjustifiable whim. Even at less fundamental levels, however, the existence of this paradox dictates the need to pay close attention to the complexities of cost and benefit in genetic manipulations of humans.

Because of the complexity we are discovering within the living world, there appears to be something like a biological version of the Heisenberg uncertainty principle. Biological systems are dynamic and interactive, and the interactions aggregate through time. Biological systems are thus inventive. Therefore, interventions in the biological system are inventions within a system which is already inventing. The outcomes of interventions are therefore difficult to control. In its applied form, this principle goes something like this: it is a lot easier to understand how biological systems work than it is to redirect them to specific alternative outcomes. Consequently, just because it becomes increasingly easy to intervene in biological systems does not mean that it will be proportionately easy to control the outcomes of those interventions. This does not mean that some measure of control will not be developed. Good bets can be placed on our eventually overcoming what appear now to be insurmountable obstacles to gene therapy. However, the good bet also holds that they will not be overcome by a single magic bullet, but rather in fits and starts with techniques adapted to particular conditions and organs. Certainly the story of the treatment of cancer reinforces this likelihood. Cancer was at one time envisioned as a single disease (though manifesting itself in different organs). "The cure" for cancer was spoken of routinely. Today, few people expect that we will discover a single cure for all cancers in all patients. Instead, we are discovering multiple approaches for affecting the disease and matching those approaches to the very distinctive biologies of different patients. It seems likely that a similar variegated approach will be employed for gene therapy. Ultimately, gene "therapy" may turn out to be much more like the personalized counseling provided in psychological therapy than like "genetic engineering" with the uniform mathematical grids once imagined for it.

There are attendant significant consequences for the image of remaking humans wholesale. Clearly, such a remaking would not be a predictable, rational, controlled exercise, with products whose quality and character we could guarantee. Some may still wish to advocate the creation of transhuman species, but they should be held accountable for advocating an open gamble with unknown odds and an unknown prize, rather than permitted to pretend that they offer clearly controlled and accurately described outcomes. The same is equally true for genetic selection and genetic engineering of children. It seems increasingly unlikely that geneticists will ever be able to offer anything like a designer child. Interventions into the genome will always be subject to unpredictable outcomes. Even if we were to isolate a gene or set of genes that increases mental processing speed, inserting that genetic configuration into an embryo would not have reliable effects. At best, the insertion would offer a reasonable probability of increased mental processing speed, for genetics is always a matter of probabilities rather than a matter of certain lawlike behavior. Such an intervention will also entail all kinds of other potentialities and impacts. Genetic medicine will not be able to escape the possibility of unwanted "side effects." This eventuality is a far cry from the "shopping cart" image promoted when critics and advocates speak of genetic selection of children. Even under very advanced genetic selection procedures, children will remain the gamble that they are today.

Philosophers who write about so-called *genetic enhancement* thus mislead the public. Perhaps enhancement will bring about a probablistically predictable result, but it is also true that each enhancement also is always likely to produce decreases in qualities that were not specified in the shopping list. *Genetic specification* might be a more realistic term than *genetic enhancement,* but even that fails to capture the fact that genetic selection is in the best case scenario only partial and probabilistic. Genetic selection will probably also be genetic roulette in important ways. What can we call it that captures this sense of risk and cost-loss trade-offs? *Genetic biasing* seems to me most technically and socially accurate, but it certainly lacks the favorable connotations preferred by those who want to apply the technologies.

Some term, however, needs to capture the fact that genetics, like any other technology, features the paradox of cost and benefit. Modern life is longer than life in earlier eras, but this is a life of sustained psychological stress and extended death. Many of us judge these costs to be worth the benefits, but it is of importance to keep in mind the costs. When philosophers such as David Resnik (1994) and William Gardner (1995) argue in favor of genetic enhancement, they stipulate that genetic enhancement will be desirable at such a time as it becomes technologically viable and economically and socially feasible.

They conclude that under such conditions there are no reasons not to practice genetic enhancement. But to frame the issue in such a fashion is to suggest that we should each climb Mt. Everest if only it weren't enormously cold, risky, and physically demanding to do so. The difficult ethical question about genetic enhancement is not whether or not to practice it if there are no problems with it. The real, and difficult, ethical questions must include asking, What particular costs are we willing to bear in order to achieve what kind of limited genetic perfection? Our discourse about genetics must become faithfully aware of both the costs and benefits of genetic choice. This does not mean that we let any possible cost paralyze us. It does mean that we need to include costs in our genetic equations. The fact that technological controls always exact their costs cannot be avoided in genetics, any more than it can be avoided in any other fields.

We are now beyond the phase of youthful enthusiasm with regard to genetics. Genetic engineering of many kinds is a reality in our lives, and it will most likely be an increasing reality throughout the twenty-first century. The public discourse that gives social placement to that reality and gives it particular meanings in our lives will no doubt also be of increasing importance. I hope this study of the discourse about the gene has helped to make it clear that the fact of biological heredity has not just one social meaning but several potential meanings. Our goal should be to think together about the implications of these ranges of meanings and to promote those conceptions of the gene that best represent the complexity and the positive potentials of the human condition, both biological and social.

A key component of the required attitude is to remember that controlling ourselves always comes at a cost and involves the paradoxical character of being human. We are glorious, seeking creatures, with grand goals and high ideals. We recognize ourselves, measured against such goals and ideals, to be imperfect creatures. We wish to be more generous, more mathematically able, more musical, more athletic—less like brutes and more like gods, or demigods at least. Yet as noble as our aspirations for shedding our failings might be, our history also suggests that, being flawed as we are, we can never blindly trust our own aspirations to reshape ourselves.

APPENDIXES

NOTES

REFERENCES

INDEX

APPENDIX 1

THEORY AND METHODOLOGY

The project reported in this volume has sought to describe the struggle over the placement of genetics within U.S. public discourse in the twentieth century. Descriptions of this sort are always the product of particular theoretical and methodological assumptions and practices. Because the practices that informed this project are a distinctive blend of several theoretical traditions and methodological camps, this section offers a description and brief justification for the choices I have made. The section begins by labeling this form of analysis as the study of a rhetorical formation, which is a particular approach to social-scale discourse studies that operate from a detotalized version of theory. It traces the inspiration for the study of rhetorical formations to a combination of the agonistic rhetorical tradition and to what can now be called the Foucauldian paradigm. Next, this appendix describes the methodological choices made in the study, including the combination of critical and quantitative methods. It provides details about the coding procedures and the data base employed and concludes with some thoughts about the relationships among the various media used in the study.

Theoretical Framework: Rhetorical Formations

Two major theoretical shifts in the study of discursive history have informed this project and necessitated the blending of critical and quantitative methodologies distinctive to it. The first shift has been the movement away from a focus on individuals toward larger-scale phenomena. In historical studies, the lives of "great men" and the actions they have taken are no longer seen as the primary fulcrums for the major shifts in history. Instead, a variety of social phenomena—economics, movements in human consciousness, demographic shifts, and so on—have come to be central modes of understanding. In rhetorical studies, the singular speeches of great orators such as Abraham Lincoln, Winston Churchill, or Frederick Douglass, while still appreciated as great art and as significant actions, are now understood not as unique instigations of world trends but as focal nodes in a larger torrent of human discourse.

Accompanying the shift away from singular sources has been a second trend toward the general detotalizing of theory. No longer do good theorists expect the world to act in exactly the same fashion in every instance. Instead, the flow of human events is understood as featuring singularities and particularities. Rather than seeking a single theory that can account for all human historical phenomena, scholars increasingly assume that historical events, in this case discursive processes, can flow in a variety of different patterns. In their particularities, these potential models of change through human history are infinite; in their major outlines they are at least multiple. Instead of searching for universals, social theorists now seek patterns that may be unique to a particular historical occurrence or that may at most represent recurring motifs which may exist in other social situations but which do not necessarily occur in every instance.[1] The search for a combined version of the particular and the general, that is, the study of motifs that are available but not obligatory, has now supplemented the search for the universal or the merely particular.

The effort to combine a focus on multiple texts with such a detotalized theory resulted in the conceptualization of this project as the study of rhetorical formations. The theory of a rhetorical formation clearly borrows from the concept of a discursive formation, a term which has substantial currency in mass media studies influenced by cultural studies. However, discursive formations are usually studied within fictional entertainment products. In overtly political discourse, the talk is more directly agonistic. Public discourse overtly addresses public policy and thus is more inclined to argue explicitly for and against particular propositions. While entertainment products clearly have political dimensions and political influence, they code these preferences more indirectly, especially with regard to particular issues of governance (i.e., law and national policy). I therefore found the rhetorical model, with its history of attention to public policy debate, to be an essential corrective to the model offered by studies of discursive formations. The need for a rhetorical corrective to the study of discursive formations can best be explained by a return to Michel Foucault.

Foucault's work presumed that the official, public, and scientific discourses of an era all participate in a "discursive formation" that has key markers that distinguish it from other eras. Foucault began by describing the underlying driving force of the unity of various eras as an episteme. An episteme was different from a logical or scientific paradigm. It had no validity outside a culture. Foucault was, in this way, attempting to escape the metaphysical visions of history offered by the elite traditions of intellectual thought in the West. However, the concept of the episteme still bore marks of that Western tradition. An episteme was still a unified, coherent body of discourse/thought that shaped all the major discourses of a particular era. Foucault later tried to loosen the ties to Western philosophy and Hegelian versions of history further by transmuting his notion of archaeology into a theory of genealogy. Yet, even in his genealogical studies, Foucault still conceived of eras as guided by singular overarching (or undergirding) movements of discourse that shaped human practices at all levels. Thus, for Foucault the rise of prisons in the nineteenth century devolved from the logic of discipline enabled by panoptic practices. Foucault traced the practices of armies, prisons, and schools as the products of a singular unified world view, all guided by the same logic (though he tried to efface the notion that this logic was rational and deductive by renaming it discourse). Even later in his career, Foucault's explanations tended toward unified, singular accounts that still carried a strong odor of the Western intellectual tradition's focus on unified principle; for example, he accounted for male Greek life as the product of a theory of self-care. Foucault's work has been undeniably fruitful, and his methodology has opened up new vistas. However, both his examples and his theoretical statements encourage focus on singular, unifying principles. These principles may not be quite philosophies, but Foucault's version of discourse ran to a preference for unity and orderliness that derived from the Western intellectual tradition's preference for unified, principled accounts.

To supplement such an approach, the rhetorical tradition is useful. From its origin, rhetoricians have assumed that the persuasiveness of discourses was built from multiple factors: topics, metaphors, and enthymemes all were orchestrated together at a particular moment to produce particular results in particular audiences. The work in social theory of the last half century has shown that classical rhetoric was insufficient because it did not at all address large-scale social forces. Since the particularities of the moment were always also embedded in large-scale social forces, classical rhetorical approaches were not sufficient even for accounting fully for the particular moment. In the past two or three decades, however, a series of scholars have attempted to meld classical rhetorical theory with the findings of contemporary social theory.

The leading theorist in the area has been Michael Calvin McGee, but critical work in this direction has been conducted by a variety of scholars, most of whom have been studying the rhetoric of the conflicts over gender and race that were at a peak of intensity in the sixties and seventies. Scholars such as Solomon (1979), Vanderford (1989), Burgess (1968), Scott and Smith (1969), and Railsback (1984) have been joined by others working on particular methodological issues (Osborne, 1967, on metaphor) and related social issues (Aune, 1994, on class) to generate a model of discourse study which, I think, is growing into a unique method and theoretical framework that can be labeled the analysis of rhetorical formations.

Studies of rhetorical formations feature three distinctive elements: they attend to conjunctures of multiple rhetorical components rather than to a singular unifying principle (such as care of the self or discipline), they view discourse processes as agonistic, and they focus on change and range.

Although not incompatible with an assumption that there are underlying logics to eras, the rhetorical approach examines the particular discursive structures—metaphors, ideographs, culture types, narratives, and so on—that get offered in discussions about specific issues within a specific era. Rather than assuming that an account of the "episteme" is sufficient for understanding practices, the rhetorical approach assumes that the particular incarnations of epistemes with respect to particular social practices are important for understanding the specificity and materiality of different practices. Both genetic counseling and automobile ownership might be said to be shaped by the episteme of our age—be that identified in terms of the era's individualism, its consumer orientation, or its technological dependence; however, one does not know enough about either genetic counseling or automobile ownership as social practices, or about the ways in which they came to be social practices with particular features, if one's analysis operates at only the more abstract level of the episteme. This is not to say that explorations of discursive formations do not repay their efforts, but merely that there are other important things to be learned by turning one's attention to the rhetorical level of discourse.

Rhetorics operate through particular formal devices—metaphors, maxims, ideographs, narratives, and so on. These devices constitute focal points for research because they drive particular formulations of social practices and define the experiences of those practices. These rhetorical devices are not independent of economic forces or spatial demographics or of any of the multiple other material inputs to a social system, but rhetorical devices are important in their own right. Experiences are always mediated through our language about them. Multiple forces shape a practice, but the experience of the practice is always defined through rhetoric about it. On this view, it matters whether one thinks of a genetic practice as "stock breeding" or whether one thinks of it as "recoding" one's genes. It is not enough to think of both metaphors as merely different versions of the same genetic paradigm. Similarly, on this account, it is not enough to focus on metaphors, because no single discursive component defines the discourse of an era. The potential meaning of the coding metaphor was severely constrained by the dominant model of gene-environment interactions. And the metaphor's meaning changed significantly when it was displaced to a commercial era and accompanied by the blueprint metaphor and revised models of gene-environment interactions.

Therefore, to study a rhetoric is not to seek for a single unifying principle. It is to seek for the important discursive units that recur in discussions of an era and to explore their interactions. Classical rhetoric offers an extensive catalogue of these devices, chief among which are metaphors, topoi, and enthymemes. Supplementing the classical list are more recently developed rhetorical components including especially narratives, ideographs, and images. In this study, preliminary readings suggested that metaphors, key words (labels for desired children), and particular special topics (such as the relationship between gene and environment) were of greatest significance, so these are given central attention in the study.

In addition to being formed by multiple related but independent components, rhetorical formations are always contested; unlike Foucault's smoothly operating panopticon, discourses about genetics have long been debated and contentious. Although particular visions may become hegemonic, they rarely become fully dominant, and the particular hegemonic forms they take are shaped as much by the contest among different visions as they are by any singular force. This agonistic conception, built into the rhetorical perspective, is based on a fundamental assumption about the nature of language. For Foucault, language is not persuasive; it commands. Foucault's children and prisoners never appear to question the mechanisms of discipline that are placed upon them. Language orders them. In contrast, rhetorical study presumes that language does not give commands to which human beings respond as automatons. Instead, language seduces, it offers, it persuades (Brockriede, 1972; Foss and Griffin, 1995). Often, language is so seductive, so naturalized, that people act in accord with it as though they were automatons, but they do not always do so, especially when it is blatantly not in their interest. Children routinely rebel against discipline, overtly and covertly. Prisoners constantly rebel against both prison and society, overtly and covertly. Even when the response to language is not rebellion, it may be active rather than passive. One can choose actively to assent to a discourse. More often, perhaps, one disarticulates a discourse, incorporating some fragments and rejecting others (McGee, 1998). Every act of speech, every production of discourse, is therefore an agonistic act. Even when a singular discourse dominates the mass media, this does not justify viewing that discourse as an uncontestable command. Instead, that discourse is still a seduction or a plea or an offering. Most of the time, however, the discourse in the public arena is faced by both external and internal contradictions. Rhetoricians study these challenges and contradictions rather than focusing solely on the uniformities.

This choice gains added force to the extent that we understand that discourse exists in time (Leff, 1988). From a rhetorical perspective, each moment is a moment of change. Each set of words is a request for assent instead of counterargument or rejection. Therefore, there is never a stable dominant ideology. The ruling classes are only as powerful as their last and next seduction. This does not, of course, mean that ruling classes have no power. It does mean that they must constantly expend power to maintain dominance. Given the multiplicity of forces operating in a society, this means that ruling classes are themselves faced with constant instability and the need to reformulate themselves constantly.

Foucault's work emphasized what he called discontinuity. But Foucault's version of discontinuity was a series of stable, uniform blocks separated by radical, unexplainable breaks. Foucault was reacting against two alternative images: the Platonic vision of a static universe in which no fundamental change ever happens, and the Hegelian universe in which change happens, but it represents a smooth progression of a unified spirit. Foucault's model of punctuated equilibria provided a productive alternative to these two prominent world views. There is, however, a fourth alternative, the Heraclitean vision of a world that has no unified order and which is constantly changing. The Heraclitean world, unlike that of Foucault, does not feature blocks of stagnant time in which Platonic miniworlds exist.

The Heraclitean world is extremely difficult to write about and even more difficult to study. There is, however, a productive cross that can be made between Foucault and Heraclitus. Foucault's vision offered no way to study the process of change. Radical change from one episteme to another simply happened. Foucault was not primarily interested in the process of change. If we adopt Heraclitus's focus on change and a modified version of Foucault's model of punctuated equilibrium (modified to recognize the multiple discursive units and challenging forms within discourse), we have a way of studying the flow of discourse through time that recognizes periods of temporary stasis (another key term from classical rhetoric). To see a period as featuring stasis rather than stability emphasizes the agonistic features of each moment of history. In periods of apparent stability (that is, where forms are reliably repeated), one is not viewing tranquillity or dominant ideologies firmly entrenched in everyone's mind. If periods were to feature true stability, then change out of such periods would not occur. Instead, at times of social stasis, one is viewing something like a balance of forces. Each side is actively seeking to seduce the public to its world view, but each seduction is countered by the opposition. Change happens when new seductions are invented by one party or the other or when changing conditions make one type of seduction more appealing than another for large or powerful constituencies. In the case of eugenics, the very success of the paradigm brought about its failure when its enactment in Nazi Germany gave greater presence and credibility to the negative features that its opponents had been asserting (with mixed success in the United States). The noun *rhetorical formation* should thus always be understood as carrying the connotations of the verb *form*. Public rhetorics are always in formation; they are never finally formed.

Each of these elements—multiple components, agonistic qualities, and change—should lead us to understand an important quality of a rhetorical formation as being the relative degree to which it features particular characteristics. If we are trying to be sensitive to change, then we need to know whether something is increasing or decreasing. Aristotle listed the topoi of "greater or lesser" as a central topic of rhetorical investigation. This topic stands in fairly stark contrast with the currently privileged topoi of presence/absence or identity/nonidentity. Western philosophy has continually asked *is* questions (or denied them, as in the case of Derrida); rhetoric instead suggests questions of degree (among others). Thus, unlike the Western intellectual tradition, rhetoricians would not ask whether, as a species, we are essentially social or selfish; survival demands both. The better question is, To what degree do we require each? I urge consideration of the possibility that a crucial failing of our times is the failure to ask questions of degree rather than questions of presence/absence and identity/nonidentity. Instead of seeking to find whether *a* is *b*, we might spend more time asking a question such as, How much of *a* is *b*? This approach leads to questions about ranges rather than questions about which components are dominant. In the social sciences, this means greater focus on standard deviations and cluster analysis and factor analysis than on means and t-tests. In the critical humanities it means asking questions like, What range of alternatives was available to articulate this practice? and How frequently were different articulations present? instead of asking, Which articulation dominated?

In this book, I have operationalized this last feature most imperfectly. In the quantitative section, I have relied on means, chi-squares, and t-tests instead of any alternatives. I am new to the world of statistics, and my access to alternatives is too limited personally, though I believe we are also too limited professionally and have not developed sufficient statistical tools for focusing on range. Even in the critical area my efforts at portraying the range of discourses have for the most part been swamped by other factors. Where possible I have tried to comment on the shifting ranges of determinism and anti-discriminatory discourse, but in retrospect I wish those sections had more development. I came to understand the importance of range too late in this project to do it justice. I hope that this will be taken as one of the crucial findings of this study and one of its implications for future research. For I think that the issues raised by genetics are centered around questions of degree. We cannot escape altogether questions about what we want for our children, or what it means to be a creature tied to other beings by heredity. It matters greatly, therefore, the degrees to which we are willing to go to reshape our children and to see the links between ourselves and our genes.

Methodology

The search for rhetorical formations on the large social scale makes advantageous the combination of critical and quantitative methodologies. Crits and quants have long squabbled at each other, each deriving their claim to superior power of explanation by highlighting the limitations of the other's tools. As discourse studies moved from single speeches to large masses of discourse, however, this bifurcation has constituted a serious methodological problem. Some studies in political science, sociology, and journalism (often going under the label *content analysis*) have sought to provide quantitative analyses of social change in discourse. From the point of view of rhetoric and communication studies, these have generally been ignored as insufficient because they have not been grounded in theories of discourse and they have not addressed appropriate units of discourse, especially narratives, metaphors, and other complex discursive forms (instead, these quantitative studies have usually focused on single words, "values," or themes).

Rhetoricians have tried to provide critical studies of these large masses of discourse. In fact, I would argue that this has been the distinctive characteristic of critical studies in the post-Black era of rhetorical criticism. However, these studies face the limits of the human agent as synthesizer of massive quantities of texts, and these are serious limitations. While exceptional critics in clear instances can overcome these limits to produce excellent studies of the flow of large-scale discourse, all critics can be assisted by quantitative methodologies when those methodologies are understood as counting tools, embedded in the critical project, rather than as overarching frameworks that constrain critical thought within a hypothesis-testing method. It is possible to use numbers in a postpositivistic fashion.

Such combinations of critical and quantitative work are rarely attempted, both because they are difficult to perform (requiring multiple methodological skills on the part of those who undertake them) and because they are rarely viewed favorably (since both quantitative and critical scholars are generally intolerant of each other, having much at stake in defending the exclusive value of their own methods and outputs). In spite of these unfavorable winds, this dual methodology was employed here because contributions from each approach were essential to achieving the type of balanced and critical assessment appropriate to an issue so momentous as the future of humankind's heredity. On the one hand, it is impossible to make reliable claims about the relative proportions of various forms of discourse in different periods without using quantitative measures. Previous analyses of public genetic discourse that have not employed quantitative methodologies have done little more than collect examples of one type of discourse and proclaim these dominant, without any attention to their role in the balance of discourse and debate (e.g., Hubbard and Wald, 1993; Nelkin and Lindee, 1995).[2] These studies have thereby provided important critical promptings, but they have also displayed the inherent weaknesses of critical readings that would claim to settle issues of dominance and proportion. Perception biases cloud the possibility of such assessments.

On the other hand, quantitative methodologies on their own are also insufficient to this task. Placed as they usually are within logical positivist frameworks, they discourage speculation, critical interpretation, and the linking of disparate correlated factors of different levels of abstraction. Traditionally, also, quantitative methodologies are dependent on qualitative precursors to determine what factors are significant enough to warrant quantitative assessment. For these reasons, the results in this study are based on both critical and quantitative assessments of a common, systematic data base of public discourse about genetics.

THE DATA BASE

As a rhetorician, I am interested in public policy, and I have therefore framed this study in terms of public discourse. It is for this reason also that I have bounded this study within the United States. However important multinationalism is becoming, and however much nation-states may dissolve in the future, during the twentieth century public policy was still made by nation-states, so that attention to these political boundaries in descriptions of discursive formations remains useful.

Public discourse about heredity in the United States has been incessant and widely dispersed, and it has arisen from multiple voices. Most of such discourse goes unrecorded, so that our assessments of the contents and meaning of public discourse can rest upon only the incomplete and inevitably biased records provided by archives of popular magazines, newspapers, television programs (where preserved), the *Congressional Record,* books, and so on. Fortunately, good arguments can be made that the discourse which has been recorded is probably also that which has been the most influential in public law and in the establishment of social structures and that which represents a substantial and important component of the public discourse on a topic (though it does not represent proportionately and accurately the opinions of the citizens and inhabitants of the nation). At the least, such recorded public discourse represents the bases for the legitimation of particular customs and public policies, regardless of whether it represents the true purposes, preferences, and intents of those who spoke it. Moreover, though the representations in these records are biased toward the most powerful groups and away from the least powerful voices, the discourse to which we have access from history is enormous, consequential, and indicative of much of the populace. Further, that discourse generally contains some signs of the preferences of less powerful groups, and critical methodologies can ferret out and attend to the concerns and preferences of these less powerful voices.

The disjuncture presumed here between the will of the people and the public is intentional. I presume that in the U.S. system of government, the people are not isomorphic with the public. All members of the population (i.e., the people as individuals) have some theoretical access to debates about public policy. However, in practice, those who are relatively literate have greater access to public arenas (to some degree as audience and to a greater degree as speaker) than those who are less literate. Similarly, those who are wealthier have more access than those who are less wealthy. Those who are organized into interest groups have more access than isolated individuals. There are many other variables that matter, but the point is that access to voice in the public arena is differentially distributed among the populace.

There is a difference, then, between the populace and the public. The public, however, is an abstraction. There is no single place or publication that one can identify as the public. In the twentieth century, the mass media (probably unfortunately) serve as stand-ins for the public. This was somewhat true in the nineteenth century, but in that era the speeches of politicians were more directly and frequently printed in newspapers, and stump speeches appear to have been more widely attended. News media were more likely to convey large quotations or descriptions of speeches by politicians than they are today. Today, reporters have stepped in between politicians and the populace. Thus, to a significant extent, journalists have become the voice of the public. Contrary to the claims of many, congressional debates are still of some importance (though less than before and less than ideal, perhaps). However, the public voice is probably best located today by reading newspapers and magazines and watching television news and documentaries. This does not mean that public policy is always and only responsive to this public voice. Much legislation is indeed gerrymandered in smoke-filled rooms. However, legislation cannot be gerrymandered in such a fashion in direct opposition to a strong public voice, at least not for an extended period.

This leaves the question of the role of fictional discourse in public discourse. There can be little doubt that fictional discourse plays a substantial part in forming the public dialogue, and recent books on genetics have been particularly interested in fictional products (Pernick, 1996; Turney, 1998; van Dijck, 1998). The heavy influence of genetics on science fiction makes this route particularly interesting and intriguing. However, there are enormous difficulties in interpreting the significance of fictional products. As Turney's study makes clear, it is difficult to know whether a particular trend toward a large number of negative science fiction stories about genetics means that genetics is unpopular or whether they are simply a sort of benign expression of repressed fears that is part and parcel of a generally positive public response.[3] Moreover, science fiction products often have a particular audience that is not isomorphic with either the populace at large or the public. Too much attention to these creative and intriguing products, therefore, can distort one's impression of the public consciousness about a subject.

If one were interested in the question of "origins," then one would need to brave these difficulties of interpretation attendant upon fictional products. However, since I am not trying to describe what books or writers originated what thoughts (in fact, such a focus is antithetical to my project), it is better to presume that to the extent that fictional products are influential, they will show up in the more general nonfiction public discourse. I have thus omitted any direct study of fictional discourse, just as I have omitted direct study of scientific discourse. While a full accounting of the history of genetics in this century will have to take all these factors into account, this study cannot aspire to be a full accounting. It is quite enough to present a comparative account of the themes of perfectionism, determinism, and discrimination in the most common public media.

There is substantial good reason, therefore, for taking as one's data base the magazines, newspapers, television programs, and *Congressional Record* if one wishes to provide an accounting of public discourse about heredity. However, neither critics nor quantitative scholars could feasibly provide a detailed and precise account of the entire body of recorded discourse on genetics. The volume of discourse is enormous, and it has not been indexed sufficiently to provide viable access to its full complement. Consequently, this study is based on structured random samples of indexed discourse in magazines, newspapers, the *Congressional Record,* and news programs. Randomization increases the chance that the researcher's biases are not guiding the contents of the study. A randomized sampling process increases the chances that what is located are the most significant discourses, because these themes will be the ones repeated sufficiently frequently to show up as a strong presence in the sample.

The sample of popular magazines provided the backbone for this study. This sample consisted of 653 magazine articles drawn from the *Reader's Guide to Periodical Literature* between 1919 and 1995. The *Guide* in this era was divided into five-year time blocks (pentades), and the headings "genetics," "gene," "heredity," "eugenics," "defectives," "sterilization" (and forms of these terms) were searched for all popular articles (articles in scientific journals and popular science journals were excluded, as were articles in periodicals from outside the United States). In pentades where there were fewer than 50 articles, all articles were included in the sample. In pentades where there were more than 50 articles, all articles were numbered, and 50 were selected by use of a random numbers table. Visual inspection of chi-square and t-tests comparing all individual year blocks to each other indicated that there were four major time periods, each having its own distinctive rhetorical profile. This account and the findings in appendix 2 are based on these four time periods; time period 1, the era of classical eugenics, ranges from 1919 to 1934 (n = 151); time period 2, the era of family genetics, ranges from 1940 to 1954 (n = 100); time period 3, the era of experimental genetics, ranges from 1960 to 1976 (n = 134); and time period 4, the era of medical genetics, ranges from 1980 to 1995 (n = 147) and probably beyond. Time blocks of approximately five years have been omitted between each of these periods because these years represented transitional periods where discourses of the closing and coming eras mingled (accounting for the additional number of articles in the total of 653). In the narrative account these periods have been defined more broadly on the basis of the article reading, so that the classical era ranges from 1900 to 1935; the era of family genetics, from 1940 to 1954; the era of genetic experimentation, from 1956 to 1976; and the era of genetic medicine, from 1980 to 1995. In most cases transitions were gradual, and these boundaries are merely convenient approximations.

The magazine sample forms the skeleton of the study because the magazine discourse is the only set of discourse for which a relatively comprehensive index exists across the full duration of this study. Without the use of such indexes, it is impossible to construct a data base that is not heavily influenced by direct researcher biases. Regrettably, the discourse of the *Congressional Record* before the nineties is very poorly indexed. Only a pittance of entries are indexed under the relevant headings. Television news programs are well indexed by the Vanderbilt Archive, but exist only for the last two time periods. Newspaper indexes before the watershed of on-line indexing technology in the late eighties are extremely spotty. Because of the importance of the newspapers, a careful sample was designed and studied both quantitatively and critically, and these results are incorporated into the study (although they are not included in the tables in appendix 2). However, the availability of indexes limits the breadth of that sample. Hence, the newspaper sample was designed not as an independent sample with internal validity, but as a means of comparison between newspapers and magazines. In periods 1 and 2, 50 articles were randomly selected from the *New York Times.* In period 3, 50 articles were randomly selected from the years 1940–1950 from indexes of the *Los Angeles Times,* the *Washington Post,* the *Chicago Tribune,* and the New Orleans *Times-Picayune.* These newspapers were selected because of their regional representation and because they had indexes at this period. In time period 4, a similar sample was collected in addition to 50 articles from the *New York Times*, but on-line reference bases were used. Unfortunately, because of an error by the student responsible, all these articles come from 1989 and 1990.[4]

The television news programs included in this study were selected by a random numbers procedure from the Vanderbilt Archive's on-line catalogue. In period 3, 21 news programs were coded. In period 4, 40 news programs were coded. Summaries of all these articles are available on line. The coding and critical analysis of these samples were supplemented by the reading of important books in various eras and by the consideration of secondary treatments of the subject, but the systematic comparisons, depth, and breadth of the sample give it primacy over other resources. The utility of that sample, of course, rests on the quality and character of the critical and quantitative assessment performed with it. Quantitative results are reported in appendix 2.

FROM CRITICAL METHODS TO QUANTITATIVE METHODS

This project began with a critical reading of the discourse that had been selected through the social scientific random sampling procedures. The focus of the reading was guided by my theoretical presuppositions (described above) and by the questions and challenges raised by critics of contemporary discourse about genetics (described in the introduction). My critical reading was in some ways at odds with claims of prior critics. While I certainly noticed examples of the kinds of discourse that they claimed dominated public discourse, I also noticed many other types of discourse components as well. I then formulated testable propositions about the relative dominance of various of these units—those mentioned by critics and the other units that I saw.

With the financial support provided by a grant from the Ethical, Legal, and Social Implications Branch of the Human Genome Project, I was able to pay three coders to count the frequencies of these components in the discourse across time. First, coders were trained, and the counting protocols I had developed were modified to make them intersubjectively verifiable. Some items were dropped because they did not generate intercoder agreement and because the instrument was too unwieldy. The original instrument had too many questions, and the coders could not successfully focus on all those issues. Intercoder reliability was then established by having all three coders read and count in nine different categories across approximately 20 percent of the magazine discourse (see appendix 3 for coding protocol). Intercoder reliability was established at 90 percent for degree of determinism, 85 percent for type of determinism, 86 percent for differentiation, 95 percent for preferences about children (perfectionism), 91 percent for chief purpose of article, 92 percent for types of discriminatory discourse, 94 percent for types of ethical issues raised, 93 percent for metaphor, and 80 percent for valence. The quantitative analyses were performed on data from one coder. For the 20 percent of the discourse where multiple coders were used, the data from the coder who had the highest consensus and quality (least likely to miss objectively present items) were used. Much of the disagreement among coders reflects the inevitability of coders' missing items because of the complexity of the coding instrument, especially among those categories designed to record the simple presence/absence of a discursive unit (e.g., of a metaphor). However, the categories vary in the degree to which they require global judgments of the article or measure simple appearances of terms. Those items that reflect global judgments, especially the categories of valence and degree of determinism, had to be significantly simplified in order to generate intercoder reliability. Coders could agree whether articles were mostly positive, mostly negative, or roughly balanced. They could not agree whether articles were somewhat positive or mostly positive. This represents a decrement in the richness of the study in order to increase its reliability and replicability. Such are the trade-offs necessitated by quantitative research.

FROM QUANTITATIVE RESEARCH TO CRITICAL INTERPRETATION

The rationales in which quantitative methods are generally embedded mitigate against comfort with extensive interpretation of findings. Quantitative methods are usually embedded in laboratory research and hypothesis-testing methodologies. In those contexts, it makes sense not to interpret much beyond the data because another study can be performed to (dis)confirm the speculations/interpretations a researcher might offer. However, in historical research some of the most important and interesting interpretations cannot be confirmed in that fashion. While we have performed some audience studies with contemporary audiences, we cannot perform audience studies with the public of 1930. Those people have escaped the grasp of the scientific investigator. We cannot, therefore, provide further research to explore whether or not the stock-breeding metaphor actually provided intellectual comfort to members of the professional class by casting them as breeders and others as the to-be-bred. We must operate from reasonable inference.

The need to operate from reasonable inference and argumentation does not invalidate this type of interpretation. All research is limited by its methodologies and by the ever-present potentiality that someone will eventually play Einstein to one's Newton (to Copernicus, to Ptolemy, etc.). All knowledge is constrained to be only the best argument we can make for a finding at a given time. Because understanding the workings of discourse in history is important to us, and because critical interpretation provides the best form of analysis and argument available for some of our historical understandings, no apology is needed for using the best available method.

Once the quantitative results were in hand, I needed to construct an interpretation of these results. Some aspects of that interpretation were self-evident. For example, contrary to the implications of several critics, the level of determinism and expressed discrimination did not increase in the most recent era of genetic research. However, other aspects of the interpretation required greater speculation. The quantitative data reveal fairly clearly which metaphors are prominent in each era. They do not, however, suggest what the significance in these changes in metaphors might be. Even the meaning of the data about degree of determinism is not as clear in some ways as it might seem at first glance. Given the simplicity of the categories (gene only, gene + environment, anti-gene), there was substantial range for change in meaning across time within those variances. Thus, I needed also to provide a critical reading of what the model of the gene-environment relationship was in various periods. I therefore returned and reread all the discourse in the study, trying to make more sense of the quantitative data by understanding them in relationship to other clues the discourse might offer. To do this, I used all the critical tools with which I have been made familiar over the past 20 years.

Critical methodologies have long been important in a variety of fields ranging from anthropology to literature to rhetoric to history. There are various emphases and points of focus in these methods. Some have been labeled structuralist; some, close reading; some, experiential; and so on. My own work is best categorized under the contemporary rhetorical tradition (e.g., Brock, Scott, and Chesebro, 1989; Burgchardt, 1995; Foss, 1989; Hart, 1990), but it has also been influenced by the various forces that have recently congealed to produce "cultural studies," and there are strong overlaps among all these schools. The particular motifs in my own repertoire are guided not only by the rhetorical emphasis but also by the theoretical presuppositions discussed above (but see also Condit, 1990; Condit and Lucaites, 1993). However systematic the gathering of data, in the end, the critic is an artist, a rhetor, a theorist of the particular, a storyteller. I thus concluded by trying to construct stories about the probable relationships among various components of discourse in each era and the differences among the eras. The story I constructed was the one that seemed to me to provide the fullest account of the most important elements in the discourse. There are, however, serious gaps in the stories I have told, the most serious of which are the gaps among the different media.

Multiple Media

This project was conceived as a study of multiple media. In the end, the magazine discourse dominates this account because it provided the most systematic sample by social scientific criteria. However, newspapers, congressional discourse, and television news were all integral to my interpretation. They were not wildly different from the magazines, so I believe that the magazine data reflect the general trends in the public discourse as a whole. However, there are some important changes in media across time that careful readers will want to consider in assessing this project.

The first change in media across time is internal to the magazines themselves. The use of a standard index across time was designed to counter any selection biases I might have as a critic. These biases are strong. Left to my own devices I would have picked a more polarized set of discourse than the one selected by the random numbers method. However, the use of a standard index carries some of its own problems. The character of mass magazines has changed over the years, and the index does not correct for, but rather accommodates, these changes. The first change is one of numbers. Far more magazines are printed and indexed today than in 1919. This makes it impossible to assess whether the coverage of genetics has increased or not through time.

More important, the norms of many magazines have changed. In the early period, the largest majority of magazines used an advocacy approach. That is, they picked a particular view and argued for it. Only a few magazines in that era presented more than one side of an issue. Today, most magazines try to incorporate multiple views into a single article. This change in reporting style has an impact on the apparent level of polarization of the discourse. If one examines tables 1 and 7 (appendix 2), it appears as if the discourse of the eugenic era was relatively highly polarized, whereas later discourse was more balanced. I think that apparent balance merely represents the movement of polarized discourses into the same article rather than its appearance in different magazines in different articles. Thus, it is difficult to assess whether the level of debate has increased or decreased through time. This particularly leads to a leveling with regard to an issue like determinism. Instead of some articles emphasizing purely genetic approaches and other articles arguing against the influence of genes, today we have articles that address the interaction of genes and environment. The mean level of determinism has not changed, but its character is certainly different when assertions of the roles of the gene and the environment are combined in one article instead of conscientiously poised in oppositional articles.

These changes within the magazines are accompanied by differences between magazines and newspapers. In general, newspaper articles are more prone to report discoveries and less prone to explore background issues and relationships. This is partly because newspaper articles are shorter and partly because newspapers see their role as the immediate presentation of "events," whereas the role of magazines is generally envisioned to include more commentary and interpretation. In addition, magazines, almost by definition, are national media, whereas there is a substantial gap between coverage of issues provided by national (and regional) newspapers and that provided by local newspapers. Coverage of issues in local newspapers is more variable in its slant and quality than that of the national and regional newspapers.

These differences have implications for the results of this study. In most cases, the general character of the discourse in the era is the same whether it appears in newspapers or magazines or on television. The metaphors, expressions of preferences for children, and emphasis on issues of discrimination are the same. However, there is some difference in slant and degree with other components of the discourse. Newspapers, for example, are somewhat more deterministic than the magazines, and they exhibit a slight increase in level of determinism through time (the mean shifts from 2.33 in period 1 to 2.18 in period 2, then moves to 2.5 in period 3 and 2.66 in period 4, a trend that is statistically significant overall at the .001 level; compare with the means shown in table 1). It is difficult to know, however, whether this reflects a true shift in the newspaper coverage or whether this reflects the change in the sample. The earliest sample includes only the *New York Times,* whereas later periods include an increasingly broad range of regional papers.

Newspapers exhibit patterns similar to those of mass magazines with regard to type of article, except that they show a dramatic increase in articles reporting physical features linked to genes in period

4. Newspapers are less likely than magazines to differentiate (perhaps in part because they are more likely to cover only one condition at a time), but the trend across time is similar to that of the magazines.

News programs on television can be tracked only in periods 3 and 4. They show generally more regressive trends than either newspapers or magazines, but the samples here are very small and therefore warrant extreme caution. The level of determinism has increased from a mean of 1.7 in period 3 to 2.57 in period 4 (statistically significant at the .001 level). Coverage of behavioral genetics has increased from 10 percent of articles to 15 percent (not a statistically significant increase). The low level of differentiation (10 percent of articles) has also decreased (to 5 percent; also not a statistically significant decrease). Like the newspapers but unlike the mass magazines, coverage of the linkage of physical traits to genes increased across these two time periods (statistically significant at the .001 level). There is a very low level of reporting of ethical issues in the news programs and infrequent use of various metaphors, and these have not changed across time.

News articles, whether on television or in newspapers, thus tend to be very similar to the magazines but somewhat less deterministic and somewhat less attentive to broader social issues. This may arise from some differences between the ideology of newspapers and news magazines (especially the national outlooks of the magazines as contraposed to the variable regional and local outlooks of some of the newspapers in the sample) and in the training of the two kinds of journalists. However, the differences are also arguably a result of the different functions served by magazines and newspapers. Whatever the cause, this study suggests that newspaper and television journalists are somewhat less likely to provide a progressive account of genetics, and this may warrant additional attention to this group of public media. It also suggests to educational personnel the importance of encouraging attention to magazine reading for the formation of an informed citizenry. There may have been some relative deemphasis of magazine reading in favor of newspaper reading in some curricula. As educational agendas are reformulated to incorporate new media such as the internet, the comparative role of newspapers and magazines in conveying broad frameworks should be considered.

What's Next?

This study was obviously a large undertaking. It sought to provide a broad overview of a broad sweep of history. Except for the period of eugenics, most of this history of public discourse had not previously received much careful attention. It is inevitable that as others begin to work in more specific subcomponents of the discourse, the stories written here will have to be modified or in some cases abandoned altogether. However, there is something to be said for broad perspectives. As more and more socially oriented disciplines (political science, history, communication studies) have become increasingly bound to the fragmented life-styles of the social sciences, broad looks at human trends have become less fashionable. Whatever their faults judged by logical positivist criteria, however, such broad looks are crucial for us to gain what used to be called perspective. I hope that this project has modeled ways in which perspective can be gained without overhomogenizing the stories about our past, without assimilating the diversity of the past into a single spirit acting through history. I hope, more important, that this project contributes to the broader conversation about genetics, for gaining perspective not only through studies such as this, but also through sustained public conversation about the ways in which we will place new technologies, seems to me to be a pressing agenda item for the next century of the gene.

APPENDIX 2

QUANTITATIVE DATA

This appendix presents the data from the mass magazine sample. Further information about the methodology and sample are available in appendix 1. Comparative data for other media are incorporated there as well. The coding sheet and instructions are available in appendix 3. Unless otherwise indicated, significance levels are based on Fisher's exact two-tailed test comparing adjacent time periods. Asterisks (single, double, or triple) indicating the level of significant change from one period to the next are located between the two time periods. Thus, if a given feature in time period 1 differs to a significant degree from that feature in time period 2 on the basis of Fischer's exact test or t-test, this is indicated by one or more asterisks placed between the raw numbers or means for each of these time periods. In many tables, columns do not add to the total number shown or to 100 percent because some articles do not include any of the measured features or because they may include more than one of these features.

Table 1. Numeric and percentage distributions of articles, by relative degree of determinism indicated

	1919–34		1940–54		1960–76		1980–95	
Articles with statements:	n	%	n	%	n	%	n	%
Opposing genetic influence (or pro-Lamarck)	7	5	1	1	0	0	0	0
Attributing influence to both gene and environment	79	54	44	61	60	61	60	66
Attributing influence to gene only (or anti-Lamarck)	45	34	27	38	39	40	31	40
Mean determinism[a]	2.29		2.36		2.40		2.34	

[a]gene only = 3, gene + environment = 2, opposition to gene = 1.

Note: There are no significant differences between adjacent time periods as calculated using Fisher's exact test or t-test.

Table 2. Numeric distribution of articles, by type of metaphor about genes

Metaphor	1919–34		1940–54		1960–76		1980–95
Stock breeding	53	***	9	*	3		0
Lottery/odds	16		13		19	*	10
Atom	3		2	*	12		5
Blueprint	1		0		9	**	25
Code	0		0	***	29		21
Command	39		24		30	**	12
Genes in control	3	***	12		7		6
Humans in control of genes	7		4	***	24	***	3
Total articles	151		100		134		147

$^{*}p < .05$
$^{**}p < .01$
$^{***}p < .001$

Table 3. Numeric distribution of articles, by main purpose

Article purpose	1919–34		1940–54		1960–76		1980–95
Protect reproductive rights	21	**	3		8		5
Espouse individual reproductive benefits	3	**	28		45	***	20
Present individual health applications	0		3	***	32		46
Control human reproduction	64	**	13		8	**	0
Discuss investment issues	0		0		0	***	14
Discuss agriculture and industrial applications	0	**	5		2	***	58
Announce discoveries	3	*	9	*	29	*	17
Total articles	151		100		134		147

$^{*}p < .05$
$^{**}p < .01$
$^{***}p < .001$

Table 4. Numeric distribution of articles, by content area of ethical issues addressed

Issue	1919–34		1940–54		1960–76		1980–95
Quality of gene pool	86	***	16		22	**	9
Anti-discrimination	2		2	*	11	*	24
Safety	2		2	*	14	*	32
Total articles	151		100		134		147

*p < .05
**p < .01
***p < .001

Table 5. Number and percentage of articles addressing ethical issues, and mean comparisons

	1919–34	1940–54		1960–76		1980–95
No. and % of articles addressing ethical issues	33 (22%)	32 (32%)		58 (43%)		66 (45%)
Mean number of issues per article, for total article base	0.245	0.37	*	0.537	*	0.776
Mean number of issues per article for articles addressing ethical issues	1.12	1.16		1.24		1.73
Total articles	151	100		134		147

*p < .05
Note: Significance based on t-test of number of ethical issues addressed per total article base.

Table 6. Numeric distribution of articles, by type of label indicating preferred qualities in children

Label	1919–34		1940–54		1960–76		1980–95
Superior	22	**	4		6	*	0
Healthy	38		18	***	53	***	28
"(Ab)normal"	10		12	*	32	***	9
"Perfect"	2		1		3		2
Total articles	151		100		134		147

*p < .05
**p < .01
***p < .001

Table 7. Numeric distribution of articles, by nature of portrayal of eugenics and genetics

Portrayal	1919–34		1940–54		1960–76	1980–95
Mostly negative	22		3		14	10
Balanced	5		2		25	28
Mostly positive	99		68		92	82
Mean favorableness	2.61	*	2.89	**	2.60	2.60

$^{*}p < .05$
$^{**}p < .01$
Note: Significance based on t-test comparing adjacent time periods.

Table 8. Number and percentage of articles expressing differentiation among the degrees of genetic causation of differing trait types

	1919–34		1940–54	1960–76		1980–95
No. and % of articles expressing differentiation	5 (3%)	*	11 (12%)	21 (19%)	*	36 (32%)
Total articles	146		89	113		111

$^{*}p < .05$

Table 9. Numeric and percentage distributions of articles, by type of characteristic attributed to genetic causation

Statements linking genes to:	1919–34			1940–54			1960–76			1980–95	
	n	%		n	%		n	%		n	%
Physical features	95	63		64	64		92	67		90	61
Mental features	99	66	***	38	38	*	69	51	**	49	33
Behaviors	65	43	***	19	19		21	16		21	14
Ambiguous qualities	26	17		26	26		29	22		52	35
Total articles	151			100			134			147	

$^{*}p < .05$
$^{**}p < .01$
$^{***}p < .001$

Table 10. Numeric distribution of articles expressing discriminatory attitudes toward specific conditions, by condition type

Condition type	1919–34		1940–54	1960–76		1980–95
Physical disease	14		6	11	**	1
Physical trait	13	**	1	3		2
Mental condition	67	***	13	8		2
Behavior	29	***	3	0		1
No discrimination	14		6	19		30
Total articles	151		100	134		147

$^{*}p < .05$ (does not apply for this table)
$^{**}p < .01$
$^{***}p < .001$

APPENDIX 3

CODING SHEET AND INSTRUCTIONS

This appendix includes the final coding sheet used by coders, including the written instructions they received. Coders also received verbal instructions, practice, and training with the coding sheets. X's signify line breaks provided for data entry.

CIRCLE ANSWER

Coder B = 1, C = 2, S = 3

X

Year 1C = 01, 2C = 02, 3C = 03, 4C = 04, 5C = 05, 6C = 06, 1A = 07, 2A = 08, 3A = 09, 4A = 10
S1 = 11, S2 = 12, S3 = 13, S4 = 14, N = 15

X

Article no. (three digit: 001 etc.)

I. Determinism TYPE X

Det type Ambig.	0 = No, 1 = Yes	NS [not stated]/hered char/nonspecific
Det type Phys.	0 = No, 1 = Yes	disease/trait/components
Det type Mental	0 = No, 1 = Yes	illness/retard/intelligence
Det type Behavior	0 = No, 1 = Yes	talent/personality/beliefs/morality/ vice/crime/additions/homsex/prosp

X

If it isn't clear what they are talking about circle ambiguous (this would mean that no specific features were referenced in the article—highly unlikely, but still possible). Otherwise list as many of the other categories as are indicated.

Distinctions:
1) physical disease is something referred to as such or one that your common sense would say is a disease
2) physical traits are things like height, eye color, weight, etc.
Three categories of mental conditions:
1) mental illness is things like Alzheimer's or manic depressive disorder
2) Mental retardation refers to references to substantially subnormal abilities—including those associated with "genetic diseases" like Down syndrome and Fragile X
3) Intelligence—refers to general intelligence, especially those references where they are talking about the desirability of manipulating one's intelligence to get geniuses, etc.

There is obviously overlap among these categories. Try to use the contextual clues to see whether they are treating the mental condition as (1) a disease—degeneration, strange responses, etc. (2) retardation—sub-sub par (not merely below average, but seriously limited, or (3) general, normal ranges of mental ability.

Things like piano playing fall under intelligence if skill is highlighted but under behavior if liking the activity or participating in it regularly is highlighted.

Addictions include things like alcoholism and drug addiction.

References to "promiscuity" fall under the category behavior.

Financial prosperity is coded under behavior as well.

II. Det degree

1 = NS/Ambig
2 = Gene only
3 = Gene +
4 = Anti-Gene
5 = Lamarck (+/–)
6 = Lamarck (–)
7 = Lamarck (+)

If there is no clear reference, file under 1.

If the reference is ambiguous "the stuff of life"—file under 1.

If the gene is the only causal force mentioned, code under 2.

If the gene is accorded influence along with explicit mentions of other factors such as environment, culture, nutrition, etc., file under 3.

If the view that the gene is powerful is attacked, file under 4.

If a Lamarckian view is discussed, but no clear pro/con or both are given, 5.

If a Lamarckian view is attacked, 6.

If a Lamarckian view is supported, 7.

III. Det differentiation 0 = Nonalloc or NS, 1 = Alloc

If the article clearly assigns differing degrees of influence on different kind of biological outcomes, put 1 (allocation = allocates different role to different conditions/traits). For example, if the article says that genes cause Huntington's disease, but that genes have influence on intelligence, that article is differentiating and thus gets a 1.

If the article makes no differentiation put 0. This applies even if the article only addresses one condition.

IV. Preferences about children

Indicate if the article indicates that parents/society should desire or may desire children of these sorts. If the word is in quotation marks on the list, it must appear in exact form (or root; e.g., perfect/perfection is OK). If the word is not in quotation marks you may list cases that mean the same thing.

"perfect"	0 = No	1 = Yes
superior	0	1
"better"	0	1
"perf healthy"	0	1
disposition	0	1
"ave/type"	0	1
like parent	0	1
appearance	0	1
diverse/unique	0	1
sex selection	0	1
specific race	0	1

healthy/not defect	0	1	not degen/not handicapped
"ab/normal"	0	1	
intellig/not feeble	0	1	

X

V. Purpose A: topic: Which of the following topics forms the MAJOR focus of the article?

repro/ind	0 = No	1 = Yes	(concern with individual reproduction; e.g., genetic counseling articles)
patient health	0	1	(concern with nonreproductive health concerns)
ag/ind	0	1	(concern with genes in agriculture or industrial applications)
stocks	0	1	(concern with stock market impact)
discoveries	0	1	(describes new genetic discoveries)
repro/protect	0	1	(concern with protecting individual from society/government/etc.)
repro/control	0	1	(concerned to control errant reproduction by individuals/families)
other	0	1	specify:

X

VI. Purpose B: ethical concerns: discrimination expressed towards:

physical disease	0	1
physical trait	0	1
mental	0	1
behavioral	0	1

anti-discrim	0	1 opposition to discrimination based on genes/heredity

X

Indicate if the article expresses negative discriminatory attitudes with those of each of these types of conditions. E.g., if the article expresses negative attitudes toward those with mental illness and a desire to eliminate those problems through selective abortion, that would be a 1 mental. If the article emphasizes explicit opposition to discrimination based on genes/heredity, score this as anti-discrimination 1.

VII. Purpose C: ethical concerns

individ rights	0 = No	1 = Yes	decide/freedom/choice/privacy/ coerc/unborn
discrimin	0	1	insurance/employ/label/equal/ genoci
divers gene pool	0	1	value of disabilities
safety/environ	0	1	ecology/world health
qual gene pool	0	1	gen load/race decl/degen/ proportion/imprv race/social welfare/burden/race mixing/ purity
secondary effects	0	1	feed poor/workers/small farmers
"play God"/spirit	0	1	
breach species integ	0	1	"sanctity of human germ-plasm"
overpopulation	0	1	
animal rights	0	1	
family members	0	1	rights

If the article lists ethical or social concerns in any of these categories score it as such. Note that the category title is to the left and additional examples that may occur are to the right.

X

VIII. Metaphors

lottery	0 = No	1 = Yes	
accident	0	1	dice/chance/luck/fate/odds
control gene	0	1	
controle (we)	0	1	
code	0	1	bk/library/message/alphabet/words
blueprint	0	1	program/computer disck
template	0	1	transmit/machine
command	0	1	direct/determi/govern/reg/rule/law
atom	0	1	space/BNW [brave new world]/ andromeda/Frankenstein/play God
stock	0	1	(i.e., breed stock, good stock, not Wall St.)
other	0	1	complexity/blood tells/mold/map/ family tree/carriers/catalyst/mystery/ bead/particle/OTHER ________

Indicate if any of the following metaphors are used TO DESCRIBE genetics or heredity.

X

IX. Valence

1 = negative
2 = relatively balanced
3 = positive
4 = pro/anti Lamarck
5 = not on genes/heredity/DNA

X = BLANK

Be sure to circle precise components when further examples exist in right-hand column (sometimes this may mean circling the item on the left side of the numbers if the general designator is also the specific designator in the instance).

NOTES

Introduction

1. The scholars listed as concerned about determinism are also generally the persons who raise issues of discrimination, but see also Martin Pernick (1996), who has raised the issues of determinism and perfectionism in a slightly different fashion.

2. Instead of using the older label "Down's syndrome," I have employed "Down syndrome," the preferred term of most organizations run by and for those with this chromosomal variation (e.g., National Down Syndrome Society, National Association for Down Syndrome). See "How to Treat People with Down Syndrome," www.downsyndrome.com/trumble.html.

3. I am aware of the strong poststructuralist objections to the epistemology I am employing here. I believe that I have provided an answer to those objections in "The Materiality of Coding" (Condit, 1999b) by incorporating materialistic and poststructuralist theories.

4. There is a debate in the discipline of rhetorical studies about the proper meaning of the term *public.* Michael Calvin McGee has sought to displace the term in favor of the concept "the people" (see Bitzer, 1968; McGee, 1975, 1987; McGee and Martin, 1983; for an overview of the issue see Lucaites, Condit, and Caudill, 1998). While McGee is correct to disparage a concept of the public that presumes it to have a kind of rationally based knowledge and an essential existence, in Anglo-American democracy his concept of the people has been an active force that works within a public sphere that has been defined by institutional processes. That public sphere is not permanent, and it too is formed in the rhetorical will of a people and that people's willingness to abide by the sphere, but the public sphere created by one instantiation of the people also then exerts force on subsequent constitutions of the people. Thus, both terms signify empirical processes in political discourse. The term *popular* does not necessarily denote "the people," since *the people* implies a collective identity for McGee, whereas the popular does not necessarily manifest itself outside individual living rooms.

5. My observations are that students of elite sources have misrepresented those sources as well and that there is not a singular focus on biology in these sources either; note the euthenic component to Galton's own definition quoted above (see also Pernick, 1996: 43). Because I have not rigorously studied these other sources, I use a sentence structure that grants these assumptions to those working in that area, even though I have my doubts about their soundness.

6. Much of the literature in "science studies" operates in a debunking spirit, seeking to delegitimate the power and authority of science in society. My own

aims are different. I seek to use rhetorical analysis as an academic tool to improve the practice of science in society in particular and targeted ways. My perspective is, as Lyne (1990: 54) puts it, "not intended to show rhetoric as countertheory or an alternative to traditional scientific practices, but rather as a part of those practices." I believe that both sides in the "science wars" share a crucial assumption that is false. Oppositional critics of science believe that if they explore the rhetorical functioning of science they have delegitimated all the practice of science. (And this is what scientists react strongly against.) Far too many scientists, however, employ the same *either/or* logic, accepting the premise that rhetorical analysis of science can only delegitimate it and arguing instead that language practices are insignificant to the outcome of science (see, for example, a thoughtful, yet wrong, response by a scientist to the rhetoric of science: McShea, 1992). Instead, some forms of rhetorical analysis can be subsumed within the practices and methodologies of science as one of the techniques that help improve rigor and avoid error. Indeed, once one understands that language has its own distinctive characteristics, and once one accepts that language is inevitably a part of science, then it is desirable (rather than threatening) that studies of the language used in science be undertaken in order to improve the science itself. I operate from and encourage a *both/and* stance that sees the practice of science as dependent both upon the features of a very real material world and upon the language practices we use in our interactions with this world, the scientific community, and the broader society. While this book has implications for encouraging science studies per se to move in this nonoppositional (and nonreductive) route, it is a study of science-society interaction that focuses on the public side, rather than on the science side, and that is a form of research that has not generally been seen as included in science studies per se.

7. These assumptions are elucidated in several textbooks of the field (e.g., Brock, Scott, and Chesebro, 1989; Foss, 1989; Hart, 1990).

8. The idea that free will is possible within the concept of original sin has been defended artfully and vigorously, but the need for such extended defense indicates that the concept of original sin has deterministic overtones and thus constitutes a rhetorical problem demanding address by theologians within the Christian religion.

9. This is the standard that seems to me to be implicit in Katz Rothman (1998), and it is one that no human endeavor can meet.

Chapter 1. Breeding the Human Stock

1. This kind of claim derived from Malthusian discourse that predated the developments of eugenics as associated with genetic research, but it became central to the eugenic argument.

2. For an argument against the significance of social Darwinism, see Bannister 1979.

3. The errors in the notion of the genes as discrete entities are discussed below.

4. As chapter 2 will note, even such mentions constituted significant difficulties for the persuasiveness of eugenic discourse.

5. For those unfamiliar with close analysis of discourse, this is an unusually high dominance. Never in any of my previous studies—of abortion, of racial equality, and of reproductive technologies—have I seen a metaphor so uniformly used in a broad range of public discourse.

6. Several social historians agree that the eugenic movement in the United States was never widely popular among the citizenry as a whole. The distinction between popular and public can be of use for clarifying this statement. Eugenics was highly prominent and may have been at least close to being a dominant public discourse, even while it was not a dominant popular discourse. This is because there is sometimes (or often, depending on one's political ideology) a difference between what is most popular with the majority of the nation's inhabitants and what dominates the public discourse (Page and Shapiro, 1992). This does not deny that strands of the eugenic discourse, such as racism, were widely popular. Nor does it deny that some applications of eugenic discourse (especially sterilization in cases of institutionalization for mental deficiency) received public acquiescence. It only argues that the core discourse of eugenics, if we understand it as focused on a major reconstitution of the genetic heritage of the population, was never widely and explicitly endorsed and agitated for by the working and farming masses. The reasons for the failure of the metaphor are discussed in chapter 2.

7. These initiatives have been associated with the Progressive Era (Pickens, 1968). I am not qualified to enter the broader debate about the motives of the progressives and the breadth and limitations of that movement. Whatever the conditions and circumstances of the broader movement, I believe that the demographic changes were important for the eugenicists, and these linkages have been fairly carefully traced by those social historians who have studied the eugenics movement in detail (especially Kevles, 1995; Larson, 1995; Paul, 1995; Pickens, 1968).

8. These two visions were not entirely discrete. The former developed into the latter, and the latter is evident in later chapters of Weismann's work.

9. This research on the appearance of the term *germ-plasm* in the sample was performed by Stephanie Palmer.

10. So-called positive eugenics entailed increasing the reproduction of the superior classes. It was almost always spoken of as a voluntary activity to be encouraged, in contrast with "negative eugenics," which was the discouragement of breeding on the part of the unfit and was more often assumed to entail involuntary methods including segregation and sterilization.

11. Pernick's study focuses on a film, *The Black Stork,* which addressed the euthanizing of physically deformed infants. He therefore emphasizes the role of physical appearance in eugenics. Both the topic matter (physical deformity) and the medium (film, which is highly sensitive to visual aesthetics) lead him to overstate the role of appearance in the American eugenic movement as a whole. Diane Paul (1998: 157–58) describes the shift away from the assumption that undesired mental characteristics were linked to physical characteristics

and the active efforts on the part of some eugenicists to separate the two. Markers of race and class were important visual elements of photographs and did signify desired qualities, but these were understood as significant in terms of their moral and intellectual linkages.

12. The discourse about this act is large, and many different opinions were expressed. Some advocates argued that cultural differences were not signs of inequality, but should be kept in their natural place of origin. This other version of the anti-immigration argument did not require that these others were absolutely less fit but merely less fit for this nation.

13. Actually, an ambiguity about the relationship between eugenics and euthenics exists along with a tendency to deploy it strategically. Most critics and historians have identified eugenics as being opposed to euthenics and focused around biological (especially genetic) factors. However, when objectionable actions are given euthenic justifications (such as in the case of sterilization), the critics continue to define these negative actions as existing under the eugenic umbrella (see, e.g., Reilly, 1991).

14. For example, Kevles (1995: 48) argues that Charles Davenport was familiar with the implications of the interactions among genes and used them in his physical studies but not in his studies of moral and intellectual traits.

Chapter 2. Challenges to Eugenics

1. In the mass magazines throughout the first half of the century, there was continued support for sterilization of those who had been institutionalized. The restriction of involuntary sterilization to those who were institutionalized seems to have provided a key element for balancing reproductive rights against sterilization interests. Sterilization gained popular acquiescence only when applied to those whose rights to individual autonomy were otherwise already discounted. This allowed preservation of the right to reproductive autonomy for the members of the community because it permitted sterilization only for those who had been removed from the community. Whether or not this was a desirable policy, it was clearly a principled rather than a capricious distinction. In operation, however, Smith has convincingly demonstrated that the practice of involuntary sterilization was capriciously applied. On the other hand, involuntary sterilization was often involuntary with regard to the patient but done under familial sanction, and, given the role of families and their position in the victims' lives, this inclusion had some significance in mitigating the power of the government and the institution.

2. Both Paul and Reilly argue that eugenics achieved more popular support than earlier accounts indicate; however, this broader popularity is documented inadequately and assumes an overly broad definition of eugenics that identifies as eugenic any racist doctrine associated with immigration restrictions, any doctrine associated with sterilization, and sometimes merely any doctrine associated with genetics. Neither of these authors demonstrates popular support for mass sterilization for eugenic purposes. Both rely heavily on a survey by *Fortune* magazine during the Depression, which showed that a

majority of readers supported compulsory sterilization of "mental defectives" and "criminals" ("Fortune Survey," 1937; Paul, 1995: 83; Reilly, 1991: 125). The readership of the business magazine *Fortune* is, of course, grossly unrepresentative (in our survey the magazines *Forbes* and *Fortune* were among the magazines most likely to have supported discriminatory eugenic discourses). Moreover, support for sterilization of these two groups is not identical with the earlier support for sterilization of the "feeble-minded," for these two later categories encompass narrower, somewhat more precisely defined (and therefore smaller) groups because of narrowing of the terms in public usage as a result of increased social scientific precision in classification. Hence, while the attitudes of the *Fortune* readers represent a mix of punitive and eugenic responses, they are not identical to support for the broad eugenic goal of improving the breed. Reilly admits (1991: 95–97) that the poll results can be better understood as a desire to avoid the need to provide social support for a limited group of children whose parents would be unable to provide care for them and/or were deemed unfit. In the classical era of eugenics, one sterilized to avoid the production of misfit children. In the Depression era, one sterilized to avoid the production of children who would have to be given substitute parents. Paul quotes Reilly on this, but she lumps these "social and genetic" rationales together.

3. Paul quotes a statement by Popenoe and Johnson indicating that everyone believed that feeble-minded and epileptic persons did not have a right to marry. Popenoe and Johnson may or may not have reflected the public opinion of the times (certainly many opponents of eugenics agreed with them), but their statement confuses the presumption of a general right to reproduction with the issue of whether or not particular persons were outside the general community in some way that abrogated that right. Part of the confusion is caused by the shifting referent *the feeble-minded.* Those who agreed that compulsory sterilization was acceptable for the feeble-minded generally tended to see that category as a relatively small one, populated by persons completely incapable of caring for themselves. The more numerous one envisioned this group to be, the less likely one was to see sterilization as an option. Very few persons thought both that the feeble-minded constituted a large proportion of the population and that involuntary sterilization should be applied to that entire population. Thus, opposition to mass sterilization was based on the presumption of a reproductive right, but this presumption was not necessarily incompatible with sterilization in "extreme" cases.

4. There were 6 pro-Lamarckian articles and only 3 opposed in this era, in contrast with the next era, which featured 12 anti-Lamarckian articles and only 1 supportive article. Debates about the correctness of Lamarckian positions have been revived from time to time. For example, the discovery that bacteria share fragments of DNA that provide antibiotic resistance in a nonreproductive fashion was described as a Lamarckian means of inheritance.

5. An intermediate version of this view was the "Baldwin effect," which suggested that learned behaviors could shift the distribution of the genes in the population (David Depew, personal communication, 1998).

6. Vernon Kellogg's defense of eugenics reflects the public atmosphere of ridicule. He said, "Why laugh down a good word for a good idea because the abuse of cranks has given eugenics over to the tender mercies of the comic press and comic stage?" (1924: 207).

7. Reilly argues that the Nazis did not have an impact on American eugenic attitudes. His proof, however, consists of statements by active eugenicists and sterilization boards, especially in California. Our evidence shows clearly, however, that though active eugenicists may not have changed their minds in response to Nazi practices, their statements became less convincing to the public (and probably the populace as a whole) as Nazi implementations of those practices became increasingly visible. Reilly has attributed actions of sterilization boards to active eugenicists, so that the lack of change in their beliefs also does not reflect broader public sentiment. He argues that the radical decline in sterilization during World War II was a product of the lack of surgeons, and this is clearly true, but it does not demonstrate that the number of sterilizations would have stayed the same or increased had a sufficient supply of surgeons been available. It merely indicates that we do not know what would have happened in that case.

8. For a discussion of the theoretical concept of presence in argument, see Perelman and Olbrechts-tyteca (1969).

9. Diane Paul (1998: chap. 7) makes a brilliant argument that the discovery of the Hardy-Weinberg principle did not dissuade eugenics-oriented scientists both (1) because they were willing to accept minimal short-term gains and to attend upon longer-term gains (which for some was true, and there are examples in the public discourse), and (2) because even under the assumptions of the Hardy-Weinberg calculations, undesired characteristics which made up as much as 10 percent of the population would still show substantial decreases in the first generation or two through selective breeding. As applied to the impact of science on public debate, however, Paul's argument ignores the fact that the scientific advances of the era were also simultaneously delegitimating the second assumption. The demonstrations of the complexity of genes made simple categories such as "feeble-mindedness" increasingly untenable. At the same time as the Hardy-Weinberg formula came about, therefore, geneticists were rapidly coming to understand that there were no undesired conditions of a prevalence of 10 percent. Thus Paul's argument that changes in attitudes did not depend upon "a fundamental mistake" corrected by the Hardy-Weinberg formula alone is correct, but this does not mean that changes in the science as a whole played no role in the changing public attitudes.

10. The use of the reductionistic, simplifying vision has been well-documented by critics of genetics operating from science studies (Doyle, 1997; Fox Keller, 1995; Kay, 1995). The use of complexity and diversity in genetics is evident to any first-year genetic student but is also deserving of further scholarly study.

11. For technical reasons, images in magazines at the turn of the last century were relatively rare, and they grew more frequent through time. Whatever the absolute numbers of images, however, there were proportionately more

images in pro-eugenic than in anti-eugenic articles. Later, anti-genetic articles would feature, at most, cartoons, while pro-eugenic articles featured a broad range of images.

12. This reductionistic view of genetics continues to be presented by scientists to the public. On 29 January 1996, Nobel Prize winner Paul Berg presented a lecture on genetics and human values at the University of Georgia. He began by using single-gene-associated traits as the paradigmatic case and then subsuming all genetic medicine and technology under this model. This does not mean that scientists do not understand and recognize the complexity of genetic interactions, but merely that they continue to treat these as "add-ons" to the fundamental model of the gene producing characteristics.

13. There were many other proposals as well, including marriage and child-bearing bounties, free sexual congress, and marriage control through certification for health.

Chapter 3. The Transformation to Genetics

1. Reilly (1991) argues that this drop-off is overstated. He focuses on continuing sterilization. Additionally, Selden's (1989) work shows that eugenic statements continued to be included in newly published high school biology textbooks until the sixties. Neither of these authors makes clear distinctions among eugenics, reform eugenics, and the new genetics. Both of these arguments, however, suggest that many professionals and some scientists did not give up on eugenics. In the public space, though isolated voices in support of eugenics continued to be heard, the movement as a serious public phenomenon died out or was transformed.

2. The model of the relationship between science and the rest of society that I believe most accurately captures that relationship portrays science and the public sphere as semiautonomous regions operating in the cottony penumbra of the society as a whole. Each sphere is subject to the constant hidden "gravitational" forces of the other and also to short-term, focused, direct incursions by the other.

3. Throughout the period there is some slight lack of uniformity in this reporting, owing perhaps to the preference for environmentalist accounts and for free inquiry by reporters on the Left who may have had either positive or negative reactions to the Soviet Union.

4. The press went even further, associating Lysenko with the Nazis, both directly and by implication. In the *Saturday Review,* Hermann Muller (1948: 65), a geneticist supportive of communist doctrines but who experienced repression of free research in the Soviet Union, explained that the "similarity between the Nazi and Communist theories" accounted for censorship of Western genetics at the 1937 International Genetics Congress. More subtly, *Commonweal* asserted that Lysenko was offering to make everyone a "superman" (Bernstein 1946: 525). In this way, the old association between the Nazis and eugenics was transformed, so that Lysenko (through anti-genetic Lamarckianism) was now seen as the source of a plan to build a hostile nation of superpeople.

5. There is no discrete physical entity one can neatly label as a gene. Chromosomes are double stranded, and only one strand contains the coding unit for a given protein. Moreover, there are noncoding sections of DNA within the coding unit itself, and there are activation sequences that are not contiguous with the coding sequence. Fifty years later, it turns out that there is a very rough association between the beads-on-a-string metaphor and groupings of transcription units around nucleosomes, but the complexities of chromatin structure still bely the simplicity of the early particulate, or "beads," model.

6. See Turney 1998: 127. The persistence of the negative view among the public and a neutral view in the scientific community was demonstrated in a study performed by Beth O'Grady, Roxane Parrott, and myself (Condit, Parrott, and O'Grady, in press).

7. It also provided Muller a public platform to disseminate his reform eugenic visions.

8. This was an ambiguous move because the "individual" is not identical with the "family," but historically the patriarchal family was identified with the interests of the father as individual. The structure of the family and assumptions about it were changing in this era, however, so the treatment of family and gender issues was ambiguous. See chapter 4.

Chapter 4. The Family Gene

1. It is worth thinking about the fact that the assignment of "heredity" to genetics and not to family nurturance required the assumption that family environments (or cultures) were not in some sense inherited. This seems to be a false assumption (though more so in eras where social change was less rapid), but it was consonant with the subsumption of family environment within broader social structures (especially class), which were also then being understood in an ahistorical sense.

2. These attitudes were supported by reports of scientific experiments in which genetic problems were shown to manifest themselves at regular and repeatable parts of the developmental process (Plumb, 1949).

3. An exception was provided in the *Saturday Review,* "The Future of Our Genes" (1953). This article argued, in the classic eugenic mode, that since there was some genetic influence, eugenic action must be taken.

Chapter 5. Recoding Humanity

1. American scientists rapidly accepted the structure, though debates continued about the sufficiency of DNA as the hereditary material (the protein model died hard). The press, however, initially turned to Pauling, who may have been less than enthusiastic, in part because of his own work in the area, which included a prior and different model. However, Pauling soon also accepted the structure, but the news window had by then already closed.

2. It is odd that these critiques attack the "accuracy" of the metaphors (van

Dijck, 1998: 177), since they tend to operate from epistemologies that would hold that metaphors cannot be "accurate." It is ironic that these critiques attack the scientists' use of a version of the coding metaphor that is monosemic (e.g., van Dijck, 1998: 129, 151), when their own critiques provide a monosemic reading of the scientists' use of the coding metaphor. While I do not deny that some of the interpretations of the coding metaphors offered by these critics are both tenable and sometimes active, in keeping with a polysemic view of discourse, I offer here and in part 4 some alternative interpretations of the ways in which these metaphors could be understood and could interact with other components of the discourse, as well as some evidence that those modes of meaning were actually active. Note that this alternate interpretation is more consistent with the interpretation of the implications of the intellectual movement about semiotics, of which the genetics discourse was arguably one offshoot.

3. Lewin's series provides the cleanest example. In the first edition of *Genes,* published in 1983, he introduces genetics with the atomistic beads-on-a-string metaphor. The coding metaphors are presented, but they come later in the text. It is not until the third edition of *Genes,* published in 1987, that the coding metaphors are predominant.

4. I base this judgment of delay in part on a study I performed at the meeting of the American Society of Human Genetics in 1995. I selected 12 posters to represent the proportional appearance of the implicit model of the gene from all the posters at the meeting and then interviewed the presenting authors, asking them about their model of the gene. A quarter of the authors were self-reflective about the tension between the functional definition and the chemical definition. Most of the others provided the now textbook truncated chemical definition, something like "a gene is a coding unit of DNA."

5. Katz Rothman (1998: 23–24) argues for the superiority of the recipe metaphor on the grounds that bread made with ingredients from different parts of the country or under different conditions ends up being different. This is true of building projects as well. The same blueprint used by different people does not produce identical products. In most ways, a recipe is simply a blueprint for baking. Katz Rothman is, however, correct in noting that, to some extent, the element of time is more apparent in a recipe, since in most cases the time for executing a recipe is measured in minutes. Recipes thus often include both ingredients and directions for combining the ingredients (though the recipes I received from my mother-in-law never did!). In contrast, blueprints are usually executed over much longer time frames, and because of their complexity, the directions for their execution are not always included. This does not mean, however, that blueprints are time-insensitive, as is evident in the building plans for most of the great cathedrals of the world. Katz Rothman's lack of familiarity with building plans and greater familiarity with baking may make these other elements more evident to her in the term *recipe* than in the term *blueprint,* but this can not be expected to be true of persons with different sets of experience, and the research report below bears this out. It is odd that critics of genetic research who interpret genetic metaphors in a reductive and

monosemic manner so often criticize geneticists for reductivism and a failure to understand the polysemy of coding systems. See also chapter 8.

6. It is difficult to assess whether this was because the persons using the phrase increasingly had not read Huxley's book, or because the phrase has a descriptive force of its own, independent of the book, or because reporters did not find Huxley's world as dystopian as some of us might find it. Probably in different cases each of these things is true.

7. The use of artificial insemination provided a partial exception to the lack of acceptable means. Artificial insemination by donor generated only mild opposition and was steadily, though not widely, used in the last quarter of the century. Because it addressed only half of a genetic legacy and because sets of genes are so enormously complex, it offered far less control than the dreamers of this era sought.

8. This fairly pervasive sense of the irreversibility of the changes also dominates the ethicists' condemnation of germ-line genetic manipulation. However, it is substantially in error with regard to the genetics. If we were able to remove a sickle-cell gene readily, we could presumably also resynthesize such a gene and stick it back into the human germ line. Instead, the fear is a displacement of a social fact: once we accept genetic engineering as a social practice, it may be difficult to eradicate it from our social fabric.

9. Decisions about public funding for various research avenues did need to be made, and some temporary bans might be desirable, given that no one knew when exactly surrogacy, cloning of a mammal or human, and chimerical gene exchange between humans and other species would occur, and given that surrogacy experiments performed on women were fairly widespread with what is arguably a lack of suitable informed-consent procedures or sufficient prior research in animal models to minimize risks (Arditti, Klein, and Minden, 1989).

10. Some additional predictions included Rorvik's suggestion that genetic surgery would occur in 10–20 years (Rorvik, 1969b) and *Harper's* looser claim that it would take "several decades" for genetic engineering to occur (Etzioni, 1973). *Psychology Today* predicted that cloning of humans would occur in 20–50 years (Ausubel, Beckwith, and Janssen, 1974), as opposed to Horace Freeland Judson's claim (1975: 40) that "no road goes there." Interestingly, Judson also claimed that sampling fetal blood cells from maternal blood was likely to be tricky, but if it could be accomplished it would be 20 years away (some versions of it became technically feasible in the late nineties, although in a narrower scope than originally thought). He also thought that gene therapy was many decades away, perhaps impossible, and that genetic counseling could achieve the same ends more rapidly. *Ms.* was the most accelerated in its forecast, predicting human cloning within 10 years (Rivers, 1976).

Chapter 6. Genetic Counseling

1. This difficulty is reflected in the response of various disability rights groups to genetics. Most groups support continued genetic research in the

hope that it will generate cures. Some groups support parental access to selective abortion. Some groups actively oppose the ideology behind selective abortion because of the potential implications about the human worth of those with varying abilities.

2. Variations included "Will the Baby Be Normal?" and "Will My Baby Be Normal?"

3. Some disability rights activists argue in favor of equating their identity with their disability because of the utility of such an approach in immediate political action and by analogy with other civil rights movements. Others wish to ensure that they not be coerced into "treatment" for a disability against their will.

4. When the council eventually lifted the moratorium, the magazine then used the incident as an example of the wonders of democracy, illustrating the usual assumption that a form of government is desirable only to the extent that its last decision favored us. For a fuller discussion see chapter 7.

Chapter 7. Ethical Challenges and Biohazards

1. These principles were not honored in the era, for embryo research and experimentation on women for in vitro fertilization went forward without sufficient prior research, informed consent, or external monitoring.

2. This arguably was the case later in the century when Richard Seed decided to attempt human cloning for profit (John Carey, 1998; "Human Cloning to Start in Russia," 1998; "Seed Vows to Continue Quest to Clone Human Beings," 1998).

Chapter 8. Blueprints for Genetic Commerce

1. José van Dijck (1998: 92–93) offers a substantially different interpretation of the discursive dynamics of this era. She traces the rise of the factory metaphor to Richard Dawkins's book *The Selfish Gene* and downplays the importance of the blueprint metaphor. Such originary tracings of metaphors to single authors are unconvincing because metaphors become pervasive and important only when they reflect broad social forces. They are not most profitably viewed as having their "origins" in the pens of individual authors.

2. Even Watson and Crick, who were heavily invested in the idea that genes are key players in human outcomes, now also explicitly endorsed the environmental perspective, though they did so in a nondevelopmental fashion. A *Time* interview quoted them as saying, "That's not to say that everything we do is determined by our genes. Heredity is modified by experience" (Jaroff, 1993).

3. See also note 5, chapter 5.

4. This article portrayed the arguably legitimate resistance of physicians to the tests as a simple ignorance on the part of primary-care physicians, thus marking the "opponents are ignorant" topoi as one that cut across not only the public but also professional alignments.

5. I argue that these criteria were only putatively scientific because accuracy

has a higher value than speed in science, whereas in the regulatory agencies speed assumes a higher role. Truly scientific answers cannot be delivered on a production timetable.

6. Scientists sought to negotiate their own interests within the matrix. Most supported patenting of the new forms of life; many scientists were entering biotechnology firms or entering into agreements to share their university-based developments with industry, so they stood to gain financial rewards from patenting. However, many scientists resolutely resisted patenting of naturally occurring biological products themselves (e.g., stem cells) or of DNA sequences for which there was no known function.

7. Lily Kay (1993) argues that this funding lineage was not only heavily influenced by medical goals but also included a deliberate focus on human perfectibility. Both medically based funding and such politically motivated funding had the same effect of focusing research on human perfectibility.

Chapter 9. Toward a Personal Health Genetics

1. Preimplantation diagnosis required a couple to undergo in vitro fertilization. Then, technicians performed genetic tests on the embryos resulting from the fertilization process, and they implanted only those embryos without the genetic defects measured by the tests they applied. Some persons saw this as a way around the moral quandaries offered by abortion. Aborting a fetus in the second trimester because of its "failure" of genetic testing was closer to selecting against adults with particular genetic conditions than was choosing to implant particular embryos. The distinction remained one of degree rather than one of absolute difference, however. Moreover, in vitro fertilization was expensive and had its own emotional roller coasters (Condit, 1994, 1996b). The new technology did not seem to solve the old moral or philosophical problems, or to erase the emotional tensions, or to reduce the costs. Thus, though the magazines tended to endorse preimplantation genetic screening and to quote its users as endorsing it enthusiastically as "all worth it" (Cohen, 1993: 273), the basic difficulties of genetic selection that made genetic counseling an autonomy-enhancing social structure continued.

2. The claim is often made that there were not enough genetic counselors available, but training in genetic counseling required only a master's program. At the same time, genetics and biotechnology were booming, and Ph.D.'s with genetic specialties were beginning to be mass produced and even overproduced. This means that the root difficulty was a failure to value genetic counseling rather than an inability to produce sufficient genetic counselors.

3. These attitudes were changing slowly within the medical profession at the close of the century, but commercial and institutional constraints continued to mitigate against real enactment of truly autonomous decision processes for patients/clients.

4. The problem was handled by treating MSAFP (maternal serum alpha fetoprotein) and triple-screen tests as though they were not the first stage in the process of genetic selection. These tests were the ones most likely to be

routinized in obstetrical and clinical practices without extended genetic counseling. The use of amniocentesis and CVS was more often accompanied by more sufficient counseling.

5. Although the motivation might also have been the hope that the delegitimation of nondirectiveness would lead to the discrediting of genetic tests in general, it seems to have had the alternative effect of delegitimating counseling while maintaining the tests.

6. Academic critics worried, for example, that sustained concern with genetic information would lead not simply to individual choices with regard to diet and exercise but also to increased and excessive medical surveillance (T. Thompson, 1991: 168). The realistic nature of this fear was suggested by attitudes expressed in the mass media itself; an article in *Good Housekeeping,* for example, referred favorably to the possibility that a woman who knew she was at increased risk for ovarian cancer might have her ovaries removed for "peace of mind" (Krause, 1994: 162).

7. The quotation was attributed to physicist Richard Seed, who, according to the article, "cofounded a research enterprise for embryo transfer experiments on female humans." Randolph Seed was implicated in the quotation through the claim that he had defined asthma as a genetic defect.

Chapter 10. Ethical Challenges for Genetic Health

1. This also contrasts with the tendencies of our readers in our audience study (Condit, 1999a; Condit and Williams, 1997).

2. Pictures in the later period generally tended to feature middle-class families interacting with high technology. Middle-class families were thus identified as in need of remediation through these technologies. In contrast, earlier images featured middle-class families as "superior" examples (see, e.g., the "fitter family" contests) who were not in need of correction or improvement but who merely needed to reproduce themselves as they were. Systematic research on the visual images was done by Holly Gooding.

3. Another version of this uneven treatment of women occurred with regard to the issue of risk. As women's magazines pointed out, when it came to incurring genetic risk associated with women, companies expelled them from workplaces with toxic chemicals on the grounds that "zero risk" was the only acceptable principle—that any risk to any fetus was too great (Kirp, 1990). Such risks, however, were not only tolerable for male workers (who might also transmit chemically inflicted genetic mutations to their offspring), but also an absolute sacred business principle in other contexts. For example, the CEO of Monsanto railed against the "zealots" who flew the "risk-free" banner, and thus "stop much technological development" (Harbison, 1990: 723).

4. The risks of prostate cancer and breast cancer are probably not equal. Prostate cancer, as far as we can currently tell, is less likely to be fatal to men before they would die of other natural causes than breast cancer is likely to be fatal to women before they would die of other natural causes. However, in this case, the women at issue do not have cancer. They have been identified

only as at a potentially increased risk for cancer because of their genetic configurations, and they are still advised to undergo massive surgery with major consequences of both cosmetic and convenience sorts, while the men, who have been identified as having cancer, are advised against the less significant surgery on the grounds of their lesser consequences of cosmetic and convenience sorts.

5. Of course, Richard Seed would soon make it evident that there were indeed sufficient investigators outside Judson's criteria of responsibility willing to undertake such human experimentation (John Carey, 1998; "Human Cloning to Start in Russia," 1998; "Seed Vows to Continue Quest to Clone Human Beings," 1998).

Conclusions and Speculations

1. Evolutionary biology thus does not tell us much of interest if it shows that there are mean differences in philandering between men and women. Ask instead, these questions: What is the range of philandering behavior among women? among men? In what kinds of environments are highly philandering women genetically favored? socially favored? or favored with personal satisfaction? Now, are the conditions in which these women are genetically favored the same as those in which they are socially favored or satisfied? And in what conditions is society favored in a variety of ways by their philandering? After answering all those questions for all categories of men and women (heterosexual and homosexual), we might begin to address social issues, such as, What kind of environment ought we collectively to construct? And which groups ought we to favor/disfavor?

2. I do not make this claim in ignorance of the substantial critiques of the Western optimist's tendency to construct a so-called whig view of history, nor do I make the claim in ignorance of postmodern theories of narrative or in ignorance of the potentiality that progressive accounts merely reflect our own comfort with our own vocabularies and discomfort with alien vocabularies. I have thought carefully and self-critically about these potentials. I have concluded that the progressive account is both the best and the most accurate (by very different epistemological criteria and value systems) story to be told about this discourse. I accept that what I tell is only a story, but I contend it is a good story, the best story for our times. I see it not as a story that guarantees that technology will make all right with the world, but rather as a story that our struggles can help us to keep from allowing technology to make all wrong with the world.

3. Cloning of animals can provide the necessary research basis for cloning of human parts for medical replacements. Cloning of human parts (except perhaps the brain) should be legal if it becomes technically feasible. Cloning of full human beings to use their body parts is obviously immoral. A cloned person has equivalent rights to the person from whom she or he was cloned.

4. Those who believe in a soul see the soul as another alternative. However, in one way or another the soul must be integrated with these other factors.

This leads us to describe how these other factors work (especially given that, by its nature as a nonempirical phenomenon, we cannot investigate how the soul interacts with these factors. If we could investigate that through anything other than religious authority, the soul would take on the characteristic of an empirical phenomenon). Thus we can know the soul only in an empirical sense by knowing the sociobiological factors in which it manifests itself.

5. In my judgment Conrad and Weinberg missummarize media discourse about genetics as much as they accuse the press of doing, and in ways similar to the press (though with reversed valence), with regard to the summarizing of scientific discourse. They argue that the frame of the mass media with regard to genes and alcohol is "genes cause alcoholism," whereas most of the evidence they quote in their article supports the frame that "genes play a role in causing alcoholism," which appears to be a substantially different and more defensible frame. In spite of this tendency to reductionism, the research is instructive.

6. See E. O. Wilson's (1998: 130–43) treatment of the difference between the two measures. Wilson errs in suggesting the two measures are in practice interchangeable, because he assumes that the only possible environments that are of interest are those de facto included in heritability measures. But advocates of social change are precisely interested in new potentials for new environments that are not included in such existing measures. See also Alper and Beckwith, 1993.

7. Critics such as Hubbard and Wald (1993) and Lippman (1993) dispute this, arguing that funds for genetic research and technologies should be redirected and would pay greater dividends elsewhere. It may well be true that there are more cost-effective expenditures than genetic research, but there are clearly many less cost-effective expenditures as well, and allocation decisions are rarely based on direct comparisons. Arguably, the cost of the Human Genome Project came out of moneys that would otherwise have gone to the superconducting supercollider rather than to vaccinations and nutrition supplements to poor children. Arguably also, we could afford both the HGP and vaccinations and nutrition supplements for poor children. I take the position that genetics is not inherently a discriminatory and exclusionist technology and that we should work to bring its benefits to everyone who needs them.

8. This does not invalidate the practice of genetic counseling or justify the argument that counselors should go ahead and be directive, since they will always be imperfectly nondirective. It is worth noting that genetic counseling is a relatively young practice which is only beginning to develop an adequate professional knowledge base to guide its practices.

Appendix 1. Theory and Methodology

1. I borrow the concept of motif from DNA studies which show that there is a series of recurring coding sequences that are associated with particular categories of functions. These coding sequences are not identical to one another but show statistically significant similarities to each other. They are combined in different patterns to produce unique organisms.

2. An exception is José van Dijck's (1998) analysis, which explores multiple aspects. However, the lack of quantitative analysis in that study leads to some conclusions that seem to me to be errant (see notes throughout this book) and to some self-contradiction. Admittedly, however, it is difficult to tell whether the problems in that analysis arise from a lack of quantitation or rather from a general commitment to a particularly anarchic version of post-modernity emphasizing fragmented presentation. Jon Turney's (1998) analysis of a broader range of discourses (British and U.S. as well as nineteenth- and twentieth-century) has fewer self-contradictions and outright errors. It is somewhat more systematic, though not quantitative, and it is more balanced, demonstrating that critical analyses do not have to be simplistically one-sided. Martin Pernick (1996) puts limited quantitation to good effect in his analysis of reception of a particular episode in the eugenic controversy.

3. The difficulty of knowing the implications of the quantitative proportions of various issues is less of a problem with nonfictional discourse, though still it must be reckoned with. For example, the increase in discourse about ethics at the end of the twentieth century cannot simply be taken to mean that we are "more ethical" now than in the second era, when there was little of such discourse. Obviously, recent discourse is indicative of the fact that we are much closer to implementing major genetic technologies, and ethical discourse is a reaction to that. Numbers must always be interpreted, and I have tried to judge these issues within my text. There is peril, therefore, in excessive reliance on the tables of numbers on their own in appendix 2.

4. Funding was no longer available to correct this error when it was discovered. Because the quantitative results of the newspaper study are not reported in detail here, this error was not corrected.

REFERENCES

Alper, Joseph S., and Jonathan Beckwith. 1993. "Genetic Fatalism and Social Policy: The Implications of Behavior Genetics Research." *Yale Journal of Biology and Medicine* 66: 511–24.

Alvarez, Walter. 1972. "Scientists Now Able to Tell if Fetus Has Heritable Disease." *Los Angeles Times,* 30 March, IV, 16 (3).

Andrews, Lori B. 1982. "Genetic Counselors: How They Can Help and How They Can't." *Parents Magazine* (November): 92–97, 178–84.

Andrews, Lori B., Jane E. Fullarton, Neil A. Holtzman, and Arno G. Motulsky, eds. 1994. *Assessing Genetic Risks: Implications for Health and Social Policy.* Washington, D.C.: National Academy Press.

"The Andromeda Fear." 1974. *Time* (29 July): 59.

Arditti, Rita, Renate Duelli Klein, and Shelley Minden. 1989. *Test-Tube Women: What Future for Motherhood?* London: Pandora Press.

Auerbach, Stuart. 1975. "NIH Scientists Find Cancer-linked Gene." *Washington Post,* 2 May, A, 7 (1).

Aune, James Arnt. 1994. *Rhetoric and Marxism.* Boulder, Colo.: Westview Press.

Ausubel, Frederick, Jon Beckwith, and Karen Janssen. 1974. "The Politics of Genetic Engineering: Who Decides Who's Defective?" *Psychology Today* (June): 30, 32, 34–39.

Avise, John C. 1998. *The Genetic Gods: Evolution and Belief in Human Affairs.* Cambridge, Mass.: Harvard University Press.

"Bad Blood." 1951. *Today's Health* (December): 16.

Bailey, Ronald. 1988. "Ministry of Fear." *Forbes* (27 June): 138–39.

Bailey, Ronald. 1990. "The Fourth Hurdle." *Forbes* (2 April): 166–67.

Baltimore, David. 1983. "Can Genetic Science Backfire? That's the Chance We Take." *U.S. News and World Report* (28 March): 52–53.

Bannister, Robert C. 1979. *Social Darwinism: Science and Myth in Anglo-American Social Thought.* Philadelphia: Temple University Press.

Bartels, Diane M., Bonnie S. Leroy, and Arthur L. Caplan, eds. 1992. *Prescribing Our Future: Ethical Challenges in Genetic Counseling.* New York: Aldine de Gruyter.

Baumiller, Robert C. 1964. "Will the Child Be Normal?" *America* (15 February): 220–21.

Begley, Sharon. 1992. "A New Genetic Code." *Newsweek* (2 November): 77–78.

Berens, Conrad. 1955. "Your Eyes and Heredity." *Today's Health* (November): 34–35.

Berg, Roland H. 1971. "Will My Baby be Normal?" *McCall's* (August): 52, 55, 137.

Bernhardt, Barbara A. 1997. "Empirical Evidence That Genetic Counseling Is

Directive: Where Do We Go from Here?" *American Journal of Human Genetics* 60: 17–20.
Bernstein, Joseph. 1946. "Science under Dictate." *Commonweal* (13 September): 522–25.
"Biotechnology Is Now Survival of the Fittest." 1982. *Business Week* (12 April): 36–37.
Bitzer, Lloyd F. 1968. "The Rhetorical Situation." *Philosophy and Rhetoric* 1: 1–14.
"Bizarre Crimes and Punishment." 1993. *USA Today* (magazine) (April): 16.
Blank, Robert H. 1990. *Regulating Reproduction.* New York: Columbia University Press.
Block, J. L., and J. Stein. 1971. "Miracle Baby of Carolyn Sinclair." *Good Housekeeping* (June): 86–87.
"The Blood Line." 1955. *Newsweek* (15 August): 50.
"The Body: From Baby Hatcheries to 'Xeroxing' Human Beings." 1971. *Time* (19 April): 37–39.
Bornstein, Paul. 1983. "What's Your Ph.D. Ratio?" *Forbes* (13 August): 112–14.
Bosk, Charles L. 1992. *All God's Mistakes: Genetic Counseling in a Pediatric Hospital.* Chicago: University of Chicago Press.
Brock, Bernard L., Robert L. Scott, and James W. Chesebro. 1989. *Methods of Rhetorical Criticism: A Twentieth-Century Perspective.* 3d ed., rev. Detroit: Wayne State University Press.
Brockriede, Wayne. 1972. "Arguers as Lovers." *Philosophy and Rhetoric* 5: 1–11.
Bronson, Gail. 1986. "Where's the Demand?" *Forbes* (20 October): 138, 142.
Brownlee, Shanon. 1990. "James Wilson: The Man Who Throws Good Genes after Bad Ones." *U.S. News and World Report* (31 December): 80.
Brownlee, Shanon. 1994. "The Narrowing of Normality." *U.S. News and World Report* (22 August): 67.
Brownlee, Shanon, with Joanne Silberner. 1990. "The Assurance of Genes." *U.S. News and World Report* (23 July): 58.
Brunger, Fern, and Abby Lippman. 1995. "Resistance and Adherence to the Norms of Genetic Counseling." *Journal of Genetic Counseling* 4: 151–68.
Bryan, William Jennings. 1925. "The Bible Is Good Enough for Me." *Collier's, the National Weekly* (1 August): 8.
Burgchardt, Carl R., ed. 1995. *Readings in Rhetorical Criticism.* State College, Pa.: Strata Publishing.
Burgess, Parke G. 1968. "The Rhetoric of Black Power: A Moral Demand?" *Quarterly Journal of Speech* 54: 122–33.
Burke, Kenneth. 1968. "Definition of Man." In *Language as Symbolic Action,* 3–24. Berkeley, Calif.: University of California Press (originally published 1966).
Bylinsky, Gene. 1974. "What Science Can Do about Hereditary Diseases." *Fortune* (September): 148–60.
Bylinsky, Gene. 1994. "Genetics: The Money Rush Is On." *Fortune* (30 May): 94–108.
Callanan, N. P., B. Cheuvront, and J. R. Sorenson. 1995. "Non-traditional Genetic Education for Potential CF Carriers: Comparison of Risk Perception,

Knowledge, and Satisfaction among CF Relatives Screened in Clinic or at Home." Abstract of a paper presented at the annual meeting of the American Society for Human Genetics, October 1995. *American Journal of Human Genetics* 57: A29.

Campbell, John Angus. 1987. "Charles Darwin: Rhetorician of Science." In *The Rhetoric of the Human Sciences,* ed. John Nelson, Allan Megill, and Donald N. McCloskey, 69–86. Madison: University of Wisconsin Press.

Campbell, John Angus. 1990. "Scientific Discovery and Rhetorical Invention: The Path to Darwin's Origin." In *The Rhetorical Turn: Invention and Persuasion in the Conduct of Inquiry,* ed. Herbert W. Simons, 58–90. Chicago: University of Chicago Press.

"Cancer and the Brave Hope." 1947. *Newsweek* (14 April): 56.

Capron, Alexander Morgan. 1985. "Technology and Ethics: Human Genetic Engineering." *Current* (January): 8–13.

Capron, Alexander M., Marc Lappé, Robert F. Murray, Jr., Tabitha M. Powledge, Sumner B. Twiss, and Daniel Bergsma, eds. 1979. *Genetic Counseling: Facts, Values, and Norms.* New York: Alan R. Liss, Inc.

Carey, John. 1990. "The Genetic Age." *Business Week* (28 May): 68–71, 74, 78, 82–83.

Carey, John. 1992. "This Genetic Map Will Lead to a Pot of Gold." *Business Week* (2 March): 74.

Carey, John. 1993. "The $600 Million Horse Race." *Business Week* (23 August): 68.

Carey, John. 1994. "Gene Therapy: Promises, Promises." *Business Week* (26 September): 92.

Carey, John. 1998. "Don't Ban Cloning, Don't Ban Cloning." *Business Week* (2 March): 44.

Carey, Joseph. 1984. "New Insight into Genes: Now the Payoff." *U.S. News and World Report* (6 August): 57.

Carey, Joseph. 1985. "Genes." *U.S. News and World Report* (11 November): 56.

Carlisle, Norman. 1951. "What Heredity Means to You." *Coronet* (April): 140–46.

Carlson, A. Cheree. 1988. "Limitations on the Comic Frame: Some Witty American Women of the Nineteenth Century." *Quarterly Journal of Speech* 74: 310–22.

Carlson, Elof Axel. 1991. "Defining the Gene: An Evolving Concept." *American Journal of Human Genetics* 49: 475–87.

Carpenter, Betsy. 1989. "The Body's Master Controls." *U.S. News and World Report* (8 May): 57–59.

Cassidy, Robert. 1974. "University Professors under Attack." *Chicago Tribune,* 26 March, I, 14 (3).

Cimons, Marlene. 1985. "Your Health, Your Heredity: How Strong Is the Link?" *Glamour* (June): 232–33, 266–68, 284.

Clark, Matt, with Mariana Gosnell, Daniel Shapiro, and Mary Hager. 1984. "Medicine: A Brave New World." *Newsweek* (5 March): 64–70.

Clayton, Ellen Wright. 1994. "What the Law Says about Reproductive Genetic Testing and What It Doesn't." In *Women and Prenatal Testing: Facing the*

Challenges of Genetic Technology, ed. Karen H. Rothenberg and Elizabeth J. Thomson, 131–60. Columbus: Ohio State University Press.
"Clinic for Ancestors." 1946. *Newsweek* (23 December): 54–55.
"Clues to Chemistry of Heredity Found." 1953. *New York Times,* 13 June, 17 (8).
Cohen, Mark. 1993. "Designer Genes." *GQ* (October): 268–75.
Condit, Celeste M. 1990. *Decoding Abortion Rhetoric: Communicating Social Change.* Urbana: University of Illinois Press.
Condit, Celeste M. 1994. "Hegemony in a Mass Mediated Society: Concordance about 'Reproductive Technologies.'" *Critical Studies in Mass Communication* 11 (September): 205–30.
Condit, Celeste M. 1996a. "How Bad Science Stays That Way: Of Brain Sex, Demarcation, and the Status of Truth in the Rhetoric of Science." *Rhetoric Society Quarterly* 26: 83–109.
Condit, Celeste M. 1996b. "Media Bias for Reproductive Technologies." In *Evaluating Women's Health Messages: A Resource Book,* ed. Roxanne Louiselle Parrott and Celeste Michelle Condit, 341–55. Thousand Oaks, Calif.: Sage.
Condit, Celeste M. 1999a. "Public Discourse about Genetic Blueprints: Audience Responses Are Not Necessarily Deterministic or Discriminatory." *Public Understandings of Science,* special issue.
Condit, Celeste M. 1999b. "The Materiality of Coding: Rhetoric, Genetics, and the Matter of Life." *Rhetorical Bodies,* ed. Jack Selzer and Sharon Crowley. Madison: University of Wisconsin Press.
Condit, Celeste M. Forthcoming. "Women's Reproductive Choices and the Genetic Model of Medicine." *Body Talk: Rhetoric, Technology, Reproduction* (working title), ed. Mary M. Lay and Laura Gurak. Madison: University of Wisconsin Press.
Condit, Celeste, and John Louis Lucaites. 1993. *Crafting Equality: America's Anglo-African Word.* Chicago: University of Chicago Press.
Condit, Celeste, and Melanie Williams. 1994–1995. "Gender Differences and Argumentation: A Positional Account of the Reception of Genetics Arguments." *Speaker and Gavel* 32: 1–12.
Condit, Celeste, and Melanie Williams. 1997. "Audience Responses to the Discourse of Medical Genetics: Evidence against the Critique of Medicalization." *Health Communication* 9: 219–36.
Condit, C. M., Roxane L. Parrott, and Beth O'Grady. In press. "Principles and Practice of Communication Processes for Genetics in Public Health." In *Genetics and Public Health: Translating Advances in Human Genetics into Disease Prevention and Health Promotion,* ed. Muin J. Khoury, Wylie Burke, and Elizabeth Thomson. New York: Oxford University Press.
Conrad, Peter. 1997. "Public Eyes and Private Genes: Historical Frames, News Constructions, and Social Problems." *Social Problems* 44: 139–54.
Conrad, Peter, and Dana Weinberg. 1996. "Has the Gene for Alcoholism Been Discovered Three Times Since 1980? A News Media Analysis." *Perspectives on Social Problems* 8: 3–25.
Cooley, Charles H. 1923. "Heredity and Instinct in Human Life." *Survey* (1 January): 454–56.

Crane Davis, Dorothy. 1968. "Predicting Tomorrow's Children." *Today's Health* (January): 32–37.
Cranor, Carl F., ed. 1994. *Are Genes Us? The Social Consequences of the New Genetics.* New Brunswick, N.J.: Rutgers University Press.
Cravens, Hamilton. 1988. *The Triumph of Evolution: The Heredity-Environment Controversy, 1900–1941.* Baltimore: Johns Hopkins University Press (originally published 1978).
Crick, Francis. 1988. *What Mad Pursuit.* New York: Basic Books.
Crooke, Stanley T. 1988. "Knowledge and Power: Biotechnology." *Vital Speeches of the Day* (15 September): 732–35.
Curran, John J. 1987. "Will Biotech's Boom Go Bust?" *Fortune* (6 July): 101–2.
Darlington, C. D. 1947. "The Retreat from Science in Russia." *Nineteenth Century* (October): 157–68.
Davidson, R. Michael. 1969. "Man's Participatory Evolution: Living in a Biological Revolution." *Current* (March): 4–10.
Davis, Bernard. 1984. "Ban Experiments in Genetic Engineering?" *U.S. News and World Report* (3 October): 44.
Davis, Maxine. 1947. "If You Want a Baby." *Good Housekeeping* (February): 42–43, 173.
Dawkins, Richard. 1976. *The Selfish Gene.* New York: Oxford University Press.
"Dean Inge Bids Church Heed Eugenics' Value: Says Advocates Are Not Fools or Cranks, Urging Churchmen Not to Be Indifferent." 1930. *New York Times,* 10 October, 19 (3).
"Detecting an Old Killer." 1971. *Time* (4 October): 57.
Dobzhansky, Theodosius. 1966. "Equality: A Geneticist's View." *Catholic World* (November): 104–12.
Doyle, Richard. 1997. *On Beyond Living: Rhetorical Transformations of the Life Sciences.* Stanford, Calif.: Stanford University Press.
Duster, Troy. 1990. *Backdoor to Eugenics.* New York: Routledge.
Eisenberg, Lucy. 1966. "Genetics and the Survival of the Unfit." *Harper's Magazine* (February): 53–58.
"The End of Insurance." 1990. *New Republic* (9 and 16 June): 26.
Entman, Robert M. 1990. "Modern Racism and the Images of Blacks in Local Television News." *Critical Studies in Mass Communication* 7: 332–45.
Etzioni, Amitai. 1973. "Genetic Engineering." *Harper's* (December): 11.
Etzioni, Amitai. 1974. "Biomedicine and Ethics: What Role for Genetic Engineering?" *Current* (January): 38–49.
"Exploring the Secrets of Life." 1963. *Newsweek* (13 May): 63–66.
Farrell, Thomas. 1990. "From the Parthenon to the Bassinet: Along the Epistemic Trail." *Quarterly Journal of Speech* 76: 78–84.
Farrell, Thomas B., and G. Thomas Goodnight. 1981. "Accidental Rhetoric: The Root Metaphors of Three Mile Island." *Communication Monographs* 48: 271–300.
"Fears Moron Types Will People Nation: Dr. Wiggam Says Democracy Will Not Outlast Century Unless the Intelligent Reproduce." 1930. *New York Times,* 22 March, 4.

"Fortune Survey: Sterilization of Criminals." 1937. *Fortune* (July): 106.

Foss, Sonja K. 1989. *Rhetorical Criticism: Exploration and Practice.* Prospect Heights, Ill.: Waveland Press.

Foss, Sonja K., and Cindy L. Griffin. 1995. "Beyond Persuasion: A Proposal for an Invitational Rhetoric." *Communication Monographs* 62: 2–18.

Foucault, Michel. 1980. *Power/Knowledge: Selected Interviews and Other Writings, 1972–1977.* Ed. Colin Gordon. New York: Pantheon.

Foucault, Michel. 1990. *Politics, Philosophy, Culture: Interviews and Other Writings 1977–1984.* Ed. Lawrence D. Kritzman. Trans. Alan Sheridan and others. New York and London: Routledge.

Fox Keller, Evelyn. 1995. *Refiguring Life: Metaphors of Twentieth-Century Biology.* New York: Columbia University Press.

Frank, Robert Edward. 1995. "A Rhetorical History of the 1989 Revolution in the German Democratic Republic: Calling Forth the People." Ph.D. diss., University of Georgia, Athens, Ga.

"Freakishness of Heredity." 1915. *Scribner's Magazine* (April): 518.

French, Kathryn J., Theresa D. Frasier, and C. Jay Frasier. 1996. "Knowing When to Say When and Why: Media Messages Aimed at Preventing Women's Alcohol Consumption." In *Evaluating Women's Health Messages,* ed. Roxanne Louiselle Parrott and Celeste Michelle Condit, 190–203. Thousand Oaks, Calif.: Sage Publications.

Fried, John J. 1975. "Is Science Creating Dangerous New Bacteria?" *Reader's Digest* (December): 133–36.

Fromer, Anne. 1956. "She Predicts What Your Baby Will Be Like." *Coronet* (November): 131–34.

"Frozen Fatherhood." 1961. *Time* (8 September): 68.

"The Future of America: A Biological Forecast." 1928. *Harper's Magazine* (April): 529–39.

"The Future of Our Genes." 1953. *Saturday Review* (5 December): 25.

"Gaining and Losing Human Power." 1919. *Literary Digest* (15 November): 25–26.

Galton, Francis. [1909] 1985. "Eugenics: Its Definition, Scope and Aims." In *The History of Hereditarian Thought,* ed. Charles Rosenberg. New York/London: Garland Publishing Co. (originally a paper read before the Sociological Society at a meeting in the School of Economics and Political Science, London University, 16 May 1904; originally published in *Essays in Eugenics, London: The Eugenics Education Society,* 1909).

Gardner, William. 1995. "Can Human Genetic Enhancement Be Prohibited?" *Journal of Medicine and Philosophy* 20: 65–84.

Geller, Lisa N., Joseph S. Alper, Paul R. Billings, Carol I. Barash, Jonathan Beckwith, and Marvin R. Natowicz. 1996. "Individual, Family, and Societal Dimensions of Genetic Discrimination: A Case Study Analysis." *Science and Engineering Ethics* 2: 71–88.

Gelman, David, with Thomas M. DeFrank and Jeff B. Copeland. 1976. "Politics and Genes." *Newsweek* (12 January): 50–54.

"The Gene Machine." 1976. *Newsweek* (13 September): 72.

"The Gene Makers." 1970. *Newsweek* (15 June): 91.
"Genes: Sliced and Pictured." 1949. *Newsweek* (24 January): 44.
"The Gene-Testing Boom Is Still Set for Someday." 1994. *Business Week* (21 November): 102–3.
"Genetic Engineering." 1972. *Newsweek* (15 May): 72.
"Genetic Engineering Stirs Debate." 1975. *Washington Post,* 23 April, A, 2 (2).
"Genetic Moratorium." 1976. *Time* (19 July): 67.
"Genetic Rosetta Stone." 1960. *Time* (23 May): 50.
"Genetics and Medicine." 1941. *New York Times,* 2 February, D5.
Gianturco, Michael. 1990. "Theme for the Nineties." *Forbes* (8 January): 304.
Gilman, William. 1966. "Tampering with Our Genetic Blueprint: This Time We're Warned." *New Republic* (12 November): 12–13.
Glick, Daniel. 1989. "A Genetic Road Map." *Newsweek* (2 October): 46.
"Gloomy Nobelman." 1946. *Time* (11 November): 96.
Goddard, Henry Herbert. 1931. *The Kallikak Family: A Study in the Heredity of Feeble-mindedness.* New York: Macmillan Co.
"Going Nature One Better." 1986. *U.S. News and World Report* (24 March): 55.
Goodman, Walter. 1965. "What Happens When Life Begins." *Redbook* (July): 55, 94–96.
Gorman, Christine. 1988. "A Mouse That Roared." *Time* (25 April): 83.
Gorman, Christine. 1995. "The Doctor's Crystal Ball." *Time* (10 April): 60–62.
Gribbin, John. 1985. *In Search of the Double Helix: Darwin, DNA, and Beyond.* Aldershot, England: Wildwood House.
Gruenberg, Benjamin C., and Joel Berg. 1956. "The Stuff That Life Is Made Of." *Parents Magazine* (May): 42–43, 114–16.
Gumpert, Martin. 1946. "The Nazis' Biological Warfare." *Nation* (18 May): 597–98.
Gur, R. C., L. H. Mozley, P. D. Mozley, S. M. Resnick, J. S. Karp, A. Alavi, S. E. Arnold, and R. E. Gur. 1995. "Sex Differences in Regional Cerebral Glucose Metabolism During a Resting State. *Science* 267: 528–31.
Haggerty, Sandra. 1974. "Of Race and Genetics." *Los Angeles Times,* 10 September, II, 7 (1).
Hall, Stuart, Chas Critcher, Tony Jefferson, John Clarke, and Brian Roberts. 1978. *Policing the Crisis: Mugging, the State, and Law and Order.* London: Macmillan.
Haller, Mark H. 1984. *Eugenics: Hereditarian Attitudes in American Thought.* New Brunswick, N.J.: Rutgers University Press (originally published 1963).
Hamilton, Joan O'C. 1990. "The Genetic Age." *Business Week* (28 May): 68–83.
Hamilton, Joan O'C. 1992. "Biotech: America's Dream Machine." *Business Week* (2 March): 66–74.
Hamilton, Joan O'C. 1994. "Biotech: An Industry Crowded with Players Faces an Ugly Reckoning." *Business Week* (26 September): 84–92.
Hamilton, Michael. 1969. "New Life for Old: Genetic Decisions." *Christian Century* (28 May): 741–44.
Hammons, Helen. 1961. "Heredity Counseling." *Parents Magazine* (January): 40–41, 78, 103, 111.

Harbison, Earle H., Jr. 1990. "Technological Innovation and Political Leadership." *Vital Speeches of the Day* (15 September): 723–26.
Harding, Sandra. 1986. *The Science Question in Feminism.* Ithaca and London: Cornell University Press.
Hart, Roderick P. 1990. *Modern Rhetorical Criticism.* Glenview, Ill.: Scott Foresman/Little, Brown Higher Education.
Hasian, Marouf A., Jr. 1996. *The Rhetoric of Eugenics in Anglo-American Thought.* Athens, Ga.: University of Georgia Press.
Hasian, Marouf A., Jr., Celeste M. Condit, and John L. Lucaites. 1996. "The Rhetorical Boundaries of 'the Law': A Consideration of the Rhetorical Culture of Legal Practice and the Case of the 'Separate But Equal' Doctrine." *Quarterly Journal of Speech* 82: 323–42.
Haskins, Caryl P. 1945. "The Changing Picture of Heredity." *American Scholar* (January): 97–105.
Hayes, Brian. 1998. "The Invention of the Genetic Code." *American Scientist* (January–February): 8–14.
Haynes, Valerie M. 1974. "Once Lovely Boy Wastes Away: Tay-Sachs Victim." *Times-Picayune* (New Orleans), 28 April, I, 23 (1).
"Heredity." 1952. *Today's Health* (June): 67.
"Heredity and Mental Illness." 1956. *Today's Health* (May): 22, 52, 53–55.
Herrnstein, Richard J., and Charles Murray. 1994. *The Bell Curve: Intelligence and Class Structure in American Life.* New York: Free Press.
Hicks, Clifford B. 1969. "George Beadle Talks about the New Genetics." *Today's Health* (July): 44–47, 61.
Hill, Shirley A. 1994. *Managing Sickle Cell Disease in Low Income Families.* Philadelphia: Temple University Press.
Hirschhorn, Kurt. 1968. "On Re-doing Man." *Commonweal* (17 May): 257–61.
Historical Statistics of the United States: Colonial Times to 1970. 1970. Washington, D.C.: U.S. Government Printing Office.
Hoffman, Roald. 1998. "Chemistry's Essential Tension: The Same and Not the Same." Charter Lecture, University of Georgia, 23 February.
Holmes, Samuel J. 1926. "Eugenics Vital to the Human Race." *Current History* (December): 348–50.
Holtzman, Neil A. 1989. *Proceed with Caution: Predicting Genetic Risks in the Recombinant DNA Era.* Baltimore: Johns Hopkins University Press.
Hooton, Earnest. 1943. "Morons into What?" *Woman's Home Companion* (August): 4, 96.
"Houses to Go to Parents of 'Faultless' Children." 1934. *New York Times,* 30 July.
"How about a Smart Pill for Investors?" 1986. *Forbes* (5 May): 146–47.
"How the Races of Man Developed." 1953. *Life* (18 May): 101–6.
Hubbard, Ruth, and Elijah Wald. 1993. *Exploding the Gene Myth: How Genetic Information Is Produced and Manipulated by Scientists, Physicians, Employers, Insurance Companies, Educators, and Law Enforcers.* Boston: Beacon Press.
Huber, Peter. 1987. "New Biotechnologies: The Uncertain Regularity Trumpet." *Vital Speeches of the Day* (1 April): 369–71.

Huber, Peter. 1991. "Property Lines." *Forbes* (13 May): 124.
"Human Cloning to Start in Russia." 1998. AAP News Service, 7 March.
Hunt, Morton M. 1954. "Doctor Kallmann's 7000 Twins." *Saturday Evening Post* (6 November): 20–23, 81–82.
"Husbanding the Nation's Manhood." 1910. *World's Work* (October): 13470–71.
Impoco, Jim. 1993. "Green Genes." *U.S. News and World Report* (18 October): 58–60.
"Improving Mankind." 1928. *New York Times*, 22 January, III, 4 (3).
"Inheritance of Acquired Characters Proved at Last." 1919. *Current Opinion* (August): 107.
"Intelligence: Is There a Racial Difference?" 1969. *Time* (11 April): 54.
James, D. C. S., L. A. Crandall, B. A. Rienzo, and R. W. Trottier. 1995. "Roles of Physicians, Genetic Counselors and Nurses in the Genetic Counseling Process." *Journal of the Florida Medical Association* 82: 403–10.
Jaroff, Leon. 1989. "The Gene Hunt." *Time* (20 March): 62–67.
Jaroff, Leon. 1992. "Seeking a Godlike Power." *Time*, special issue (Fall): 58.
Jaroff, Leon. 1993. "Happy Birthday Double Helix." *Time* (15 March): 56–59.
Johnson, Julie. 1992. "Patenting Life." *Ms.* (November/December): 82–83.
Jones, Stanton L. 1995. "My Genes Made Me Do It." *Christianity Today* (24 April): 14–18.
Jordan, Winthrop D. 1977. *White over Black: American Attitudes toward the Negro, 1550–1812.* New York: W. W. Norton (originally published 1968).
Judson, Horace Freeland. 1975. "Fearful of Science." *Harper's Magazine* (March): 32–41.
Judson, Horace Freeland. 1979. *The Eighth Day of Creation: Makers of the Revolution in Biology.* New York: Simon and Schuster.
Judson, Horace Freeland. 1983. "Thumbprints in Our Clay." *New Republic* (19–26 September): 12–17.
Kaempffert, Waldemar. 1944. "Papers Read before Chemical Society Show How Chemists Are Shaping Our Lives." *New York Times*, 17 September, E9.
Kaempffert, Waldemar. 1947. "Experiments of a Geneticists to Discover the Connection between Cancer and Heredity." *New York Times*, 3 August, E7.
Katz Rothman, Barbara. [1986] 1993. *The Tentative Pregnancy: How Amniocentesis Changes the Experience of Motherhood.* New York: W. W. Norton.
Katz Rothman, Barbara. 1998. *Genetic Maps and Human Imaginations: The Limits of Science in Understanding Who We Are.* New York: W. W. Norton.
Kay, Lily E. 1993. *The Molecular Vision of Life.* New York/Oxford: Oxford University Press.
Kay, Lily E. 1995. "Who Wrote the Book of Life? Information and the Transformation of Molecular Biology, 1945–55." *Science in Context* 8 (4): 609–34.
Kaye, Howard L. 1997. *The Social Meaning of Modern Biology: From Social Darwinism to Sociobiology.* New Brunswick, N.J.: Transaction Publishers (originally published 1986).
Keehn, Joel. 1992. "Mean Green." *Buzzworm* (January/February): 33–37.
Kellogg, Vernon. 1924. "The Human Future." *World's Work* (June): 204–7.
Kevles, Daniel J. 1995. *In the Name of Eugenics: Genetics and the Uses of Human*

Heredity. Cambridge, Mass.: Harvard University Press (originally published 1985).

Kevles, Daniel J., and LeRoy Hood, eds. 1992. *The Code of Codes: Scientific and Social Issues in the Human Genome Project.* Cambridge, Mass.: Harvard University Press.

Khoury, M. J. 1997. "Relationship between Medical Genetics and Public Health: Changing the Paradigm of Disease Prevention and the Definition of a Genetic Disease." *American Journal of Medical Genetics* 71: 289–91.

Kirp, David. 1990. "Health: Some Women Workers Are Already Facing This Choice: Quit or Be Sterilized." *Vogue* (May): 198, 203–4.

Kitcher, Philip. 1996. *The Lives to Come: The Genetic Revolution and Human Possibilities.* New York: Simon and Schuster.

Kleegman, Sophia J., and Mildred Gilman. 1947. "Why Can't You Have a Baby?" *Parents Magazine* (December): 31, 68.

Kline, Kimberly N. 1996. "The Drama of in Utero Drug Exposure: Fetus Takes First Billing." In *Evaluating Women's Health Messages,* ed. Roxanne Louiselle Parrott and Celeste Michelle Condit, 61–75. Thousand Oaks, Calif.: Sage Publications.

"A Knockout for the 'Perfect Man.'" 1932. *Literary Digest* (1 October): 20.

Kolker, Aliza, and B. Meredith Burke. 1994. *Prenatal Testing: A Sociological Perspective.* Westport, Conn.: Bergin & Garvey.

Kornberg, Arthur. 1981. "Genetic Engineering: 'Revolution in Medicine.'" *U.S. News and World Report* (16 March): 74.

Kornhauser, Arthur. 1946. "Does Heredity Make You What You Are? Poll of Experts." *American Magazine* (May): 52–53.

Korzybski, Alfred. 1958. *Science and Sanity: An Introduction to Non-Aristotelian Systems and General Semantics.* Clinton, Mass.: Colonial Press (originally published 1924).

Kotulak, Ronald. 1975. "Manipulation." *Chicago Tribune,* 21 September, 23–24, 26, 28, 30.

Krause, Carol. 1994. "How to Research Your Medical Family Tree." *Good Housekeeping* (August): 161–62.

Krimsky, Sheldon. 1982. "Social Responsibility in an Age of Synthetic Biology." *Environment* (July/August): 2–11.

Laird, Donald A. 1952. "Intelligence and Heredity." *Today's Health* (August): 23, 31.

Lamberg, Lynne. 1976. "Genetic Screening: Learning What You Never Wanted to Know." *Today's Health* (March): 28–31, 53–54.

Larson, Edward J. 1995. *Sex, Race, and Science: Eugenics in the Deep South.* Baltimore and London: Johns Hopkins University Press.

Lasagna, Louis. 1962. "Heredity Control: Dream or Nightmare?" *New York Times Magazine* (5 August): 7, 58–61.

Lash, Joseph P. 1949. "Russian Science and Heresy." *New Republic* (3 January): 12.

Latour, Bruno, and Steve Woolgar. 1979. *Laboratory Life: The Social Construction of Scientific Facts.* Beverly Hills, Calif.: Sage.

Laurence, William. 1949. "Studies Hint Cause of Alcoholism May Lie in Inherited Food 'Block.'" *New York Times,* 28 April, 33 (5).

Lederberg, Joshua. 1969. "Genetic Engineering: Controlling Man's Building Blocks." *Today's Health* (November): 24–27, 80, 83.

Lederberg, Joshua. 1970. "Participatory Evolution: What Controls for Genetic Engineering?" *Current* (September): 48–51.

Leff, Michael. 1988. "Dimensions of Temporality in Lincoln's Second Inaugural." *Communication Reports* 1: 26–31.

Leo, John. 1984. "Are Criminals Born, Not Made?" *Time* (21 October): 94.

Leo, John. 1989. "Genetic Advances, Ethical Risks." *U.S. News and World Report* (25 September): 59.

Lessing, Lawrence. 1966. "Into the Core of Life Itself." *Fortune* (March): 146–51, 174–76.

Levine, Jonathan. 1990. "Cutting the Heart Out of European Biotech?" *Business Week* (18 June): 177–78.

Lewin, Benjamin. 1983. *Genes.* New York: Wiley.

Lewin, Benjamin, 1987. *Genes.* 3d ed. New York: Wiley.

Lippman, Abby. 1992. "Led (Astray) by Genetic Maps: The Cartography of the Human Genome and Health Care." *Social Science and Medicine* 35: 1469–76.

Lippman, Abby. 1993. "Prenatal Genetic Testing and Geneticization: Mother Matters for All." *Reproductive Genetic Testing: Impact upon Women, Fetal Diagnosis and Therapy* 8 (suppl. 1): 64–79.

"Lottery of the Genes." 1943. *Newsweek* (13 September): 86–87.

Lucaites, John L., Celeste M. Condit, and Sally Caudill. 1998. *Contemporary Rhetorical Theory: A Reader.* New York: Guilford Press.

Ludmerer, Kenneth M. 1972. *Genetics and American Society: A Historical Appraisal.* Baltimore and London: Johns Hopkins University Press.

Luria, S. E. 1969. "Modern Biology: A Terrifying Power." *Nation* (20 October): 406–9.

Lyne, John. 1990. "Bio-Rhetorics: Moralizing the Life Sciences." In *The Rhetorical Turn: Invention and Persuasion in the Conduct of Inquiry,* ed. Herbert W. Simons, 35–57. Chicago: University of Chicago Press.

Lyne, John, and Henry F. Howe. 1990. "The Rhetoric of Expertise: E. O. Wilson and Sociobiology." *Quarterly Journal of Speech* 76: 134–51.

Lyne, John, and Henry F. Howe. 1992. "Gene Talk in Sociobiology." *Social Epistemology* 6: 109–64.

Lyon, E., S. Hogan, J. McDonald, N. Crow, K. Ward, M. Leppert, and R. Burt. 1995. "Alternative Models for Return of Genetic Information in Presymptomatic Testing: Study of a Family with Attenuated Adenomatous Polyposis Coli." Abstract of a paper presented at the annual meeting of the American Society for Human Genetics, October. *American Journal of Human Genetics* 57: A28.

McAuliffe, Kathleen. 1987a. "Genes That Predict Disease." *Reader's Digest* (September): 17–24.

McAuliffe, Kathleen. 1987b. "Predicting Diseases." *U.S. News and World Report* (25 May): 64–69.

McCormick, Richard A. 1985. "Genetic Technology and Our Common Future." *America* (27 April): 337–42.
McDougall, Kenneth J. 1976. "Genetic Engineering: Hazard or Blessing?" *Intellect* (April): 528–30.
McGee, Michael Calvin. 1975. "In Search of 'The People': A Rhetorical Alternative." *Quarterly Journal of Speech* 61: 235–59.
McGee, Michael Calvin. 1987. "Power to the <People>." *Critical Studies in Mass Communication* 4: 432–37.
McGee, Michael Calvin. 1998. *Rhetoric in Postmodern America: Conversations with Michael Calvin McGee.* Ed. Carol Corbin. New York: Guilford Press.
McGee, Michael C., and Martha A. Martin. 1983. "Public Knowledge and Ideological Argumentation." *Communication Monographs* 50: 47–65.
McShea, Daniel W. 1992. "Gene-talk Talk about Sociobiology." *Social Epistemology* 6: 183–92.
McWethy, Jack. 1975. "A Move to Protect Mankind." *U.S. News and World Report* (7 April): 66.
McWethy, Jack. 1976. "Science's Newest 'Magic'—Blessing or Curse?" *U.S. News and World Report* (12 July): 34–35.
Mahoney, Richard. 1993. "Biotechnology and U.S. Policy: Policy Is What You Do, Not What You Say." *Vital Speeches of the Day* (15 April): 400–403.
Marshall, Carolyn. 1987. "Our Bodies, Our World." *Vogue* (October): 404, 489.
Marteau, Theresa, and Martin Richards, eds. 1996. *The Troubled Helix: Social and Psychological Implications of the New Human Genetics.* Cambridge: Cambridge University Press.
Medawar, Peter Brian. 1960. "The Future of Man." *Saturday Evening Post* (1 October): 32–33, 91–93.
Mendel, Gregor. [1865] 1959. "Experiments in Plant Hybridization." In *Classic Papers in Genetics,* ed. James A. Peters. Englewood Cliffs, N.J.: Prentice-Hall.
Mereson, Amy. 1988. " 'Will My Baby Be Okay?' New Genetic Tests Predict Diseases." *Glamour* (December): 258.
Meyer, Frank S. 1967. "The Heredity-Environment Problem." *National Review* (3 October): 1074.
Michie, Susan, Faye Bron, Martin Bobrow, and Theresa M. Marteau. 1997. "Nondirectiveness in Genetic Counseling: An Empirical Study." *American Journal of Human Genetics* 60: 40–47.
"Mighty Mice." 1982. *Time* (27 December): 79.
Miller, Judith, and Mark Miller. 1983. "Designer Genes." *USA Today* (magazine) (July): 49–51.
Milunsky, Aubrey, and George Annas. 1984. *Genetics and the Law.* Vol. 3. New York: Plenum.
"Minimizing the Risks in Genetic Engineering." 1976. *Business Week* (9 August): 66–67.
Mohr, Otto L. 1934. "Heredity and Human Affairs." *Forum* (August): 118–21.
Monmaney, Terence. 1995. "Genetic Testing: Kids' Latest Rite of Passage." *Health* (January): 46–49.

Mosedale, Laura. 1989. "Looking for Mr. Good Genes?" *Glamour* (October): 152.
Muller, H. J. 1948. "The Destruction of Science in the USSR." *Saturday Review of Literature* (4 December): 13–15, 63–65.
Murphy, Thomas P. 1984. "Things That Go Bump in the Night." *Forbes* (2 January): 312–13.
Murphy, Timothy F., and Marc A. Lappé, eds. 1994. *Justice and the Human Genome Project.* Berkeley: University of California Press.
Murray, Robert F., Jr. 1979. "Genetic Health: A Dangerous, Probably Erroneous, and Perhaps Meaningless Concept." In *Genetic Counseling: Facts, Values, and Norms,* ed. Alexander M. Capron, Marc Lappe, Robert F. Murray, Jr., Tabitha M. Powledge, Sumner B. Twiss, and Daniel Bergsma, 71–80. New York: Alan R. Liss, Inc.
Neel, James V. 1951. "Heredity and Health." *Today's Health* (May): 39, 56.
Neel, James V., and Murray Teigh Bloom. 1953. "Brown-eyed Parents, Blue-eyed Child, WHY?" *Collier's* (7 March): 11–14.
Neimark, Jill. 1993. "Family Ties: The Health and Heredity Link." *Mademoiselle* (May): 96–98.
Nelkin, Dorothy. 1992. "The Social Power of Genetic Information." In *The Code of Codes: Scientific and Social Issues in the Human Genome Project,* ed. Daniel J. Kevles and LeRoy Hood, 177–90. Cambridge, Mass.: Harvard University Press.
Nelkin, Dorothy, and Susan Lindee. 1995. *The DNA Mystique: The Gene as Cultural Icon.* New York: W. H. Freeman.
"The New Heredity as a Tissue of Absurdities." 1919. *Current Opinion* (February): 104–5.
Newton, Michael. 1976. "Will Our Baby Be 'Normal'?" *Family Health/Today's Health* (July): 18, 62–63.
Olby, Robert. 1994. *The Path to the Double Helix: The Discovery of DNA.* New York: Dover Publications (originally published 1974).
Osborn, Michael. 1967. "Archetypal Metaphor in Rhetoric: The Light-Dark Family." *Quarterly Journal of Speech* 53 (April): 115–26.
"Our Drift toward Degeneracy." 1921. *Literary Digest* (5 March): 21.
Page, Benjamin, and Robert Y. Shapiro. 1992. *The Rational Public: Fifty Years of Trends in Americans' Policy Preferences.* Chicago: University of Chicago Press.
Paul, Diane B. 1994. "Eugenic Anxieties, Social Realities, and Political Choices." In *Are Genes Us? The Social Consequences of the New Genetics,* ed. Carl F. Cranor, 142–54. New Brunswick, N.J.: Rutgers University Press.
Paul, Diane B. 1995. *Controlling Human Heredity: 1865 to the Present.* Atlantic Highlands, N.J.: Humanities Press.
Paul, Diane B. 1998. *The Politics of Heredity: Essays on Eugenics, Biomedicine, and the Nature-Nurture Debate.* Albany: State University of New York Press.
"Pause in Test-Tube Baby Research Urged." 1972. *Los Angeles Times,* 1 May, I, 5 (1).
Pearl, Raymond. 1908. "Breeding Better Men." *World's Work* (January): 9818–24.
Pease, Daniel C., and Richard F. Baker. 1949. "Preliminary Investigations of

Chromosomes and Genes with the Electron Microscope." *Science* 109 (7 January): 8–10.
Pellegrino, Edmund D., John C. Harvey, and John P. Langan, eds. 1990. *Gift of Life: Catholic Scholars Respond to the Vatican Instruction.* Washington, D.C.: Georgetown University Press.
Perelman, Chaim, and Lucy Olbrechts-tyteca. 1969. *The New Rhetoric: A Treatise on Argumentation.* Trans. John Wilkinson and Purcell Weaver. London/Notre Dame: University of Notre Dame Press.
"'Perfect Man' Aim of Nazi Eugenists." 1935. *New York Times,* 15 September.
Pernick, Martin S. 1996. *The Black Stork: Eugenics and the Death of 'Defective' Babies in American Medicine and Motion Pictures Since 1915.* New York: Oxford University Press.
Peters, Ted. 1997. *Playing God: Genetic Determinism and Human Freedom.* New York/London: Routledge.
Pickens, Donald K. 1968. *Eugenics and the Progressives.* Nashville: Vanderbilt University Press.
Plumb, Robert K. 1949. "How a Life Grows in Mice Detailed." *New York Times,* 21 August, 57 (1).
"Polydactylism in Georgia." 1945. *Life* (17 December): 40.
Portugal, Franklin H., and Jack S. Cohen. 1977. *A Century of DNA: A History of the Discovery of the Structure and Function of the Genetic Substance.* Cambridge Mass.: MIT Press.
Potter Dagget, Mabel. 1912. "Women: Building a Better Race." *World's Work* (December): 228–34.
"The Power of Inheritance." 1947. *Newsweek* (28 April): 60.
Puner, Helen. 1959. "Heredity." *Parents Magazine* (October): 71, 92–94, 96.
Railsback, Celeste C. 1984. "The Contemporary American Abortion Controversy: Stages in the Argument." *Quarterly Journal of Speech* 70: 410–24.
Rapp, Rayna. 1988. "Chromosomes and Communication: The Discourse of Genetic Counseling." *Medical Anthropology Quarterly* 2: 143–57.
Ravage, Barbara. 1993. "Medical Genetics: A Brave New World." *Current Health* (February): 6–12.
Reilly, Philip R. 1991. *The Surgical Solution: A History of Involuntary Sterilization in the United States.* Baltimore: Johns Hopkins University Press.
Rensberger, Boyce. 1973. "Born Free?" *Mademoiselle* (January): 64–65, 119–20.
Resnick, David. 1994. "Debunking the Slippery Slope Argument against Human Germ-line Gene Therapy." *Journal of Medicine and Philosophy* 19: 23–40.
Rhein, Reginald, Jr. 1985. "Splicing Together a Regulatory Body for Biotechnology." *Business Week* (14 January): 69.
Rifkin, Jeremy. 1984. "Ban Experiments in Genetic Engineering?" *U.S. News and World Report* (8 October): 44.
"Risks Involved in the Effort to Breed Superior People." 1920. *Current Opinion* (October): 499–500.
Rivers, Caryl. 1976. "Genetic Engineers: Now That They've Gone Too Far, Can They Stop?" *Ms.* (June): 49–51.

Rogers, Michael. 1983. "Gene Splicing Leaves the Lab." *Newsweek* (15 August): 63.
Rorvik, David M. 1969a. "The Brave New World of the Unborn." *Look* (November): 79–83.
Rorvik, David M. 1969b. "Making Men and Women without Men and Women." *Esquire* (April): 108–15.
Rosenfeld, Albert. 1959. "The Futuristic Riddle of Reproduction." *Coronet* (February): 121–26.
Rosenfeld, Albert. 1974. "Medicine's Mighty Molecules." *Saturday Review/World* (14 December): 50–54.
Rosenfeld, Albert. 1975. "Should We Tamper with Heredity?" *Saturday Review* (26 July): 44–45.
Rosner, Mary, and T. R. Johnson. 1995. "Telling Stories: Metaphors of the Human Genome Project." *Hypatia* 10: 104–29.
Rostand, Jean. 1959. "Can Man Be Modified?" *Saturday Evening Post* (2 May): 25, 97–98, 101.
Rothenberg, Karen H., and Elizabeth J. Thomson. 1994. *Women and Prenatal Testing: Facing the Challenges of Genetic Technology.* Columbus: Ohio State University Press.
Rusk, Howard A., Chester A. Swinyard, and Michael R. Swift. 1969. "Solving the Mystery of Birth Defects." *Parents Magazine* (November): 61–62, 145–46.
Sandroff, Ronni. 1989. "The Baby Shoppers." *Vogue* (May): 246, 248, 256.
Sarto, Gloria E. 1970. "The Expectant Mother: Genetic Counseling." *Redbook* (December): 13, 14, 21.
"Says Cancerous Plants Show Sign of Swastika." 1940. *New York Times,* 24 April, 3 (2).
Schaeffer, Charles. 1987. "Will the Baby Be Okay?" *Changing Times* (March): 97–103.
Scheinfeld, Amram. 1953. "You, Your Child, and Heredity." *Parents Magazine* (June): 40–42, 78–80.
Schulman, Roger. 1985. "Reading the DNA Spiral for Lurking Hereditary Risks." *Business Week* (21 October): 72.
"Scientists Urged to Go Slow on Test-Tube 'Superhumans.' " 1972. *Los Angeles Times,* 29 December, I, 1 (5).
Scott, John Paul, and John L. Fuller. 1965. "What Dogs Tell Us about Man's Future." *Saturday Review* (6 March): 47–51.
Scott, Robert L., and D. K. Smith. 1969. "The Rhetoric of Confrontation." *Quarterly Journal of Speech* 55: 1–8.
"The Secret of Life." 1958. *Time* (14 July): 50–54.
"The Secrets of the Cell." 1970. *Time* (15 June): 43–44.
"Seed Vows to Continue Quest to Clone Human Beings." 1998. *Transplant News* (14 March).
Sefcovic, Enid. 1997. "Forming 'Workers Rights': The Discourse of the Wagner Act." Ph.D. diss., University of Georgia, Athens, Ga.
Selden, Steven. 1989. "The Use of Biology to Legitimate Inequality: The Eu-

genics Movement within the High School Biology Textbook, 1914–1949." In *Equity in Education,* ed. Walter G. Secada, 118–45. New York: Falmer Press.
Seligman, Jean. 1975. "Jewish Diseases." *Newsweek* (26 May): 57.
Senn, Milton J. E. 1956. "Will Your Child Inherit Your Emotions?" *Woman's Home Companion* (October): 31, 103.
Shamel, Roger E. 1985. "Biotechnology: The New Growth Industry." *USA Today* (magazine) (March): 38–41.
Sheerin, John B. 1972. "If You Think Things Are Complicated Now." *New Catholic World* (September): 212–13, 238.
"Shockley Asks Black Aid in Test." 1973. *Washington Post,* 7 July, A, 6 (1).
Singer, Eleanor, Amy Corning, and Mark Lamias. 1998. "The Polls—Trends: Genetic Testing, Engineering, and Therapy." *Public Opinion Quarterly* 62: 633–64.
Smith, J. David. 1985. *Minds Made Feeble: The Myth and Legacy of the Kallikaks.* Rockville, Md.: Aspen Systems Corp.
Smith, Kelly. 1992. "The New Problem of Genetics: A Response to Gifford." *Biology and Philosophy* 7: 331–48.
Smith, Samuel George. 1912. "The New Science." *Atlantic Monthly* (December): 801–9.
Solomon, Martha. 1979. "The 'Positive Woman's' Journey: A Mythic Analysis of the Rhetoric of STOP ERA." *Quarterly Journal of Speech* 65: 262–74.
Spallone, Patricia. 1992. *Generation Games: Genetic Engineering and the Future for Our Lives.* Philadelphia: Temple University Press.
"Speeding up Evolution." 1946. *Newsweek* (11 November): 71.
Spencer, Steven M. 1971. "New Strides in the Battle against Birth Defects." *Reader's Digest* (May): 159–64.
Springer, John L. 1958. "Small Wonder Called the Gene." *New York Times Magazine* (23 November): 15, 24–27.
Steinmann, M. 1971. "Fighting the Genetic Odds." *Life* (6 August): 18–25.
Stock, Robert W. 1969a. "The Mouse Stage of the New Biology." *New York Times Magazine* (21 December): 8–9, 51–54.
Stock, Robert W. 1969b. "Will the Baby Be Normal?" *New York Times Magazine* (23 March): 25–27, 29.
Streshinsky, Shirley G. 1966. "Does It Run in the Family?" *Parents Magazine* (November): 61, 63, 164–67.
Strother, French. 1924. "Crime and Eugenics." *World's Work* (December): 168–74.
Strunsky, Simeon. 1912. "Race-Culture." *Atlantic Monthly* (December): 850–53.
Sturtevant, A. H. 1965. *A History of Genetics.* New York: Harper and Row.
"Success Has Come to Eugenic Village." 1931. *New York Times,* 27 September, III, 4 (5).
Taylor, Charles Alan. 1996. *Defining Science: A Rhetoric of Demarcation.* Madison: University of Wisconsin Press.
Taylor, Charles Alan, and Celeste Michelle Condit. 1988. "Objectivity in Mediation: Coverage of Creation/Science." *Critical Studies in Mass Communication* 5: 293–312.

Teitelman, Robert. 1984. "Dream Delayed, Dream Denied." *Forbes* (17 December): 36–37.
Thompson, Dorothy. 1956. "Radioactivity and the Human Race." *Ladies Home Journal* (September): 11.
Thompson, Margaret W. 1967. "Genetic Counseling." *PTA Magazine* (February): 12–15.
Thompson, Trisha. 1991. "Medical Fortune-Telling." *Harper's Bazaar* (November): 165–68.
Thorndike, Joseph J. 1973. "Genetics and the Future of Man." *Horizon* (Autumn): 56–63.
"Three Men and a Messenger." 1965. *Time* (22 October): 101.
"Tinkering with Life." 1977. *Time* (18 April): 32–45.
Turney, Jon. 1998. *Frankenstein's Footsteps: Science, Genetics and Popular Culture.* New Haven and London: Yale University Press.
U.S. Department of Commerce. 1975. *Historical Statistics of the United States: Colonial Times to 1970.* Washington, D.C.: U.S. Government Printing Office.
Vanderford, Marsha L. 1989. "Vilification and Social Movements: A Case Study of Pro-Life and Pro-Choice Rhetoric." *Quarterly Journal of Speech* 75: 166–82.
van Dijck, José. 1998. *Imagenation: Popular Images of Genetics.* Washington Square, N.Y.: New York University Press.
Vogel, Gretchen. 1997. "From Science Fiction to Ethics Quandary." *Science* (19 September): 1753–54.
Waddington, H. C. 1956. "Atoms and Genes." *Nation* (18 August): 137–40.
Wallis, Claudia. 1987. "Should Animals Be Patented?" *Time* (4 May): 110.
Ward, Robert De C. 1910. "National Eugenics in Relation to Immigration." *North American Review* (July): 56–57.
Wassersug, Joseph D. 1947. "More Help for Childless Couples." *Hygeia* (November-December): 834–35+.
Watson, James. 1977. "In Defense of DNA." *New Republic* (25 June): 11–14.
Watson, James. 1980. *The Double Helix.* Ed. Gunther Stent. New York: W. W. Norton.
Weir, Robert F., Susan C. Lawrence, and Evan Fales, eds. 1997. *Genes and Human Self-Understanding: Historical and Philosophical Reflections on Modern Genetics.* Iowa City: University of Iowa Press.
Weismann, August. 1889. *Essays upon Heredity and Kindred Biological Problems.* Oxford: Clarendon Press.
Weismann, August. 1893. *The Germ-Plasm: A Theory of Heredity.* Trans. W. Newton Parker and Harriet Ronnfeldt. New York: Charles Scribner's Sons.
Weitzen, Hyman G. 1966. "The Programmed Child." *Mademoiselle* (January): 70–71, 141–43.
Wertz, D., and C. Fletcher. 1993. "A Critique of Some Feminist Challenges to Prenatal Diagnosis." *Journal of Women's Health* 2: 173–88.
Wertz, D. C., and J. R. Sorenson. 1989. "Sociologic Implications." In *Fetal Diagnosis and Therapy: Science, Ethics, and the Law,* ed. M. I. Evans, J. C. Fletcher, A. O. Dixler, and J. D. Schulman, 554–65. Philadelphia: J. B. Lippincott Co.

Wexler, Nancy. 1985. "'An Array of New Tools' against Inherited Diseases." *U.S. News and World Report* (22 April): 75–76.
"What's Big in Tomorrow's Barnyard?" 1988. *USA Today* (magazine) (June): 3–4.
"What Will a Baby Look Like?" 1943. *Saturday Evening Post* (11 September): 108.
"When Will Some Companies Stop Treating Us Like the Weaker Sex?" 1988. *Glamour* (March): 102.
"Where Did You Get Those Eyes?" 1968. *Seventeen* (June): 102–3, 164–65.
Williams, Roger. 1971. "The Biology of Behavior." *Saturday Review* (30 January): 17–19, 61.
Wilson, Edward O. 1998. *Consilience: The Unity of Knowledge.* New York: Alfred O. Knopf.
Wilson, James Q. 1985. "Genetic Traits Predispose Some to Criminality." *U.S. News and World Report* (30 September): 54.
"With Billions in the Balance, Many Moral Questions Remain." 1988. *World Press Review* (August): 28–30.
Wohlner, Grace. 1973. "Unlocking the Mysteries of Inheritance." *Parents Magazine* (November): 43–45, 116–18.
Wolfe, Anthony. 1975. "Top Secret." *Saturday Review* (17 May): 8.
Wolff, Gerhard, and Christine Jung. 1995. "Nondirectiveness and Genetic Counseling." *Journal of Genetic Counseling* 4: 3–25.
Wright, Robert. 1994a. "Feminists, Meet Mr. Darwin." *New Republic* (28 November): 34–46.
Wright, Robert. 1994b. "Our Cheating Hearts." *Time* (15 August): 44–52.
Wright, Robert. 1995a. "The Evolution of Despair." *Time* (28 August): 50–57.
Wright, Robert. 1995b. "Up from Gorilla Land: The Hidden Logic of Love and Lust." *Psychology Today* (March): 26–29.
Young, Frank E. 1988. "The Gene Revolution." *Saturday Evening Post* (October): 14, 16.
Yuncker, Barbara. 1976. "Ethnic Diseases: An Up-to-Date Report." *Good Housekeeping* (January): 22–23.
Zuckerman, Diana. 1985. "Can Genes Help Helping?" *Psychology Today* (March): 80.

INDEX

abortion: ethics of, 191, 192, 198, 203; and experimental genetics, 19, 113, 123; and family genetics, 79; and gender, 198; and genetic counseling, 124, 125–28, 129, 130, 131, 136, 137, 139, 178; legalization of, 19, 125; and science-social values relationship, 221
African Americans, 93, 196, 215. *See also* race
allele, 216, 233, 235
American Board of Medical Genetics, 184
American College of Obstetricians and Gynecologists, 181
Americans for Disabilities Act, 176
amniocentesis, 19, 146, 198, 221; and genetic counseling, 126, 127, 128, 129, 130, 138, 139, 178
Andromeda Strain, The (Crichton), 110, 149, 150
animals: cloning of, 121; and environmentalism, 202–3, 207; genetic research on, 147; human beings as, 30, 38, 40, 58, 218. *See also* plant and animal genetics; stock-breeding metaphor
artifical reproduction, 19, 111, 121
atomic research, 67, 70, 134. *See also* radiation/radioactivity
atoms, as analogy for gene(s), 70–73, 96, 99, 100, 101, 102, 105–6, 107, 155, 162, 210–11, 214, 218–19

"baby prediction," 74, 76–80
Baltimore, David, 191, 203, 204, 206
Beadle, George, 54, 118, 134
behavior: and blueprint metaphor, 165, 205; and determinism, 211, 212, 234, 235; and ethics, 147; and eugenics, 16, 28–29, 39–40, 41–42, 44, 89, 211, 213; and experimental genetics, 107; and family genetics, 63, 82, 89, 90; future discourse about, 223, 234, 235; and genetic counseling, 131, 133; and individual rights, 147; and reform eugenics, 63; and regulation, 176
behaviorism, 8, 47
biohazards, 97, 140, 147–56
biotechnology: boom-bust cycles in, 171–72; costs of, 171–72; definition of, 170; emergence of, 157, 169–72; and ethics, 195, 204; and medical genetics, 170–77; potential products of, 171, 177, 195, 204, 223; regulation of, 172–77. *See also specific application, e.g.,* disease
birth control, 28, 66, 73–74, 95. *See also type of control*
blueprint metaphor: audience for, 167–69; and behavior, 165, 205; and commercialization of genetics, 157, 159, 160–69, 187; and determinism, 163, 164–69; development of, 160, 218; and disease, 163, 219; and ethics, 205–6; and gene-environment interaction, 162–64, 168–69, 177, 205; and genome, 159, 160–69, 177, 205, 219; holistic aspects of, 160, 161, 169, 177, 205; and intelligence, 165; and manipulation, 167–69, 205; overview of, 20–21, 160–69, 205–6, 219; and physical characteristics, 165
boundary model, 84–88, 95–96, 105, 106, 211
Bryan, William Jennings, 51
building blocks metaphor, 108
business community, 172, 173–77

Cambridge, Massachusetts: and biohazard issues, 149, 153; genetic counseling in, 134
cancer, 90, 139, 162, 164, 166, 183, 200–201, 206, 213, 233–34, 243
Catholic Church, 51, 101, 105, 113, 115, 117, 118–19, 125
cells, 161–62, 206, 211, 227
Center for Disease Control, 223
chaos theory, 228, 229, 232, 233
chorionic villus sampling (CVS), 178, 198

class issues: and determinism, 212, 226; and discrimination, 215, 222–23; and ethics, 193–94; and eugenics, 34–35, 36, 43; and family genetics, 80; future discourse about, 222–23, 226; and genetic counseling, 137; and problems for genetics, 7; and reform eugenics, 63
cloning: benefits of, 224; and commercialization of genetics, 186, 224–25; and diversity, 117; ethics of, 204; and experimental genetics, 19, 109, 111, 117, 119–20, 121; future discourse about, 223, 224–25, 226; inevitability of, 3; list of supermen for, 119–20, 145; and parents, 224–25; predictions about, 121; and race, 144; regulation of, 176
coevolution, 227–29
Cold War, 69–70, 75, 95
commercialization of genetics: and cloning, 224–25; and consumer guides, 183–84; and determinism, 186–87, 206; and discrimination, 187, 206; and disease, 20–21, 187; and egalitarianism, 187, 189; and ethics, 188–207; future discourse about, 224–26; and gene-based therapies, 185–86; and genetic counseling/testing, 178–81, 183–84, 217; and genetic research, 206–7; and genetics-heredity relationship, 177, 206, 207; and individual, 207; and manipulation, 159, 169–72, 187, 207; and normality, 187; overview of, 20–21, 186–87, 206–7; and perfectionism, 177, 187, 206, 217; and regulation, 172–77; and science-social values relationship, 222; and social structure, 207. *See also* biotechnology; blueprint metaphor; personal health genetics
computer metaphor, 161, 218
Congress, U.S., 41, 149, 177
consumer guides, 157, 183–84
contingent outcomes, 232–33
cost-benefits: of biohazards, 148–51; paradox of, 244–45
Crick, Francis, 19, 99, 100, 109, 140, 155
crime/criminals: and determinism, 212; and discrimination, 215; and eugenics, 16, 36, 40, 41, 213; and family genetics, 86; and gene-environment interaction, 86; and medical genetics, 21; and race, 142, 196–97, 215
culture: and blueprint metaphor, 169; and challenges to eugenics, 47; and determinism, 211, 212, 213, 227, 228–29; diversity in, 196–97; and egalitarianism, 236, 238–39; and experimental genetics, 211; and multiculturalism, 241–42; and perfectionism, 241–42. *See also* environment
cyborgs, 19, 111, 112

Darwin, Charles, 29, 30, 70–71, 119, 145
Darwinism, 22, 32, 38, 189, 218, 230
Davenport, Charles, 41
defect/defective: and commercialization of genetics, 187; definition of, 131; and discrimination, 214, 216; and ethics, 189, 190, 192; and eugenics, 89–90; and family genetics, 92; and genetic counseling, 128–33. *See also* disabilities; intelligence; mutants/mutations; prenatal screening
degenerates, 68, 71–72
depersonalization, 32–34
determinism: and behavior, 211, 212, 235; and blueprint metaphor, 163, 164–69; and challenges to eugenics, 51, 52; changing relationship of genetics and, 209–14; and class issues, 212, 226; and commercialization of genetics, 186–87, 206; and culture, 211, 212, 213, 227, 228–29; definition of genetic, 210; and disease, 211, 213, 231; and egalitarianism, 231; and environment, 227, 230, 231; and ethics, 142, 206; and eugenics, 8–9, 43, 44–45, 51–52, 210–11, 213; and experimental genetics, 123, 156, 211; and family genetics, 17–18, 82, 85, 86, 95, 96, 169; and fitness, 213; future discourse about, 226–36; and gender, 213, 226; and gene-environment interaction, 163, 210, 211, 212–13, 232–35; and genetic code metaphor, 104–8, 211, 214; and genetic counseling, 134; and genetic manipulation, 226; and genetic research, 211, 228; and intelligence, 211, 212, 213, 234, 235; and Mendelian genetics, 227; and methodology, 13;

determinism (*continued*)
and multipotentiality of genetic configurations, 228–30, 231; overviews about, 6–10, 22, 164–69, 209–14, 226–36; and parents, 211; and personal health genetics, 182; and physical characteristics, 211, 213; and prediction, 227, 229, 230; as problem for genetics, 6–10; and race, 142, 212, 226; and religion, 22; and social policies, 212, 214; and social structure, 211, 214, 226; and variability in genetic configurations, 228, 229–31, 233; and vocabulary of genetics, 232–35
disabilities: discrimination based on, 7, 189–92, 214; and ethics, 146, 188, 189–93; and experimental genetics, 156; and genetic counseling, 128–33; and genetic engineering, 128–33; and individual rights, 146; as malleable traits, 190; and problems for genetics, 7
discrimination: and commercialization of genetics, 187, 206; and environment, 216; and ethics, 140–44, 188, 189–201, 206, 216; and eugenics, 43, 44–45, 60; and experimental genetics, 97, 156; future discourse about, 222–23; and genetic manipulation, 192; heredity's relationship with, 214–16; and individualism, 192–95; and methodology, 13; and normality, 214; overviews about, 7–10, 140–44, 189–201, 214–16; and parents, 192–93; and problems for genetics, 7–10; and regulation, 176; and social structure, 192–95; variations in, 214–16. *See also type of discrimination*
discursive formations, definition of, 14
disease: and biotechnology, 171–72; and blueprint metaphor, 163, 219; and commercialization of genetics, 20–21, 187; and determinism, 211, 213, 231; discrimination based on, 7, 189–92, 214, 215–16; and ethics, 146, 188, 189–92; and eugenics, 211; and experimental genetics, 112–13, 122, 123, 155, 211; and family genetics, 18, 63, 89, 92, 95; future discourse about, 223, 224, 239; and gene-environment interaction, 163; and genetic counseling, 124, 131–32, 138–39; and genetic testing, 224; as malleable trait, 190; and perfectionism, 217, 239; and personal health genetics, 181–82, 206; and rewriting humanity, 155; susceptibility to, 164. *See also* medical genetics; *specific disease*
DNA: and biohazards, 147–48; double-helix structure of, 19, 96, 97, 99, 100–104, 109, 140, 155, 219; and experimental genetics, 20, 99, 100–104, 109, 111; and family genetics, 96; and genetic counseling, 129, 139; and genetic "defects," 129; as malleable, 164; and medical genetics, 20; pacification of, 161; physical existence of, 227; and privacy issues, 176; "recombinant," 148; and reform eugenics, 71; and rewriting humanity, 111; as "ultimate drug," 138. *See also specific metaphor*
Dobzhansky, Theodosius, 105, 118–19, 140, 195, 215
Dor Yeshorim program, 197
Down syndrome, 7, 91, 128, 129, 130, 137, 179, 191, 192–93
drugs, 170, 171, 172, 185, 194, 203, 207, 223

ectogenesis, 109, 111
education: about eugenics, 28, 51–52, 59; about genetics, 241
egalitarianism: and commercialization of genetics, 187, 189; and culture, 236, 238–39; and determinism, 231; and ethics, 141, 146, 189–95, 200; and experimental genetics, 20, 119; and family genetics, 17, 82, 90–95, 96; future discourse about, 236–39; and gender, 94, 200; and genetic counseling, 124, 131, 132, 139; heredity's relationship with, 209–10, 214–16; and individual, 146, 236–39; overviews about, 90–95, 189–95, 209–10, 214–16, 236–39; and personal health genetics, 184; and problems for genetics, 9; and race, 141, 196; variations in, 214–16
Ellis, Havelock, 39
employment discrimination, 201–2
Energy Department, U.S., 73
environment: definition/connotations of, 83, 88, 211; and determinism, 227, 230, 231; and discrimination, 216; and ethics, 141, 192; and eugenics, 16, 39–43, 44, 51–52, 54–55; and experimental

genetics, 19–20, 105, 113, 118–19, 211; and family genetics, 17–18, 79, 84–88, 89, 93; and genetic code metaphor, 105; manipulation of, 89, 169; and medical genetics, 20–21; and negative views of heredity, 22; as nurture, 88; and personal health genetics, 182, 184; and problems for genetics, 9; and race, 93, 141; and range of environments, 230; and rewriting humanity, 118–19. *See also* environmentalism; gene-environment interaction
environmentalism, 188, 195, 201, 202–3, 207
Environmental Protection Agency (EPA), 173
epigenetic rules, 228–29
ethics: and biohazards, 147–56; and blueprint metaphor, 205–6; and commercialization of genetics, 188–207; and determinism, 142, 206; and discrimination, 140–44, 188, 189–201, 206, 216; and egalitarianism, 141, 146, 189–95, 200; and environment, 141, 192; and eugenics, 80; and experimental genetics, 20, 113; and genetic code metaphor, 205; increase in concerns about, 157; and individual, 140, 146–47, 192–95, 197; and medical genetics, 21, 190; and normality, 190, 191, 192; and perfectionism, 206; and personal health genetics, 197; pragmatic issues in, 201–5; and problems for genetics, 9; and rewriting humanity, 113; and role of government, 192–93; and science, 221–22; and social structure, 192–95. *See also specific issue*
ethnicity, 43, 143–44, 196, 197. *See also* immigration
eugenics: and behavior, 16, 28–29, 39–40, 41–42, 44, 89, 211, 213; and behaviorism, 8, 47; central questions for, 49–50, 60; challenges to, 25, 46–61, 80; and class issues, 34–35, 36, 43; coining of term, 5–6; control lever in, 25, 30–34, 35–36, 43, 44, 55, 58, 60, 83, 220; and crime, 213; and Darwin/Darwinism, 29, 30, 32, 38; definition of, 5, 28; demise of, 6, 16–17, 52–56, 61, 63, 65; depersonalization in, 32–34; and determinism, 8–9, 43, 44–45, 51–52, 210–11, 213; discrimination in, 43, 44–45, 60; and disease, 211; diversity in discourse about, 27–28; emergence of, 3; and environment, 16, 39–43, 44, 51–52, 54–55; and ethics, 80; and experimental genetics, 116; and formation of public eugenics, 27–34; and gene-environment interaction, 210–11; as genetic manipulation, 4, 5–6, 30–34, 44, 56; genetics' relationship to, 4–6, 11, 13, 27, 29–30, 44, 65, 66–67; goals/outcomes of, 5, 16, 27, 28, 39, 42–43, 44; government role in, 28; historical aspects of, 4–5; impact on public of, 46–47; implementation of programs of, 46–48, 111; and individual, 5, 25, 32–33, 239; and intelligence, 16, 28–29, 33, 39–42, 44, 48, 63, 89–90, 211; and Mendelian genetics, 6, 29, 31, 32, 44, 50, 54, 55–56; as "mom and pop operation," 193; and natural selection, 29–30; overviews about, 4–6, 15–17, 25, 29–30; and parents, 4, 16, 38–39, 49; and perfectionism, 36–37, 42, 44–45, 53, 54–55, 60, 239, 240; and physical characteristics, 39–40, 89, 211; and politics, 9, 41, 220; "positive" and "negative," 59–60; and race, 28–29, 36, 43, 47, 48, 65, 142; and research, 30–34, 53–55, 60, 65, 66; rhetorical features in, 56–60; and science, 51, 52–53; and science–social values relationship, 220; and social Darwinism, 29–30, 34, 35–36; and social policies/issues, 5, 15–16, 25, 35–36, 39, 44, 58; statistics/prediction in, 25, 31, 33, 41–42, 44; and sterilization, 3, 4, 27, 28, 41, 47–48, 49, 53, 57, 59, 60; and superiority, 28, 29, 36–37, 38, 43, 54–55, 57, 60, 90, 91, 217, 218. *See also* "fitness"; germ-plasm; reform eugenics; stock-breeding metaphor
evolution, 71–72, 118, 124, 198–200, 212, 213, 228–29, 230
"evolutionary anxieties," 38
evolutionary biology, 195, 199, 200
evolutionary psychology, 21, 195, 198–200
experimental genetics: and abortion, 113, 123; and behavior, 107; challenges to, 116–22; and culture, 211; and determinism, 123, 156, 211; and disabilities, 156; and discrimination, 97, 156;

experimental genetics (*continued*)
and disease, 112–13, 122, 123, 155, 211; and DNA, 99, 100–104, 109, 111; and egalitarianism, 20, 119; and environment, 19–20, 105, 113, 118–19, 211; and family/parents, 120–21, 211; and "fitness," 118, 122; and gene-environment interaction, 100, 105–7, 113, 211; and genetic counseling, 19, 97, 123, 124–39; and genetic engineering, 109–22; and genetic manipulation, 108, 109–22, 123, 157; and genetic research, 211; and human consciousness–gene relationship, 100; implementation of results of, 111–22; and individual rights/liberties, 119, 121; and intelligence, 105, 106, 107, 108, 117, 211; and medical professionals, 155; and mutations, 102, 104, 117; and natural selection, 112, 117, 118, 119; and normality, 20, 122; and optimism about human nature, 117–18; outcomes of era of, 156, 157; overviews about, 19–20, 97, 109–22; and perfectionism, 117, 123, 156; and physical characteristics, 211; predictions about, 121–22; and public policies, 121; and race, 20, 113, 156; and rewriting humanity, 97, 109–22, 155, 211; and role of government, 112, 116, 121; and sexism, 156; and social norms, 155, 211; and social policies, 112; and social structure, 211; and superiority, 111, 112; undecidability about, 113–15; and who decides question, 119. *See also* biohazards; ethics; genetic code metaphor

family: demise of extended, 96; and experimental genetics, 120–21. *See also* family genetics; medical family histories; parents

family genetics: and abortion, 79; and "baby prediction," 74, 76–80; and behavior, 63, 82, 89, 90; and birth control, 95; and class issues, 80; and determinism, 17–18, 82, 85, 86, 95, 96, 169; and disease, 18, 63, 89, 92, 95; and DNA, 96; and egalitarianism, 17, 82, 90–95, 96; emergence of, 17–18, 75–80, 95; end of era of, 95–96; and environment, 17–18, 79, 84–88, 89, 93; and fertility, 94, 95; and gender, 90, 94–95; gene-control metaphor of, 17–18, 83–84; and gene-environment interaction, 17–18, 82–90, 95–96; and genetic counseling, 18, 63, 77, 78, 79, 95; and genetic manipulation, 83, 84, 89; and genetic research, 75, 83–84, 90, 93; and human control, 95; and individuals, 18, 96; and intelligence, 63, 78, 82, 86, 87, 89, 92, 93; and medical family histories, 95; and medical profession, 17, 18, 77–78, 81, 95, 96; and Mendelian genetics, 77, 86; and normality, 18, 91, 92, 93, 96; overviews about, 17–18, 63, 80–81, 82; and parents, 77–78, 80; and perfectionism, 95; and physical characteristics, 82; and prediction, 86; and race, 80, 82, 86, 90, 92, 93–94, 96, 142; and role of government, 63; and science, 75–76; and sexism, 82, 96; and social policies, 18, 80, 82, 89–90; and statistics, 86–87; and sterilization, 63, 91; and superiority, 18, 95

feeblemindedness, 16, 33, 36, 40–41, 48, 49, 58, 117, 129

feminism, 145, 146, 198, 236

fertility, 73–75, 94, 95, 224

"fitness": and challenges to eugenics, 52, 53, 54–55, 57–58, 59–60; and determinism, 213; and equalization of genes, 91; and ethics, 189, 192, 202; and experimental genetics, 118, 122; and gene-environment interaction, 54–55; and germ-plasm concept, 38; and goals of eugenics, 16, 28, 43, 217, 218; in Nazi Germany, 27, 53; and reform eugenics, 66; and reproductive rights, 38–39; and stock-breeding metaphor, 36–37, 38–39, 57–58, 59–60, 218

Food and Drug Administration (FDA), 173, 174, 176

Franklin, Rosalind, 155

free inquiry, 18, 63, 69–70, 81, 141–42, 149–50, 153, 154

"free will," 51, 60

French Anderson, W., 185, 205

Galton, Francis, 5, 6, 10, 11, 50, 58–59

gender: and abortion, 198; and determinism, 212, 213, 226; discrimination based on, 195, 197–201, 215; and egalitarian-

ism, 94, 200; and ethics, 140, 144–46, 188, 195–201; and family genetics, 90, 94–95; future discourse about, 226; and genetic counseling, 145–46; and genetic research, 200; and genetic testing, 200–201; and problems for genetics, 7; and reproductive rights, 146
gene(s): atoms as analogy of, 70–73, 96, 99, 100, 101, 102, 105–6, 107, 155, 162, 210–11, 214, 218–19; and blueprint metaphor, 160–69, 205; and cells, 206, 211; complexity of, 54–56, 76, 212; configurations of, 227–31; contextualization of, 160–69, 213; as controller of human destiny, 17–18, 30–34, 35–36, 43, 44, 55, 58, 60, 83–84, 109–10, 161, 162, 210–11, 220; defining equal, 90–95; definition of, 102–3; differentiation among, 164; as discrete, 214; and DNA, 102–4; as "fixed," 85, 88, 109–10, 155; interactions among, 211; learning human history from, 209; as malleable, 167–69, 205, 213, 214, 238; meanings of, 21, 207; as nature, 88; powers of, 17–18, 83–84, 86, 89, 95, 109–10, 164, 165, 166, 169, 186–87, 219; public use of term, 6; and radiation/radioactivity, 67, 70–73, 75, 81, 95; rewriting of, 106–7, 108, 109–22, 128; and science–social values relationship, 219–22; as servant, 162; "susceptibility" of, 87–88, 105, 125, 164, 211; 3-D structure of, 162; total control over, 19–20. *See also specific metaphor or topic, e.g.* germ-plasm; gene-environment interaction
gene-based therapies: and blue-print metaphor, 219; and commercialization of genetics, 170, 185–86; concerns about, 203, 204; and experimental genetics, 121, 123; and family genetics, 90; and medical genetics, 170, 185, 186; overview about, 185–86; and personal health genetics, 206; and potential of genetics, 3, 185–86, 206, 219, 221, 243
gene-environment interaction: and blueprint metaphor, 162–64, 168–69, 177, 205; complexity of, 213; and criminals, 86; and determinism, 163, 210, 211, 212–13, 232–35; and disease, 87, 163; and eugenics, 210–11; and experimental genetics, 100, 105–7, 113, 211; and family genetics, 17–18, 82–90, 95–96; future discourse about, 232–34; and genetic code metaphor, 105–7; and intelligence, 86, 87, 213; models for, 84–88, 95–96, 105–7, 163–64, 205, 210, 211; normatively based, 164, 205, 212; and prediction/statistics, 86–87; and race, 86; and vocabulary of genetics, 232–35
gene-splicing, 170
genetic biasing, 244
genetic code metaphor: and biohazards, 151–52; criticisms of, 160; and determining the code, 104–8; and determinism, 104–8, 211, 214; development/emergence of, 96, 100–104, 218; dominance of, 155–56; and ethics, 205; and family genetics, 88; and gene-environment interaction, 105–7; and genetic counseling, 129, 132–33, 139; individualization of, 205; and intelligence, 105, 106, 107, 108, 211; literalization of, 160, 205; and mutations, 102; overviews about, 19–20, 205, 219; and rewriting humanity, 109–22, 219; and science–social values relationship, 221. *See also* experimental genetics
genetic counseling: and behavior, 131, 133; certification for, 176, 179, 222; and class issues, 137; and commercialization of genetics, 178–81, 183–84; contradictory messages in, 135–37; costs of, 183–84; and determinism, 134; and disabilities, 128–33; and disease, 124, 131–32, 138–39; and DNA, 129, 139; and egalitarianism, 124, 131, 132, 139; emergence of, 18; and ethics, 145, 197; and experimental genetics, 19, 97, 123, 124–39; and family genetics, 18, 63, 77, 78, 79, 95; functions of, 124, 125–26; funding for, 225; future discourse about, 222, 225–26, 240–41; and gender, 145–46; and genetic code metaphor, 129, 132–33, 139; and genetic "defects," 128–33; and genetic engineering, 125, 128–33, 138; and "healthy children," 126, 127, 132; and hierarchy of judgments, 128–33, 139; and individual rights, 132; influence of counselors of, 155; and intelligence, 133; limiting of, 178–81;

genetic counseling (*continued*)
and medical profession, 133–39, 179–81; nondirectiveness of, 133–39, 179–80, 222; and normality, 126, 127, 128, 129, 132, 139; and paradoxes of information, 133–39; and perfectionism, 217, 240–41; and personal health genetics, 177, 183–84; and physical characteristics, 133; and prenatal screening, 124, 125–28, 129, 130, 131, 133–39, 178, 179–80, 240; and race, 197; and rewriting humanity, 155; and role of government, 134; and social norms, 137–38, 139; and who decides question, 133–39. *See also* genetic testing
genetic engineering: of animals, 203; and biohazards, 152; and disabilities, 128–33; and disease, 122; and ethics, 203; and experimental genetics, 109–22; future discourse about, 243, 244, 245; and genetic counseling, 125, 128–33, 138; and perfectionism, 243, 244; predictions about, 121–22; public policies about, 121; and rewriting humanity, 109–22. *See also* manipulation, genetic; *specific technology*
genetic enhancement, 3, 185–86, 204, 244, 245
genetic research: on animals, 147; benefits of, 206–7, 235–36; biohazards of, 147–56; and challenges to eugenics, 53–55; and Cold War, 69–70; and commercialization of genetics, 206–7; and determinism, 211, 228; emergence of, 65; Energy Department's responsibilities in, 73; and eugenics, 30–34, 53–55, 60, 65, 66; and experimental genetics, 211; and family genetics, 75, 83–84, 90, 93; free inquiry in, 18, 69–70, 81; funding for, 90, 177, 194; future discourse about, 231; and gender, 200; goal of, 124; on humans, 147; and individual rights, 147; misrepresentation about findings of, 234–26; moratorium on, 148–51; priorities for, 194–95; questions addressed by, 231; and race, 93; regulation of, 194; and science–social values relationship, 220; statistics in, 86–87; success of, 157. *See also* experimental genetics; prediction; statistics
genetics: attitudes toward, 22–23; coining of term, 29; complexity of, 54–57, 60, 61, 63, 82, 83–84, 95; definition of, 92, 93; eugenics' relationship to, 4–6, 11, 13, 27, 29–30, 44, 65, 66–67; future discourse about, 222–45; goal of, 242; heredity's relationship with, 21, 177, 179, 186, 206, 207, 211–12; historical aspects of, 4; holistic perspective of, 159, 160, 162–63, 213–14; laissez-faire development of, 154, 155–56; multipotentiality of configurations, 228–30, 231; plant and animal, 67, 75, 90, 107, 202–3; potential of, 3, 185–86, 206; reasons for studying, 235–36; for social control, 223; three problems for, 6–10; transformation to, 65–81; in twentieth century, 209–19; variety in configurations in, 228, 229–31, 233; vocabulary of, 232–35. *See also specific topic*
genetic selection, 19, 138, 183, 197, 206, 218, 221, 222, 223–24, 244. *See also* abortion; amniocentesis
genetic specification, 244
genetic surgery, 121, 122
genetic testing: acceptance of, 3–4; and biotechnology, 170, 171–72; and commercialization of genetics, 217; costs of, 183–84, 224, 225; and disease, 224; and ethics, 143, 146, 191, 192, 200–201, 202–3; funding for, 225–26; future discourse about, 222, 223–24, 225–26, 240; and gender, 200–201; and medical profession, 217, 223–24; and perfectionism, 217, 240; and personal health genetics, 179, 181–83, 185. *See also* amniocentesis; genetic counseling; prenatal screening
genome: as blueprint, 159, 160–69, 177, 205, 219; and cell, 161–62, 206, 211, 227; complexities of, 162; and determinism, 214; holism of, 159, 160, 162–63, 212, 213–14; and medical genetics, 20; and perfectionism, 244
germ-plasm: and challenges to eugenics, 50, 54, 60, 61; and Darwinism, 32; and determinism, 210, 213; and environment, 42; and eugenics-genetics relationship, 6; and eugenics in Nazi Germany, 27; and genes, 17, 37–38, 44, 54, 61, 83, 210; and goals of eugen-

ics, 25, 43, 83; and shift from focus on individual, 32–33; and statistical relationships, 86; and stock-breeding metaphor, 33, 37–38, 44, 60, 218; as unified, 17, 37, 54, 60, 61, 129
Goddard, Henry, 48
government, role of: and ethics, 192–93; and eugenics, 28; and experimental genetics, 112, 116, 121; and family genetics, 17, 63; and genetic counseling, 134; and reform eugenics, 63, 75, 77; and regulation, 157, 172–77, 194. *See also* Nazi Germany, eugenics in
Great Britain, eugenics in, 53, 68
Great Depression, 52
Griswold v. *Connecticut* (1965), 49
growth, and blueprint metaphor, 165

Haldane, J.B.S., 109, 112
Harvard University, 149, 153
health and safety. *See* biohazards
healthiness, 18, 126, 127, 132, 214, 217
heart disease, 164, 183, 202
heredity: and commercialization of genetics, 177; genetics' relationship with, 21, 177, 179, 186, 206, 207, 211–12; and religion, 22; shifts in public discourse about, 15–21, 22; in twentieth century, 209–19. *See also specific topic*
Herrnstein, Richard, 141, 196
heterotic vigor, 116–17
homosexuality, 200, 234, 235
human beings: as animals, 30, 38, 40, 58, 218; hierarchy of, 7, 29, 191–92; tragicomic spirit of, 23
human consciousness–gene relationship, 100
Human Genome Project, 12, 73, 160, 165, 201
humanity, rewriting, 109–22, 155, 211, 219. *See also* genetic engineering; manipulation, genetic; *specific technology*
human nature: optimism about, 117–18; pessimism about, 57–58, 60
Huntington's disease, 7, 78, 122, 190, 191
Hurler's disease, 126
Huxley, Aldous, 110, 116, 150

idiocy, 33, 91, 129, 130, 189
immigration, 4, 27, 28, 35, 36, 41, 47
implantation, 3, 178
individual(s): and behavior, 147; and blueprint metaphor, 168; and commercialization of genetics, 207; differences among, 236–39; and disabilities, 146; and discrimination, 192–95; and egalitarianism, 146, 236–39; and ethics, 140, 146–47, 192–95, 197, 202; and eugenics, 5, 25, 32–33, 239; and experimental genetics, 119, 121; and family genetics, 18, 96; and genetic counseling, 132; and perfectionism, 239–40, 241–42; and race, 146, 197; rights/liberties of, 119, 121, 132, 140, 146–47, 197, 202; and society, 18, 239–40; worth of, 130, 131, 132–33, 146, 224. *See also* reproductive rights
insurance: discrimination in, 188, 201–2; payment for genetic counseling and testing by, 225–26
intelligence: and blueprint metaphor, 165; debate about, 141–42; and determinism, 211, 212, 213, 234, 235; and ethics, 141–42, 189, 190, 196; and eugenics, 16, 28–29, 33, 39–42, 44, 48, 63, 89–90, 211; and experimental genetics, 105, 106, 107, 108, 117, 211; and family genetics, 63, 78, 82, 86, 87, 89, 92, 93; and gene-environment interaction, 86, 87, 213; and genetic code metaphor, 105, 106, 107, 108, 211; and genetic counseling, 133; and race, 93, 141–42, 196
intermarriage, 93
in vitro fertilization, 109

Jensen, Arthur, 141, 142
Jews, 127–28, 143, 144, 197
journalism, 231–32
Judson, Horace Freeland, 4, 115, 161, 204, 233

Kennedy, Edward, 149
"knowledge-based" standards, 175–76

Lamarckianism, 50–51, 60, 80, 220
Lysenko, Trofim, 18, 69–70, 71

magazines, 11, 232
manipulation, environmental, 89, 169

manipulation, genetic: acceptance of, 187; of animals, 203; and biohazards, 151; and blueprint metaphor, 167–69, 205; boundaries of, 112–13; and commercialization of genetics, 159, 169–72, 187, 207; and determinism, 226; and discrimination, 192; and environmentalism, 203; and ethics, 188–89, 190, 192, 203; and eugenics, 4, 5–6, 30–34, 44, 56; and experimental genetics, 108, 109–22, 123, 157; and family genetics, 83, 84, 89; future discourse about, 224–25, 243; and genetic counseling, 133; and perfectionism, 240, 243; and personal health genetics, 184; and reform eugenics, 68; self-, 240; and stock-breeding metaphor, 218. *See also type of manipulation, e.g.* cloning
Marfan's syndrome, 119
media, characteristics of, 11–12
medical family histories, 18, 95, 181–83, 184
medical genetics: applications of, 89, 157; and biotechnology, 170–77; and blueprint metaphor, 20–21, 159–77; and commercialization of genetics, 20–21; consequences of, 186; emergence of, 90; and ethics, 190; and family genetics, 89; future discourse about, 244; and genetic counseling, 138; overviews about, 20–21, 157, 170–72; and perfectionism, 244; and personal health genetics, 184; and reform eugenics, 67; regulation of, 172–77; "side effects" of, 244. *See also* disease
medical profession: and experimental genetics, 155; and family genetics, 17, 18, 77–78, 81, 95, 96; and fertility, 73–75; future discourse about, 223–24, 240–41; and genetic counseling, 133–39, 179–81; and genetic testing, 217, 223–24; and perfectionism, 240–41; and reform eugenics, 73–75, 77; and selection of physician, 184; as sexist, 146
Mendelian genetics: and blueprint metaphor, 162; and determinism, 227; and eugenics, 6, 29, 31, 32, 44, 50, 54, 55–56; and family genetics, 77, 86; rediscovery of, 3, 6, 29–30; and science–social values relationship, 220
mental abilities. *See* feeblemindedness; intelligence
mental illness. *See* feeblemindedness; "fitness"; intelligence
metaphors, 9, 218–19. *See also specific metaphor*
methodology, 10–15
mongoloids, 91, 93, 130, 131
moral issues. *See* behavior
Muller, Hermann J., 65, 69, 71, 72, 109, 112, 119
musical tape metaphor, 108
mutants/mutations, 54, 71, 72, 102, 104, 117, 150, 216, 233

National Academy of Sciences, 67–68, 173
National Institutes of Health, 148, 149, 176–77
National Society of Genetic Counselors, 222, 225
natural selection, 29–30, 68, 72, 112, 117, 118, 119
nature, 25, 88, 96, 127, 210
Nazi Germany, eugenics in: and biohazards, 150; and challenges to eugenics, 25, 46, 52–53, 55, 57, 60–61; and demise of eugenics, 65; and germplasm concept, 210; and reform eugenics, 67–69, 73, 77; and sterilization, 3, 27, 53, 57; and stock-breeding metaphor, 16, 57; and superiority, 18
newspapers, 11, 232
New York Association of Biology Teachers, 28
normality: and commercialization of genetics, 187; and discrimination, 214, 216; and ethics, 190, 191, 192; and experimental genetics, 20, 122; and family genetics, 18, 91, 92, 93, 96; and gene-environment interaction, 205; and genetic counseling, 126, 127, 128, 129, 132, 139; and perfectionism, 217; and personal health genetics, 184; and problems for genetics, 7; and race, 93
norms, and gene-environment interaction, 84–88, 95–96, 105–7, 163–64, 205, 210, 211
Northern Europeans, 16, 34–35, 36, 39, 41
nurture, 25, 88, 96, 210, 211

pangenesis theory (Darwin), 70–71
parents: child relationship with, 120–21; and cloning, 224–25; and determinism, 211; and discrimination, 192–93; and ethics, 146, 192–93, 202; and eugenics, 4, 16, 38–39, 49; and experimental genetics, 120–21, 211; and family genetics, 78; and genetic counseling, 124, 125–28, 129, 130, 131, 133–39, 178–79; and perfectionism, 7–8, 217. *See also* family genetics; prenatal screening
particularism, 209, 229, 233
patents, 176–77, 203
perfectionism: and commercialization of genetics, 177, 187, 206, 217; and culture, 241–42; and disease, 217, 239; dualism of, 3; and ethics, 206; and eugenics, 36–37, 42, 44–45, 53, 54–55, 60, 239, 240; and experimental genetics, 117, 123, 156; and family genetics, 95; future discourse about, 239–45; and genetic counseling/testing, 217, 240–41; and genetic engineering/manipulation, 240, 243, 244; genetics' relationship with, 209–10; and genome, 244; and healthiness, 217; heredity's relationship with, 216–18; and individualism, 239–40, 241–42; and medical genetics, 244; and medical profession, 240–41; and methodology, 13; and normality, 217; overviews about, 7–10, 216–18, 239–45; paradoxes about, 239–45; and parents, 7–8, 217; and prediction, 244; and problems for genetics, 7–10; and social norms, 7
personal health genetics, 21, 157, 177, 178–87, 197, 206
physical characteristics, 39–40, 82, 89, 133, 165, 186, 189, 211, 213
physicians. *See* medical profession
plant and animal genetics, 31, 34, 67, 75, 90, 107, 202–3. *See also* stock-breeding metaphor
plasmids, 111
politics, 9, 41, 220, 231
population control movement, 66
postmodernism, 232
prediction: and "baby prediction," 74, 76–80; and determinism, 227, 229, 230; about experimental genetics, 121–22; and family genetics, 86; future discourse about, 223, 244; and gene-environment interaction, 86; about genetic engineering, 121–22; inability to make, 209; and perfectionism, 244; and science–social values relationship, 222. *See also* statistics
prenatal screening: acceptance of, 3–4; and commercialization of genetics, 183; concerns about, 202, 206; and experimental genetics, 19, 123; future discourse about, 223–24, 240; gene-based therapies as alternative to, 90; and genetic counseling, 124, 125–28, 129, 130, 131, 133–39, 178, 179–80, 240. *See also* amniocentesis
privacy issues, 176, 188, 201, 202
Progressive Era, 16, 20
proteins, 54, 161, 162, 190
public: definition of, 10; emotionality of, 152–53, 174; impact of eugenics on, 46–47; rejection of reform eugenics by, 63
public discourse, 12, 13, 14–21, 22
public rhetoric, definition of, 10, 12–13

race: and crime, 196–97, 215; and determinism, 142, 212, 226; discrimination based on, 196–97, 215; and egalitarianism, 141, 196; and environment, 93, 141; and ethics, 140–44, 146, 188, 195–201; and eugenics, 28–29, 36, 43, 47, 48, 65, 142; and experimental genetics, 20, 113, 156; and family genetics, 17, 80, 82, 86, 90, 92, 93–94, 96, 142; future discourse about, 222–23, 226; and gene-environment interaction, 86; and genetic counseling, 197; and genetic research, 93; and genetic testing, 143; hierarchies of, 93–94; and individual rights, 146, 197; and intelligence, 93, 141–42, 196; and normality, 93; and personal health genetics, 197; and problems for genetics, 7; and reform eugenics, 63, 68–69, 74; and sterilization, 141
radiation/radioactivity, 63, 67, 70–73, 75, 81, 85, 95, 107
Rand Corporation, 121
recipe metaphor, 108, 218

reform eugenics: and atomic research, 67–70, 75; and behavior, 63; and birth control, 66, 73–74; and class issues, 63; and degenerates, 71–72; end of era of, 95–96; and eugenics-genetics relationship, 65, 66–67; and evolution, 71–72; and "fitness," 66; and genetic manipulation, 68; and genetic research, 66; goals of, 63, 77; and intelligence, 63; and medical genetics, 67; and medical profession, 73–75, 77; and mutation, 72; and natural selection, 72; overview about, 63; and population control movement, 66; public rejection of, 63; and race, 63, 68–69, 74; and reproductive rights, 80; and role of government, 63, 75, 77; and science–social values relationship, 221; and sterilization, 74; and who decides question, 77
regulation, 148–54, 157, 172–77, 194, 224
religion, 22, 30, 51, 60. *See also* Catholic Church
reproductive rights, 16, 38–39, 48–49, 60, 77–78, 80, 146, 181
restriction enzymes, 111, 147–48
rewriting humanity, 97, 109–22, 155, 211, 219. *See also* genetic engineering; manipulation, genetic; *specific technology*
rhetorical formations, 14–15, 219–22
RNA, 75, 101, 102, 109
Roe v. *Wade* (1973), 49, 125
Rostand, Jean, 112, 115, 151

Sanger, Margaret, 28
"science-based" criteria, 75–76, 174–76
science fiction, 110–11, 149, 185 science/scientists: domination by, 155; and ethics, 221–22; and eugenics, 51, 52–53; and family genetics, 75–76; as heroes, 134; as ideological, 231; and metaphors, 219; power of, 221; and public discourse, 12; and rhetorical formations, 219–22; self-regulation of, 148–54, 172–77; and social values, 219–22; as source of information, 153–54. *See also* biohazards
Seed, Randolph, 186
Seed, Richard, 186
"self," 227
semiology, 101
sexism, 82, 96, 145, 146, 156
sexuality: and ethical challenges, 144–46; homo-, 200, 234, 235
Shockley, William, 141, 142, 196
shopping metaphor, 120–21, 244
sickle-cell anemia, 88, 104, 116, 129, 142–43, 215, 216
Sinsheimer, Robert, 109
social Darwinism, 29–30, 34, 35–36
social issues/policies: and attitudes about genetics, 22; and determinism, 212, 214; and development of genetics, 4; and eugenics, 5, 15–16, 25, 35–36, 39, 44, 58; and experimental genetics, 112, 121, 155, 211; and family genetics, 18, 80, 82, 89–90; and genetic counseling, 137–38, 139; and genetic engineering, 121; and perfectionism, 7; and regulation, 176; and science, 219–22. *See also* social structure; society
social structure, 7, 169, 192–95, 207, 211, 214, 226
society: individual and, 18, 239–40; and rhetorical formations, 219–22. *See also* social issues/policies; social structure
Southern Europeans, 36
Soviet Union, 18, 67, 69–70, 75, 95
Spencer, Herbert, 29–30
sperm banks, 111, 119–20
statistics: and eugenics, 25, 31, 33, 41–42, 44; and family genetics, 86–87; and gene-environment interaction, 86–87; and personal health genetics, 182. *See also* prediction
sterilization: and ethics, 141; and eugenics, 3, 4, 27, 28, 41, 47–48, 49, 53, 57, 59, 60; and experimental genetics, 122; and family genetics, 63, 91; and race, 141; and reform eugenics, 68, 74
stigmatization, 188, 201, 202, 217
stock-breeding metaphor: advantages of, 31–32, 33, 36–39; and behavior, 41; and challenges to eugenics, 50, 53, 57, 58, 60; and class issues, 34–35; credibility of, 25; as dominant eugenics metaphor, 15, 31, 34–39, 83, 108; and germ-plasm concept, 33, 37–38, 44, 60, 218; and goals of eugenics, 34, 43, 218; and Nazi Germany, 16; overview of, 34–39, 218; polarized character of, 37–39

Storm family, 126
superiority: and discrimination, 7; and ethics, 191; and eugenics, 28, 29, 36–37, 38, 43, 54–55, 57, 60, 90, 91, 217, 218; and experimental genetics, 111, 112; and family genetics, 18, 95
supermen, lists of, 119–20, 145
Supreme Court, U.S., 49, 176–77
surrogacy, 19, 109, 111
"susceptibility" of genes, 87–88, 105, 125, 187, 211
systems theory, 232

Tay-Sachs disease, 122, 127, 131–32, 137, 144, 197
Technology, access to, 146, 168, 193–94, 223, 225. *See also specific technology*
television, 11
thread of life metaphor, 108
tissue regeneration, 111
transhuman species, 244
tuberculosis, 90
twin studies, 75, 235

uncertainty principle (Heisenberg), 243
undecidability, 113–15, 154, 221
U.S. Department of Agriculture (USDA), 173

van Dijck, José, 6, 13, 66, 100, 148, 152, 161, 173
variability, genetic, 228, 229–31, 233
viral vectors, 111
vocabulary of genetics, 232–35

Watson, James: and biohazards, 148, 150, 154; and discovery of DNA structure, 19, 99, 100, 109, 140, 155; *The Double Helix* by, 13, 66; and experimental genetics, 19; and methodology, 13
Weismann, August, 10, 32, 33, 37, 70–71
who decides question: and biohazards, 151–55; and eugenics, 49–50, 60; and experimental genetics, 119; and genetic counseling, 133–39; and reform eugenics, 77
Wilson, Edward O., 212, 222, 228–29, 230, 235
women, 21, 145, 197–201. *See also* feminism; gender
World War II, 67–70. *See also* Nazi Germany, eugenics in

RHETORIC OF THE HUMAN SCIENCES

Lying Down Together: Law, Metaphor, and Theology
Milner S. Ball

Shaping Written Knowledge: The Genre and Activity of the Experimental Article in Science
Charles Bazerman

Textual Dynamics of the Professions: Historical and Contemporary Studies of Writing in Professional Communities
Charles Bazerman and James Paradis, editors

The Meanings of the Gene: Public Debates about Human Heredity
Celeste Michelle Condit

Politics and Ambiguity
William E. Connolly

The Rhetoric of Reason: Writing and the Attractions of Argument
James Crosswhite

Democracy and Punishment: Disciplinary Origins of the United States
Thomas L. Dumm

Philosophy, Rhetoric, and the End of Knowledge: The Coming of Science and Technology Studies
Steve Fuller

Machiavelli and the History of Prudence
Eugene Garver

Language and Historical Representation: Getting the Story Crooked
Hans Kellner

The Rhetoric of Economics
Donald N. McCloskey

The Rhetoric of Economics, Second Edition
Deirdre N. McCloskey

Therapeutic Discourse and Socratic Dialogue: A Cultural Critique
Tullio Maranhão

Tropes of Politics: Science, Theory, Rhetoric, Action
John S. Nelson

The Rhetoric of the Human Sciences: Language and Argument in Scholarship and Public Affairs
John S. Nelson, Allan Megill, and Donald N. McCloskey, editors

What's Left? The Ecole Normale Supérieure and the Right
Diane Rubenstein

Understanding Scientific Prose
Jack Selzer, editor

The Politics of Representation: Writing Practices in Biography, Photography, and Policy Analysis
Michael J. Shapiro

The Legacy of Kenneth Burke
Herbert Simons and Trevor Melia, editors

Defining Science: A Rhetoric of Demarcation
Charles Alan Taylor

The Unspeakable: Discourse, Dialogue, and Rhetoric in the Postmodern World
Stephen A. Tyler

Heracles' Bow: Essays on the Rhetoric and the Poetics of the Law
James Boyd White